Problem Books in Mathematics

Edited by P. R. Halmos

Problem Books in Mathematics

Series Editor: P.R. Halmos

Unsolved Problems in Intuitive Mathematics, Volume I: Unsolved Problems in Number Theory
by *Richard K. Guy*
1981. xviii, 161 pages. 17 illus.

Theorems and Problems in Functional Analysis
by *A.A. Kirillov* and *A.D. Gvishiani*
1982. ix, 347 pages. 6 illus.

Problems in Analysis
by *Bernard Gelbaum*
1982. vii, 228 pages. 9 illus.

A Problem Seminar
by *Donald J. Newman*
1982. viii, 113 pages.

Problem-Solving Through Problems
by *Loren C. Larson*
1983. xi, 344 pages. 104 illus.

Demography Through Problems
by *N. Keyfitz* and *J.A. Beekman*
1984. viii, 141 pages. 22 illus.

Problem Book for First Year Calculus
by *George W. Bluman*
1984. xvi. 384 pages. 384 illus.

Exercises in Integration
by *Claude George*
1984. x. 550 pages. 6 illus.

Exercises in Number Theory
by *D.P. Parent*
1984. x. 541 pages.

Problems in Geometry
by *Marcel Berger et al.*
1984. viii. 266 pages. 244 illus.

Claude George

Exercises in Integration

With 6 Illustrations

Springer-Verlag
New York Berlin Heidelberg Tokyo

Claude George
Université de Nancy I
UER Sciences Mathematiques
Boite Postale 239
54506 Vandoeuvre les Nancy Cedex
France

Translator
J.M. Cole
17 St. Mary's Mount
Leyburn, North Yorkshire DL8 5JB
U.K.

Editor
Paul R. Halmos
Department of Mathematics
Indiana University
Bloomington, IN 47405
U.S.A.

AMS Classifications: 00A07, 26-01, 28-01

Library of Congress Cataloging in Publication Data
George, Claude.
Exercises in integration.
(Problem books in mathematics)
Translation of: Exercices et problèmes d'intégration.
Bibliography: p.
Includes indexes.
1. Integrals, Generalized–Problems, exercises, etc. I. Title. II. Series.
QA312.G39513 1984 515.4 84-14036

Title of the original French edition: *Exercices et problèmes d'intégration,*

Printed and bound by R.R. Donnelley & Sons, Harrisonburg, Virginia.
Printed in the United States of America.

9 8 7 6 5 4 3 2 1

ISBN 0-387-96060-0 Springer-Verlag New York Berlin Heidelberg Tokyo
ISBN 3-540-96060-0 Springer-Verlag Berlin Heidelberg New York Tokyo

Introduction

Having taught the theory of integration for several years at the University of Nancy I, then at the École des Mines of the same city, I had followed the custom of the times of writing up detailed solutions of exercises and problems, which I used to distribute to the students every week. Some colleagues who had had occasion to use these solutions have persuaded me that this work would be interesting to many students, teachers and researchers. The majority of these exercises are at the master's level; to them I have added a number directed to those who would wish to tackle greater difficulties or complete their knowledge on various points of the theory (third year students, diploma of education students, researchers, etc.).

This book, I hope, will render to students the services that this kind of book brings them in general, with the reservation that can always be made in this case: that certain of them will be tempted to look at the solution to the exercises which are put to them without any personal effort. There is hardly any need to emphasize that such a use of this book would be no benefit. On the other hand, the student who after having worked seriously upon a problem, seeks some pointers from the solution, or compares it with his own, will be using this work in the optimal way.

Teachers will find this book to be an important, if not exhaustive, list of exercises, certain of which are more or less standard, and others which may seem new.

I have also noted (and this is what led me to edit these sheets) that from one year to another one sometimes forgets the solution of an exercise and that one has to lose precious time in rediscovering it. This is particularly true for those solutions of which one remembers the heuristic form but of which the writing up is delicate if one wishes to be clear and precise at the same time. Now, if one requires, quite rightly, that students write their homework up correctly, then it is befitting to submit impeccable corrections to them, where the notations are judiciously chosen, phrases of the kind "it is clear that ... " used wittingly, and where the telegraphic style gives way to conciseness. It is often the incorporation of these corrections which demands the most work; I have therefore striven to take pains with the preparation of the proposed solutions, always remaining persuaded that perfection in this domain is never attained. If this book encourages those who have to present (either orally or in writing) correct versions of problems to improve the version they submit, the object I have set myself will be partly realised.

In this book researchers will find some results that are not always treated in courses on integration; they are either properties whose use is not as universal as those which usually appear and which are therefore found scattered about in appendices in various works, or are results that correspond to some technical lemmas which I have picked up in recent articles on a variety of subjects: group theory, differential games, control theory, probability, etc., ...

In presenting such a work it is just as well to make explicit those points of the theory that are assumed to be known. This is the object of the brief outline which precedes the eleven chapters of exercises.

In view of the origin of this book, it is evident that I took as a reference point the course that I gave at the time. After having taught abstract measure theory one year, I opted the next for a course expounding only the Lebesgue integral. This is not the place to discuss the advantages and inconveniences of each of the two points of view for the first year of a master's programme. I will say only that I have always considered the course that I gave to be more a course in analysis in which it is decided to use the Lebesgue integral than as a dogmatic exposition of a particular theory of integration. The choice of exercise reflects this attitude, especially in the emphasis given to trigonometric series, thereby paying the hommage due to the theory which is the starting point of the works of Cantor, Jordan, Peano, Borel, and Lebesgue. From this it results that, except for the seven exercises of Chapter 2 concerning σ-algebras, all the others deal with Lebesgue measure on $\mathbb{R}^n$. The advantage that has to be conceded to this point of view is that it avoids the vocabulary of abstract measure theory, which constitutes an artificial obstacle for those readers who might not yet be well versed in this theory. As for students who might have followed a more sophisticated course, I can assure them that by substituting $d\mu$ for dx and $\mu(E)$ fore meas(E) they will essentially rediscover the problems as they are commonly put to them, except for pathological examples about measures that are not σ-finite and the applications of the Radon-Nikodym Theorem. Furthermore, on this latter point the more perspicacious amongst them will not fail to see that the chapter treating the relationships between differentiation and integration is not foreign to this theorem. Truthfully, there is another point that is not tackled in this book, namely the matter of Fourier transforms of finite positive measures and Stone's Theorem, which to my mind is better suited to a course on probability.

As was mentioned above, numerous exercises are devoted to trigonometric series, which provides an important set of applications of Lebesgue's theory. This has led me to include some exercises on series, summation processes, and trigonometric polynomials.

Other exercises use the theory of holomorphic functions. In particular, some results of the Phrägmen-Lindelöf type arise on two occasions; in each instance I have given its proof under the hypotheses that appear in the exercise. Quite generally, I have included in the solutions, or in an appendix to them, the proofs of certain points of analyis, topology, or algebra which students may not know.

I have chosen to make each solution follow immediately after the corresponding problem. The other method, which consists of regrouping the former in a second part of the work, seemed to me (from memories I have retained from my student days) much less manageable, especially when the problem is long, for it then becomes necessary to return often to the back of the book in order to follow the solution.

I find it difficult to cite the origin of these exercises. Many are part of a common pool of knowledge, handed down, one might say, in the public domain. Others are drawn from different classic works where they are proposed without proof or followed by more or less summary indications (in this respect it is interesting to note that in forcing oneself to write down the solutions one discovers a certain number of errors—just as many in the questions as in the suggestions offered). Certain of the exercises in this book were communicated to me orally by colleagues; I would thank them for their help here. Lastly, others are, as I have already said, lemmas found here and there, and which I have sometimes adapted.

Table of Contents

CHAPTER 0

Outline of the Course

0·1 σ-ALGEBRAS AND MEASURES

DEFINITION: *A family* A *of subsets of a set* X *is called a* σ-ALGEBRA *("sigma algebra") if* $\emptyset \in$ A, *and if* A *is closed under complementation and countable union.*

From this it follows that the set X itself belongs to the σ-algebra A, and that the σ-algebra A is closed under countable intersection. For two sets $A, B \in$ A let us denote $A - B = \{x : x \in A, x \notin B\}$; then we have $(A - B) \in$ A.

The smallest σ-algebra containing the open sets of $\mathbb{R}^p$ is the σ-algebra of *BOREL SETS* of $\mathbb{R}^p$; this σ-algebra is also the smallest σ-algebra which contains the closed (*resp.* open) rectangles of $\mathbb{R}^p$.

DEFINITION: *A (positive)* MEASURE *on a σ-algebra* A *is a mapping* μ *of* A *into* $[0,\infty]$ *such that if* E *is the* disjoint *union of a sequence of sets* $E_n \in$ A, *then* $\mu(E) = \sum \mu(E_n)$.

It follows that $\mu(\emptyset) = 0$, and then, if E is the union (not necessarily disjoint) of the sets E_n, $\mu(E) \leq \sum \mu(E_n)$. An equivalent definition is the following: If E is the union of a finite number of E_i's, each of which is in A and which are pairwise disjoint, then $\mu(E) = \mu(E_1) + \cdots + \mu(E_p)$; and furthermore $\mu(A) = \lim\mu(A_n)$ when A is the union of an *increasing* sequence of sets A_n of $\mathcal{A}$. If μ is a measure and A is the intersection of a *decreasing* sequence of sets $A_n \in \mathcal{A}$ and if $\mu(A_1) < \infty$, then $\mu(A) = \lim\mu(A_n)$.

There exists one and only one positive measure ν on the σ-algebra of Borel sets of $\mathbb{R}^p$ such that, for every rectangle P, its measure $\nu(P)$ is equal to the volume of P.

DEFINITION: *A set E of $\mathbb{R}^p$ is called a* NEGLIGEABLE SET *if there exists a Borel set A such that $E \subset A$ and $\nu(A) = 0$.*

This definition is equivalent to the existence, for every $\varepsilon > 0$, of a sequence of rectangles covering E, the sum of the volumes of which is less than ε. A countable union of negligeable sets is negligeable, and every affine sub-manifold of $\mathbb{R}^p$ that is of dimension less that p is negligeable.

DEFINITION: *A set of $\mathbb{R}^p$ is called a* LEBESGUE MEASURABLE SET *(or simply a* MEASURABLE SET*) if it belong to the smallest σ-algebra containing the Borel sets and negligeable sets of $\mathbb{R}^p$.*

In order that $E \subset \mathbb{R}^p$ be measurable it is necessary and sufficient that there exist the Borel sets A and B such that $A \subset E \subset B$ and $\nu(B - A) = 0$; upon then setting $\text{meas}(E) = \nu(A)$ one unambiguously defines a positive measure on the σ-algebra of Lebesgue measurable sets of $\mathbb{R}^p$. This measure is called the *LEBESGUE MEASURE ON $\mathbb{R}^p$*. A set is negligeable if it is measurable and of (Lebesgue) measure zero. This is why one also uses the expression *SET OF MEASURE ZERO* to denote a negligeable set.

The Lebesgue measure is invariant under translation as well as under unimodular linear transformations (i.e., those with determinant equal to ± 1). A homothety of ratio λ multiplies the Lebesgue measure by $|\lambda|^p$ (where p is the dimension of the space).

DEFINITION: *If* $\mathcal{A}$ *is a* σ*-algebra of subsets of* X *and* $\mathcal{B}$ *is a* σ*-algebra of subsets of* Y*, a mapping* $f:X \to Y$ *is said to be an* $\mathcal{A}$ - $\mathcal{B}$-*MEASURABLE MAPPING if* $f^{-1}(B) \in \mathcal{A}$ *for all* $B \in \mathcal{B}$. *When* $Y = \mathbb{R}^p$ *and* $\mathcal{B}$ *is the* σ*-algebra of Borel sets, one says, simply, that* f *is an* $\mathcal{A}$-*MEASURABLE MAPPING. In this case the definition is equivalent to requiring* $f^{-1}(V) \in \mathcal{A}$ *for every open set* V *of* Y. *Furthermore, when* $X = \mathbb{R}^q$*, the mapping* f *is said to be a BOREL MAPPING or a LEBESGUE-MEASURABLE MAPPING according as* $\mathcal{A}$ *is the* σ*-algebra of Borel sets or the Lebesgue-measurable sets of* X.

If $f:X \to \mathbb{R}$, in order that f be $\mathcal{A}$-measurable it is sufficient that $(f < a) = \{x:f(x) < a\} \in \mathcal{A}$ for all $a \in \mathbb{R}$ (and even for $a \in \mathbb{Q}$). This condition is taken as the definition of the $\mathcal{A}$-*measurability of an ARITHMETIC FUNCTION*, that is to say of a mapping of X into $[-\infty,+\infty] = \bar{\mathbb{R}}$. If f is an $\mathcal{A}$-measurable mapping of X into $\mathbb{R}^p$ and g a Borel mapping of $\mathbb{R}^p$ into $\mathbb{R}^q$, then $g \circ f$ is $\mathcal{A}$-measurable. Let us note that every continuous mapping of $\mathbb{R}^p$ into $\mathbb{R}^q$ is Borel. If (f_n) is a sequence of $\mathcal{A}$-measurable arithmetic functions, the functions $\sup f_n, \inf f_n, \limsup f_n, \liminf f_n$ are also $\mathcal{A}$-measurable.

DEFINITION: *A function is called a SIMPLE FUNCTION (with respect to the* σ*-algebra* $\mathcal{A}$*) if it is a linear combination of characteristic functions of sets of the* σ*-algebra* $\mathcal{A}$.

For every $\mathcal{A}$-measurable positive arithmetic function f there exists an increasing sequence of positive simple functions which converges to f at every point of X.

DEFINITION: *A property holding on the points of a set* A *of* $\mathbb{R}^p$ *is said to be true ALMOST EVERYWHERE ON A SET* A *if the set of points of* A *for which this property is not satisfied has measure zero.*

If f and g are two mappings from $\mathbb{R}^p$ into $\mathbb{R}^q$ (or $\bar{\mathbb{R}}$) such that f is measurable and $f = g$ almost everywhere, then g is measurable.

This allows the notion of measurability to be extended to functions that are defined only almost everywhere.

DEFINITION: *A function defined on* $\mathbb{R}^p$ *is called a* **STEP FUNCTION** *if it is a linear combination of characteristic functions of rectangles of* $\mathbb{R}^p$.

Every measurable arithmetic function on $\mathbb{R}^p$ is the limit almost everywhere of a sequence of step functions.

THEOREM: (Regularity of the Lebesgue Measure): *For every measurable set* E *of* $\mathbb{R}^p$ *one has:*

$$\text{meas}(E) = \begin{cases} \sup\{\text{meas}(K) : K \text{ compact } K \subset E\}; \\ \inf\{\text{meas}(V) : V \text{ open } V \supset E\}. \end{cases}$$

THEOREM: (Egoroff): *Let* X *be a measurable set of* $\mathbb{R}^p$ *such that* $\text{meas}(X) < \infty$ *and* (f_n) *a sequence of measurable functions such that* $f_n \to f$ *almost everywhere on* X. *For every* $\varepsilon > 0$ *there exists a measurable set* $A \subset X$ *such that:*

(i): $\text{meas}(X - A) < \varepsilon$;

(ii): $f_n \to f$ *uniformly on* A.

0·2 INTEGRATION OF MEASURABLE POSITIVE FUNCTIONS

NOTATION: If φ is a simple function on $\mathbb{R}^n$ that takes the positive values $\alpha_1,\ldots,\alpha_p$ on the (disjoint) measurable sets $A_1,\ldots,A_p$, we set

$$\int_{\mathbb{R}^n} \varphi(x)dx = \int \varphi = \sum_{i=1}^{p} \alpha_i \text{meas}(A_i),$$

with the convention that $a.(+\infty) = +\infty$ or 0 according as $a > 0$ or $a = 0$.

DEFINITION. *With the above notation, if f is a positive measurable arithmetic function on $\mathbb{R}^n$ there exists an increasing sequence (φ_i) of positive simple functions which tends towards f at every point. One then sets:*

$$\int_{\mathbb{R}^n} f(x)\mathrm{d}x = \int f = \lim_i \int \varphi_i.$$

This element of $[0,+\infty] = \bar{\mathbb{R}}_+$, which does not depend upon the sequence (φ_i) selected, is called the (LEBESGUE) INTEGRAL of f on $\mathbb{R}^n$.

This (Lebesgue) integral possesses the following properties (where f and g denote measurable positive arithmetic functions):

PROPERTY (1): *If $f = g$ almost everywhere, then $\int f = \int g$;*

PROPERTY (2): *$\int f = 0$ if and only if $f = 0$ almost everywhere;*

PROPERTY (3): *$\int f < \infty$ implies $f < \infty$ almost everywhere;*

PROPERTY (4): *$f \leqslant g$ almost everywhere implies $\int f \leqslant \int g$;*

PROPERTY (5): *$\int (f + g) = \int f + \int g$;*

PROPERTY (6) *If $\lambda \in \bar{\mathbb{R}}_+$, then $\int \lambda f = \lambda \int f$.*

One can prove the following two fundamental results:

THEOREM: (Lebesgue's Monotonic Convergence): *If (f_n) is an* increasing *sequence of measurable positive arithmetic functions, then*

$$\int \lim_n f_n = \lim_n \int f_n.$$

LEMMA: (Fatou): *If* (f_n) *is a sequence of positive arithmetic functions, then*

$$\int \liminf_n f_n \leqslant \liminf_n \int f_n.$$

These two essential properties of the Lebesgue integral are equivalent to the following statements: Let (f_n) be a sequence of positive arithmetic functions:

$$\text{(a)}: \quad \int (\sum_n f_n) = \sum_n (\int f_n);$$

(b): If $f_n \to f$ almost everywhere, and if there exists $A \in \mathbb{R}_+$ such that

$$\int f_n \leqslant A \quad \text{for all } n,$$

then

$$\int f \leqslant A.$$

Property (1) allows the definition of the integral to be extended to measurable arithmetic functions that are defined only almost everywhere.

And lastly:

NOTATION: If f is a measurable positive arithmetic function, and if E is a measurable set of $\mathbb{R}^n$, we *set*

$$\int_E f(x)dx = \int_E f = \int \mathbb{1}_E f,$$

where $\mathbb{1}_E$ *denotes* the *CHARACTERISTIC FUNCTION* of E.

The mapping $E \to \int_E f$ defines a positive measure on the σ-algebra of (Lebesgue) measurable sets.

0·3 INTEGRATION OF COMPLEX MEASURABLE FUNCTIONS

NOTATION: For every real function u, we set:

$$u_+ = \sup(u,0) = \tfrac{1}{2}(|u| + u),$$

$$u_- = \sup(-u,0) = \tfrac{1}{2}(|u| - u),$$

so that

$$u = u_+ - u_-,$$

$$|u| = u_+ + u_-,$$

$$u_+u_- = 0.$$

Let f be a complex measurable function on $\mathbb{R}^n$, and let u and v be its real and imaginary parts. The function f is measurable if and only if u_+,u_-,v_+,v_- are measurable.

DEFINITION: *f is (LEBESGUE) INTEGRABLE ON $\mathbb{R}^n$ if it is measurable and if*

$$\int |f| < \infty.$$

The INTEGRAL *is then defined by setting*

$$\int f = \int u_+ - \int u_- + i\int v_+ - i\int v_-,$$

(which has a meaning, because u_+,u_-,v_+,v_- are majorised by $|f|$).

The Lebesgue integral possesses Properties (1) and (5) of the preceding Section; it also possesses Property (4) when f and g are real, as well as Property (6) with $\lambda \in \mathbb{C}$. The sum, and the pointwise maximum and minimum of a *finite number* of integrable functions are integrable. If f is integrable, then

$$\left|\int f\right| < \int |f|,$$

with equality holding only if there exists $\alpha \in \mathbb{C}$ such that $f = \alpha|f|$ almost everywhere.

The two essential properties of the Lebesgue integral are the following:

THEOREM: (Lebesgue's Dominated Convergence): *Let (f_n) be a sequence of integrable functions that converges to f almost everywhere. If there exists a measurable positve arithmetic function g such that*

$$|f_n| \leqslant g \quad \textit{for all } n, \qquad \textit{and} \int g < \infty,$$

then f is integrable, and

$$\int f = \lim_n \int f_n.$$

THEOREM: (Term by Term Integration of Series of Functions): *If (u_n) is a sequence of integrable functions such that*

$$\sum_n \int |u_n| < \infty,$$

then the series

$$u(x) = \sum_n u_n(x)$$

is absolutely convergent for almost all x, the function u (which is defined almost everywhere) is integrable, and

$$\int u = \sum_n \int u_n.$$

If f is integrable and if E is a measurable set, we again set

$$\int_E f = \int \mathbb{1}_E f$$

(the integral of f over E). In fact this integral depends only upon the values of f on E and can be defined whenever f itself is only defined on the set E; it suffices, for example, that the function obtained by extending f by zero outside E be integrable over $\mathbb{R}^n$. In this case one says that f is *INTEGRABLE OVER E*. All the properties of the Lebesgue integral over $\mathbb{R}^n$ extend to this case. Furthermore, if f is integrable over E then it is integrable over every measurable set contained in E, and if E is the *disjoint union* of a sequence (E_n) of measurable sets,

$$\int_E f = \sum_n \int_{E_n} f.$$

Similarly, if E is the *union of an increasing sequence* (E_n) of measurable sets,

$$\int_E f = \lim_n \int_{E_n} f.$$

This formula is still valid if E is the *intersection of a decreasing sequence* (E_n) of measurable sets and if *f is integrable* over E_1.

In the case of integration on $\mathbb{R}$, it is convention to write

$$\int_I f = \int_a^b f$$

when I is an interval with endpoints a and b $(-\infty \leqslant a \leqslant b \leqslant +\infty)$, and f is either a measurable positive arithmetic function on I, or a complex integrable function on this interval. When $a > b$

we write

$$\int_a^b f = -\int_b^a f$$

when f is integrable over $[a,b]$, and a "Chasles' Formula" can then be written. If $-\infty < a < b < +\infty$, then f is Lebesgue-integrable over $[a,b]$ whenever it is Riemann-integrable over this interval, and the two integrals are equal. However, when the interval is infinite the Lebesgue integral only constitutes an extension of the notion of *absolutely convergent* (generalized) Riemann integral.

The *GENERALIZED* (or *SEMI-CONVERGENT*) *LEBESGUE INTEGRALS* can be defined in the following way. For example, let us assume that f is (Lebesgue) integrable over every interval $[0,M]$, $0 \leqslant M < \infty$; one then sets

$$\int_0^\infty f = \lim_{M\to\infty}\int_0^M f$$

when this limit exists. In this respect let us note the following Proposition:

PROPOSITION: (Second Mean Value Formula): *If f is a decreasing positive function on $[a,b]$, and g an integrable function on this interval, then*

$$\left|\int_a^b fg\right| \leqslant f(a) \sup_{a\leqslant x\leqslant b}\left|\int_a^x g\right|.$$

To close this Section let us indicate that if f is a mapping from $\mathbb{R}^p$ into $\mathbb{R}^q$ of which the q coordinates are integrable, the integral of f is the element of $\mathbb{R}^q$ of which each component is equal to the integral of the corresponding component of f. Everything that has been said above remains valid when the absolute value is replaced by a norm on $\mathbb{R}^q$.

0·4 FUBINI'S THEOREM

THEOREM: (Fubini): *Let* $X = \mathbb{R}^p$, $Y = \mathbb{R}^q$; *then the formula*

$$\iint_{X\times Y} f(x,y)\,dx\,dy = \int_X dx\int_Y f(x,y)\,dy = \int_Y dy\int_X f(x,y)\,dx$$

is valid in each of the following two cases:

(1): *f is a measurable positive arithmetic function on $X\times Y$;*

(2): *f is an integrable function over $X\times Y$.*

Amongst other things the validity of the formula means (according to the case) that for almost all $x\in X$ the function $y\mapsto f(x,y)$ is measurable (*resp.*, integrable) on Y, and that the function $x\mapsto \int_X f(x,y)\,dy$, defined almost everywhere on X, is measurable (*resp.*, integrable) on X.

In particular, in order to have the rule of "interchangeability of the order of integration" it suffices to be assured, when f is a measurable complex function on $X\times Y$, that

$$\int_X dx\int_Y |f(x,y)|\,dy < \infty.$$

Whenever f is positive and measurable this interchange is always legitimate.

Certain proofs of Fubini's Theorem use the following Lemma, which is interesting in its own right:

LEMMA: *In order that a set have measure zero it is necessary and sufficient that there exist an increasing sequence* (q_n) *of positive step functions such that*

$$\sup_n\int\varphi_n < \infty,\qquad \lim_n\varphi_n(x) = +\infty,\qquad \textit{if } x\in E.$$

In this statement one can replace "step functions" by "compactly supported continuous functions".

0·5 CHANGE OF VARIABLES

Let V be an open set of $\mathbb{R}^p$ and u a diffeomorphism of V onto $u(V)$, that is to say a bijection of V onto $u(V)$ such that u and u^{-1} are continuously differentiable. The *JACOBIAN* of u is *denoted* $J(u)$.

PROPOSITION: *The formula*

$$\int_{u(V)} f(x)\mathrm{d}x = \int_V f(u(x))|J(x)|\mathrm{d}x$$

is valid in each of the following two cases:

(1): *f is a measurable positive arithmetic function on $u(V)$;*

(2): *f is an integrable function over V.*

Amongst other things the validity of the formula means that fou is measurable (*resp.* $(fou)|J|$ is integrable) on V.

Spherical Coordinates in $\mathbb{R}^n$

$\mathbb{R}^n$ with the half-hyperplane defined by $x_1 \leqslant 0$, $x_2 = 0$ removed, possesses a proper parametric representation:

$$\begin{cases} x_1 = r\sin\vartheta_{n-2}\cdots\sin\vartheta_2\sin\vartheta_1\cos\varphi, \\ x_2 = r\sin\vartheta_{n-2}\cdots\sin\vartheta_2\sin\vartheta_1\sin\varphi, \\ x_3 = r\sin\vartheta_{n-2}\cdots\sin\vartheta_2\cos\vartheta_1, \\ x_4 = r\sin\vartheta_{n-2}\cdots\cos\vartheta_2, \\ \cdots\cdots\cdots \\ x_n = r\cos\vartheta_{n-2}, \end{cases}$$

where $r > 0$, $0 < \vartheta_i < \pi$, $|\varphi| < \pi$. The formula for the change of variables in this case comes down to replacing the x_i by their expression as a function of r, the ϑ_i's, and φ, and $dx = dx_1 dx_2 \dots dx_n$ by

$$r^{n-1}\sin^{n-2}\vartheta_{n-2} \cdots \sin^2\vartheta_2 \sin\vartheta_1 \, dr \, d\vartheta_{n-2} \cdots d\vartheta_2 \, d\vartheta_1 \, d\varphi.$$

Let us also recall that the volume of the ball $x_1^2 + \cdots + x_n^2 \leqslant 1$ is

$$V_n = \frac{\pi^{n/2}}{\Gamma\left(\frac{n}{2} + 1\right)}.$$

0·6 THE L^p SPACES

In what follows we make the convention of setting $1/\infty = 0$, $a.\infty = 0$ or ∞ according as $a = 0$ or $a > 0$, and $\infty^a = \infty$ if $a > 0$.

HÖLDER'S INEQUALITY: *Let* $1 < p, q < \infty$ *be such that*

$$\frac{1}{p} + \frac{1}{q} = 1,$$

and let f *and* g *be measurable positive arithmetic functions on* $\mathbb{R}^n$*; then:*

$$\int fg \leqslant \left(\int f^p\right)^{1/p}\left(\int g^q\right)^{1/q}.$$

MINKOWSKI'S INEQUALITY: *Let* $1 \leqslant p < \infty$*, and let* f *and* g *be measurable positive arithmetic functions on* $\mathbb{R}^n$*; then*

$$\left(\int (f + g)^p\right)^{1/p} \leqslant \left(\int f^p\right)^{1/p} + \left(\int g^p\right)^{1/p}.$$

When $p = q = 2$, Hölder's Inequality is known as the (*CAUCHY-*

SCHWARZ INEQUALITY. For $0 < p < 1$ one still has an inequality which is obtained by replacing $\leqslant$ with $\geqslant$ in Minkowski's Inequality (it is sometimes called *MINKOWSKI'S SECOND INEQUALITY*).

In order to have equality in Hölder's Inequality it is necessary and sufficient that there exist $\alpha \geqslant 0$ such that $f^p = \alpha g^q$ almost everywhere. For Minkowski's Inequality when $1 < p < \infty$, equality is only obtained if $f = \alpha g$ almost everywhere.

NOTATION: For every measurable function f on $\mathbb{R}^n$ one sets

$$\|f\|_p = \left(\int |f|^p\right)^{1/p}, \quad \text{where } 0 < p < \infty.$$

Also, there exists $M \in \bar{\mathbb{R}}_+$ such that $|f| \leqslant M$ almost everywhere, and such that this does not hold for any $M' < M$. The element M is called the *ESSENTIAL SUPREMUM* of $|f|$ and one defines

$$\|f\|_\infty = \text{ess sup } |f|.$$

DEFINITION: *With the preceding notations, for* $0 < p \leqslant \infty$ *one defines the following functional spaces:*

$$L^p(\mathbb{R}^n) \equiv \{f : f \text{ is measurable on } \mathbb{R}^n ; \|f\|_p < \infty\}.$$

These spaces are vector subspaces of the space of functions on $\mathbb{R}$.

When $1 \leqslant p \leqslant \infty$ the mapping $f \mapsto \|f\|_p$ is a semi-norm on L^p (where we write L^p for $L^p(\mathbb{R}^n)$); the kernel of this semi-norm coincides with the vector subspace N of functions zero almost everywhere, so that L^p/N is canonically provided with a norm. *In this work we shall make not distinction between* L^p *and* L^p/N, *which amounts to identifying two functions that are equal almost everywhere.* Let us point out that certain authors denote by $\mathcal{L}^p$ that which we denote by L^p, and they keep the latter notation for $\mathcal{L}^p/N$. Having taken into account the convention we have just indicated, in the remainder we shall consider $f \mapsto \|f\|_p$ as a norm on L^p $(1 \leqslant p \leqslant \infty)$.

THEOREM: *Let* $1 \leqslant p \leqslant \infty$ *and let* (u_n) *be a sequence of elements of* L^p *such that*

$$\sum_n \|u_n\|_p < \infty.$$

Then for almost all x *the series*

$$u(x) = \sum_n u_n(x)$$

is absolutely convergent; the function u *thus defined almost everywhere belongs to* L^p *and one has*

$$u = \sum_n u_n$$

in the sense of convergence in the norm of L^p.

COROLLARY: *For* $1 \leqslant p \leqslant \infty$ *the spaces* L^p *are complete. Furthermore, if* $f_n \to f$ *in* L^p *there exists a sub-sequence* f_{n_i} *such that* $f_{n_i} \to f$ *almost everywhere.*

It will be noted that if $f_n \to f$ in L^∞, then $f_n \to f$ almost everywhere. This property may fail if $f_n \to f$ in L^p, $1 \leqslant p < \infty$.

When $f \in L^p$, $g \in L^q$, $1 \leqslant p,q \leqslant \infty$, $\dfrac{1}{p} + \dfrac{1}{q} = 1$, the function fg is integrable and

$$\left|\int fg\right| \leqslant \|f\|_p \|g\|_q \qquad \text{(Hölder's Inequality)}.$$

The spaces L^p $(1 \leqslant p \leqslant \infty)$ are *BANACH SPACES*; L^2 is even a Hilbert space if the scalar product is defined by

$$(f|g) = \int f\bar{g}, \qquad f,g \in L^2.$$

For every measurable set E of $\mathbb{R}^n$ one similarly defines the spaces $L^p(E)$ by everywhere replacing the integrals taken over $\mathbb{R}^n$ by integrals over E. The space $L^p(E)$ is identified with the closed vector subspace of $L^p(\mathbb{R}^n)$ formed of the functions that vanish outside E. When $\operatorname{meas}(E) < \infty$ one has the inclusions $L^q(E) \subset L^p(E)$ if $0 < p < q < \infty$.

DENSITY THEOREM NO. 1: *Let E be a measurable set of $\mathbb{R}^n$; then the characteristic functions of the measurable sets $A \subset E$ are total for $L^p(E)$ for $1 \leqslant p \leqslant \infty$.*

DENSITY THEOREM NO. 2: *Let V be an open set of $\mathbb{R}^n$, then the characteristic functions of the rectangles P such that $P \subset V$ are total in $L^p(V)$ for $1 \leqslant p < \infty$. The set of continuous functions compactly supported in V are dense in $L^p(V)$, $1 \leqslant p < \infty$.*

Let us recall that a set A of a metric space E is said to be *DENSE* in E if $\bar{A} = E$; if E is a normed space, A is said to be *TOTAL* in E if the linear combinations of elements of A are dense in E.

0·7 CONVOLUTION

DEFINITION: *Two measurable functions f, g on $\mathbb{R}^n$ are said to be CONVOLVABLE if the function $y \mapsto f(x - y)g(y)$ is integrable for almost all x; in this latter case one can define almost everywhere the CONVOLUTION PRODUCT OF TWO FUNCTIONS f and g by*

$$(f * g)(x) = \int f(x - y)g(y)dy.$$

We have:

$$f * g = g * f.$$

If f is convolvable with g and h, then it is with $\lambda g + \mu h$ ($\lambda, \mu \in \mathbb{C}$), and

$$f*(\lambda g + \mu h) = \lambda(f*g) + \mu(f*h).$$

If $|f| \leqslant f_1$, $|g| \leqslant g_1$, f and g are convolvable whenever f_1 and g_1 are; from this it follows that if $f = u + iv$, $g = r + is$ (u,v, r,s real), f and g are convolvable if and only if each of the functions u_+,u_-,v_+,v_- is convolvable with each of the functions r_+,r_-,s_+,s_-.

Whenever f and g are convolvable, and are zero respectively outside measurable sets A and B, $f*g$ is zero outside $A + B$. In particular, if f and g are compactly supported, $f*g$ is also compactly supported.

In addition to the spaces L^p (i.e., $L^p(\mathbb{R}^n)$) introduced in the preceding Section it is useful to define the following functional spaces:

DEFINITION: *If* $1 \leqslant p < \infty$ *we define* L^p_{loc} *to be the vector space of measurable functions* f *such that for every compact set* K *of* $\mathbb{R}^n$ *one has*

$$\|f\|_{p,k} - \left(\int_K |f|^p\right)^{1/p} < \infty.$$

The functions belonging to L^p_{loc} *are said to be* **LOCALLY p-INTEGRABLE**. *If* $p = \infty$, L^∞_{loc} *is the space of* **LOCALLY BOUNDED MEASURABLE FUNCTIONS**, *that is to say that for every compact set* K *one has*

$$\|f\|_{\infty,K} = \operatorname*{ess\,sup}_{x\in K} |f(x)| < \infty.$$

One can be satisfied with looking at compact sets of the type $K_N = \{x : |x| \leqslant N\}$, where $N \geqslant 1$ is an integer and $x \to |x|$ a norm on $\mathbb{R}^n$. We then *set*

$$\|f\|_{p,K_n} \equiv \|f\|_{p,N},$$

and the formula

$$d(f,g) = \sum_{N=1}^{\infty} 2^{-N} \frac{\|f - g\|_{p,N}}{1 + \|f - g\|_{p,N}}$$

defines a metric on L^p_{loc} (with the usual condition of identifying two functions which are equal almost everywhere). This metric is compatible with the vector space structure on L^p_{loc}, $f_i \to f$ if and only if $\|f - f_i\|_{p,K} \to 0$ for every compact set K, and L^p_{loc} is complete in this metric.

DEFINITION: *The space of p-INTEGRABLE COMPACTLY SUPPORTED MEASURABLE FUNCTIONS is written L^p_c for $1 \leqslant p \leqslant \infty$; it is a vector subspace of L^p. We say that $f_i \to f$ in L^p_i if $f_i \to f$ in L^p and if, furthermore, there exists $A > 0$ such that for all i one has $f_i(x) = 0$ whenever $|x| > A$. This latter condition is expressed by saying that the f_i have their SUPPORT contained in a fixed compact set.*

DEFINITION: *The SPACE OF k-FOLD CONTINUOUSLY DIFFERENTIABLE FUNCTIONS on $\mathbb{R}^n$ is written E^k $(0 \leqslant k \leqslant \infty)$. (For $k = 0$, E^0 is defined to be the space of continuous functions on $\mathbb{R}^n$; in this case we may also write C instead of E^0).*

For every n-tuple $s = (s_1,\ldots,s_n)$ of integers greater than or equal to zero one sets

$$|s| = s_1 + \cdots + s_n,$$

and we define

$$D^s f \equiv \frac{\partial^{|s|} f}{\partial x_1^{s_1} \cdots \partial x_n^{s_n}}, \qquad f \in E^k, \quad |s| \leqslant k.$$

If $0 \leqslant k < \infty$, for every integer $N \geqslant 1$ and $f \in E^k$ we set:

$$p_N(f) \equiv \sup\{|D^s f(x)| : |s| \leqslant k, |x| \leqslant N\},$$

and if $k = \infty$,

$$p_N(f) = \sup\{|D^s f(x)| : |s| \leqslant N, |x| \leqslant N\}.$$

With these notations the formula

$$d(f,g) = \sum_{N=1}^{\infty} 2^{-N} \frac{p_N(f-g)}{1 + p_N(f-g)}$$

defines a metric on E^k compatible with the vector space structure, and for which E^k is complete. Furthermore, $f_i \to f$ in E^k if for every s such that $|s| \leqslant k$ one has $\lim_i D^s f_i = D^s f$ uniformly on every compact set of $\mathbb{R}^n$.

DEFINITION: *The vector space* C^∞ *is the vector* SPACE OF BOUNDED CONTINUOUS FUNCTIONS *on* $\mathbb{R}^n$ *provided with the* NORM

$$\|f\|_\infty = \sup_x |f(x)|.$$

It is a Banach space.

DEFINITION: *The* closed *vector subspace of* C^∞ *formed of* UNIFORMLY CONTINUOUS BOUNDED FUNCTIONS *on* $\mathbb{R}^n$ *is denoted* UC^∞.

DEFINITION: *The space of* CONTINUOUS FUNCTIONS *of* $\mathbb{R}^n$ WHICH TEND TO ZERO AT INFINITY *is denoted* C_0. *It is a* closed *vector subspace of* UC^∞.

DEFINITION: *The vector subspace of* E^k *consisting of* COMPACTLY SUPPORTED k-FOLD CONTINUOUSLY DIFFERENTIABLE FUNCTIONS *is written* D^k $(0 \leqslant k \leqslant \infty)$. *If* $k = 0$ *the space* D^0, *that is to say the space of compactly supported continuous functions, is also written* K.

We shall say that $f_i \to f$ *in* D^k *if for all* s $(|s| \leqslant k)$ *one has* $\lim_i D^s f_i = D^s f$ *uniformly on* $\mathbb{R}^n$ *and if the* f_i *have their support contained in a fixed compact set.*

If E and F are two of the spaces that have just been defined we shall *write* $E \to F$ *if* $E \subset F$ *and if* $f_i \to f$ *in* E *implies that* $f_i \to f$ *in* F. It is clear that $E \to F$ and $F \to G$ implies $E \to G$.

We have the following diagram, where $0 < k < \infty$ and $1 < p < \infty$:

$$\begin{array}{ccccccccccc}
\mathcal{E}^\infty & \to & \mathcal{E}^k & \to & C & \to & L^\infty_{loc} & \to & L^p_{loc} & \to & L^1_{loc}\\
\uparrow & & \uparrow & & \uparrow & & \uparrow & & \uparrow & & \uparrow\\
| & & | & & C^\infty & & | & & | & & |\\
| & & | & & \uparrow & & | & & | & & |\\
| & & | & & UC^\infty & & L^\infty & & L^p & & L^1\\
| & & | & & \uparrow & & \uparrow\;\uparrow & & \uparrow & & \uparrow\\
| & & | & & C_0 & & | & & | & & |\\
| & & | & & \uparrow & & | & & | & & |\\
\mathcal{D}^\infty & \to & \mathcal{D}^k & \to & K & \to & L^\infty_c & \to & L^p_c & \to & L^1_c
\end{array}$$

It will be noted that $\mathcal{D}^\infty \to E$ for any space E from the table above, and also $E \to L^1_{loc}$ for every E; note also that the L^p spaces are not mutually comparable.

NOTATION: For every function f and every $a \in \mathbb{R}^n$, we *define* the *TRANSLATION OF f BY a* by

$$f_a(x) = f(x - a).$$

Sometimes the notation $\tau_a f$ is used to designate f_a. If f and g are convolvable, then

$$(f*g)_a = f_a*g = f*g_a.$$

DEFINITION: *If E is one of the function spaces defined above, one says that E is INVARIANT UNDER TRANSLATION if $f \in E$ implies that $f_a \in E$ for every $a \in \mathbb{R}^n$. Furthermore, if $a_i \to a$ implies that $f_{a_i} \to f$ in E, one says that the TRANSLATIONS OPERATE CONTINUOUSLY ON E.*

THEOREM: (i): *The spaces* L^p, L^p_{loc}, L^p_c $(1 \leqslant p \leqslant \infty)$, $C^\infty, UC^\infty, C_0, E^k$ *and* D^k $(0 \leqslant k \leqslant \infty)$ *are invariant under translation;*

(ii): *The translations operate continuously on all these spaces, except upon* $L^\infty, L^\infty_{loc}, L^\infty_c$ *and* C^∞.

NOTATION-DEFINITION: *If* F, G, H *are three of the function spaces defined above, the notation* $F*G \subset H$ *expresses that if* $f \in F$, $g \in G$, *then* f *and* g *are convolvable and* $f*g \in H$. *Furthermore, if* $f_i*g_i \to f*g$ *in* H *whenever* $f_i \to f$ *in* F *and* $g_i \to g$ *in* G, *one writes* $F*G \subset H$ *(continuously). In the case where* $F = G = H = A$ *it is said that* A *is a* CONVOLUTION ALGEBRA. *Lastly, if* A *is a convolution algebra and* $A*E \subset E$, *one says that* A OPERATES IN E; A *is said to* OPERATE CONTINUOUSLY IN E *if* $A*E \subset E$ *continuously.*

THEOREM: (i): $L^1*L^\infty \subset UC^\infty$ *(continuously); furthermore,*

$$\|f*g\|_\infty < \|f\|_1\|g\|_\infty;$$

(ii): $L^p*L^q \subset C_0$ *(continuously) if* $1 < p,q < \infty$, $\frac{1}{p} + \frac{1}{q} = 1$; *furthermore,*

$$\|f*g\|_\infty \leqslant \|f\|_p\|g\|_q;$$

(iii): $L^1*L^p \subset L^p$ *(continuously) if* $1 \leqslant p \leqslant \infty$; *furthermore,*

$$\|f*g\|_p \leqslant \|f\|_1\|g\|_p.$$

In particular, L^1 *is a convolution algebra.*

(iv): L^1 *operates continuously in* L^p $(1 \leqslant p \leqslant \infty)$, C^∞, UC^∞ *and* C_0;

(v): $D^k*L^1_{loc} \subset E^k$ *(continuously)*, $0 \leqslant k \leqslant \infty$;

(vi): L^1_c *is a convolution algebra;*

(vii): *L_c^1 operates continuously in L^p_{loc}, L^p_c $(1 \leq p \leq \infty)$ $\mathcal{E}^k$ and $\mathcal{D}^k$;*

(viii): *$\mathcal{D}^k$ is a convolution algebra.*

Lastly, we have the formula

$$D^s(f*g) = f*D^s g,$$

whenever $f \in L^1_{loc}$, $g \in \mathcal{D}^k$ or $f \in L^1_c$, $g \in \mathcal{E}^k$ and $|s| \leq k$.

REMARK: Since $\mathcal{D}^\infty \to L^1_c \to L^1$ (continuously), the convolution algebras $\mathcal{D}^\infty$ and L^1_c operate continuously in all the function spaces which have previously been defined.

NOTATION-PROPOSITION: *For every function f defined on $\mathbb{R}^n$ we set:*

$$\check{f}(x) \equiv f(-x), \qquad \tilde{f}(x) \equiv \overline{f(-x)}.$$

When f and g are convolvable, so are $\check{f}$ and $\check{g}$ (resp. $\tilde{f}$ and $\tilde{g}$), and

$$(f*g)^\vee = \check{f}*\check{g}, \qquad (f*g)^\sim = \tilde{f}*\tilde{g}.$$

Furthermore:

NOTATION-PROPOSITION: *If $g \in L^p$, $h \in L^q$, $\frac{1}{p} + \frac{1}{q} = 1$, we set*

$$(g,h) \equiv \int gh = (g*\check{h})(0)$$

$$(g|h) \equiv \int g\bar{h} = (g*\tilde{h})(0).$$

When $f \in L^1$, $g \in L^p$, $h \in L^q$, $\frac{1}{p} + \frac{1}{q} = 1$,

$$(f*g,h) = (g,\check{f}*h), \qquad (f*g|h) = (g|\tilde{f}*h).$$

NOTATION: We *denote* by $L^p(\mathbb{T})$ the set of measurable functions on $\mathbb{R}$ that have period 2π and are such that

$$\|f\|_p = \left(\frac{1}{2\pi}\int_0^{2\pi} |f|^p\right)^{1/p} < \infty, \qquad 1 \leqslant p < \infty,$$

$$\|f\|_\infty = \operatorname*{ess\,sup}_{0\leqslant x\leqslant 2\pi} |f(x)| < \infty.$$

The set of k-fold continuously differentiable functions on $\mathbb{R}$ with period 2π is *denoted* $E^k(\mathbb{T})$ (here $0 \leqslant k \leqslant \infty$, and $E^0(\mathbb{T})$ is also denoted $C(\mathbb{T})$).

DEFINITION: *The* CONVOLUTION PRODUCT OF TWO MEASURABLE FUNCTIONS *f and g* WITH PERIOD 2π *is defined by the formula:*

$$(f*g)(x) = \frac{1}{2\pi}\int_0^{2\pi} f(x - y)g(y)dy.$$

Defining $f_i \to f$ in $E^k(\mathbb{T})$ to mean that $f_i^{(s)} \to f^{(s)}$ uniformly for every integer s with $0 \leqslant s \leqslant k$, the following hold:

$$L^p(\mathbb{T})*L^q(\mathbb{T}) \subset C(\mathbb{T}) \quad \text{(continuously), if } \frac{1}{p} + \frac{1}{q} = 1,$$

$$L^1(\mathbb{T})*L^p(\mathbb{T}) \subset L^p(\mathbb{T}) \quad \text{(continuously)},$$

$$L'(\mathbb{T})*E^k(\mathbb{T}) \subset E^k(\mathbb{T}) \quad \text{(continuously)},$$

as do the inequalities:

$$\|f*g\|_\infty \leqslant \|f\|_p\|g\|_q, \quad \text{if } f \in L^p(\mathbb{T}),\ g \in L^q(\mathbb{T}),\ \frac{1}{p} + \frac{1}{q} = 1,$$

$$\|f*g\|_p \leqslant \|f\|_1\|g\|_p, \quad \text{if } f \in L'(\mathbb{T}),\ g \in L^p(\mathbb{T}).$$

By defining $\check{f}, \tilde{f}$ as above, and

$$(f,g) \equiv \frac{1}{2\pi}\int_0^{2\pi} fg, \qquad (f|g) \equiv \frac{1}{2\pi}\int_0^{2\pi} f\bar{g},$$

one obtains the same formulae as above. Let us note that in the formulae defining $f*g$, (f,g) and $(f|g)$, one can replace the range of integration $(0,2\pi)$ by any interval of length 2π.

Clearly one can consider functions having an arbitrary period $T > 0$, it is then just a matter of replacing π by $T/2$ everywhere.

0·8 REGULARISATION OF FUNCTIONS

DEFINITION: *One calls an approximate identity in* L^1 *every sequence* (φ_i) *of integrable functions that satisfies the following conditions:*

(i): *There exists a constant* M *such that* $\|\varphi_i\|_1 \leqslant M$ *for all* i;

(ii): $\lim_i \int \varphi_i = 1$;

(iii): *For every* $a > 0$, $\lim_i \int_{|x| \geqslant a} \varphi_i = 0$.

An approximate identity (φ_i) *is said to be compact if all the functions* φ_i *vanish outside the same compact set of* $\mathbb{R}^n$.

In L^1 *there exist compact approximate identities consisting of functions belonging to* $\mathcal{D}^\infty$; *these are called* REGULARISING SEQUENCES.

THEOREM: *Let* (φ_i) *be an approximate identity in* L^1. *If* E *is one of the spaces* L^p $(1 \leqslant k < \infty)$ *or* $\mathcal{UC}$, *then for every function* $f \in E$ *one has* $\varphi_i * f \to f$ *in* E.

If the approximate identity (φ_i) *is compact this property extends to the spaces* L^p_{loc}, L^p_c $(1 \leqslant p < \infty)$, $\mathcal{E}^k, \mathcal{D}^k$ $(0 \leqslant k \leqslant \infty)$.

From this one deduces the following corollaries:

THEOREM: (Density): $\mathcal{D}^\infty$ *is dense in each of the spaces* L^p, L^p_{loc}, L^p_c $(1 \leqslant p < \infty)$, $\mathcal{E}^k, \mathcal{D}^k$.

LEMMA: (Calculus of Variations): *If* $f \in L^1_{\text{loc}}$ *is such that* $\int f\varphi = 0$ *for any* $\varphi \in \mathcal{D}^\infty$, *then* $f = 0$ *almost everywhere.*

DEFINITION: *An approximate identity in* $L^1(\mathbb{T})$ *is a sequence* (φ_i) *of integrable functions with period* 2π, *such that*

(i): $\|\varphi_i\|_1 \leqslant M$;

(ii): $\lim_i \dfrac{1}{2\pi}\displaystyle\int_{-\pi}^{\pi} \varphi_i = 1$;

(iii): *For all* a, $0 < a < \pi$, $\lim_i \displaystyle\int_{a \leqslant |x| \leqslant \pi} \varphi_i(x)dx = 0$.

If E denotes one of the spaces $L^p(\mathbb{T})$ $(1 \leqslant p < \infty)$ or $E^k(\mathbb{T})$ $(0 \leqslant k \leqslant \infty)$, then $\varphi_i * f \to f$ in E for every function $f \in E$. In $L'(\mathbb{T})$ there exist approximate identities consisting of functions belonging to $E^\infty(\mathbb{T})$; from this it follows that $E^\infty(\mathbb{T})$ is dense in each of the spaces $L^p(\mathbb{T})$, $1 \leqslant p < \infty$ and $E^k(\mathbb{T})$.

0·9 FOURIER TRANSFORMATIONS

DEFINITION: *For every* $f \in L^1$ $(= L^1(\mathbb{R}))$ *we give the name* FOURIER TRANSFORM *of* f *to the function*

$$\mathcal{F}f(y) \equiv \int_{-\infty}^{+\infty} e^{-2\pi ixy} f(x)dx, \qquad y \in \mathbb{R}.$$

The symbol $\hat{f}$ *is also employed to designate* $\mathcal{F}f$. *The function*

$$\bar{\mathcal{F}}f(y) \equiv \int_{-\infty}^{+\infty} e^{2\pi ixy} f(x)dx, \qquad y \in \mathbb{R},$$

is called the ADJOINT FOURIER TRANSFORM *of* f.

Upon setting

$$e_a(x) = e^{2\pi iax}, \qquad a \in \mathbb{R},\ x \in \mathbb{R}.$$

one obtains the following Formulas, where f and g belong to L^1:

$$f*e_a = \hat{f}(a)e_a, \tag{0 9.1}$$

$$F(\tilde{f}) = \overline{Ff}, \qquad \bar{F}(\tilde{f}) = \overline{\bar{F}(f)}, \tag{0 9.2}$$

$$F(\bar{f}) = (Ff)^{\sim}, \qquad \bar{F}(\bar{f}) = (\bar{F}f)^{\sim}, \tag{0·9.3}$$

$$F(f*g) = Ff.Fg, \qquad \bar{F}(f*g) = \bar{F}f.\bar{F}g, \tag{0·9.4}$$

$$F(f_a) = e_{-a}F(f), \qquad \bar{F}(f_a) = e_a\bar{F}(f), \tag{0·9.5}$$

$$F(e_af) = (Ff)_a, \qquad \bar{F}(e_af) = (\bar{F}f)_{-a}, \tag{0·9.6}$$

$$(Ff|g) = (f|\bar{F}g), \tag{0·9.7}$$

$$f*\bar{F}g = \bar{F}(Ff.g). \tag{0·9.8}$$

Furthermore, if f is piecewise continuously differentiable, and if f and f' belong to L^1, then

$$\hat{f'}(y) = 2\pi iy\hat{f}(y).$$

Whenever f and $x \mapsto g(x) = xf(x)$ belong to L^1, $\hat{f}$ is continuously differentiable and

$$\hat{g}(y) = -\frac{1}{2\pi i}\hat{f}'(y).$$

THEOREM: (Riemann-Lebesgue): F *is a linear mapping of* L^1 *into* C_0 *such that* $\|Ff\|_\infty \leqslant \|f\|_1$.

THEOREM: (Fourier's Inversion): *If* f *and* Ff *belong to* L^1, *then* $f = \bar{F}(Ff)$ *almost everywhere. In particular, if* $Ff = 0$, *then* $f = 0$ *almost everywhere.*

THEOREM: (Fourier-Plancherel): *If* $f \in L^1 \cap L^2$, *then* $\|Ff\|_2 = \|f\|_2$.

It follows from the Fourier and Fourier-Plancherel Theorems that the restriction of F and $\bar{F}$ to $L^1 \cap L^2$ extend in an unique way into two isometries of L^2 onto itself, which are inverses of each other; these will still be denoted F and $\bar{F}$. If $f \in L^2$ and $\hat{f} = Ff$, then

$$\hat{f}(y) = \lim_{M\to\infty} \int_{-M}^{M} e^{-2\pi ixy} f(x)dx,$$

$$f(x) = \lim_{M\to-\infty} \int_{-M}^{M} e^{2\pi ixy} \hat{f}(y)dy,$$

in the sense of convergence in L^2. Formulas (2),(3),(5),(6),(7) are still valid for $f,g \in L^2$; Formula (4) generalises to the case where $f \in L^1$ and $g \in L^2$; and Formula (8) can be replaced by

$$F(fg) = Ff * Fg$$

whenever $f,g \in L^2$. Lastly, let us note that if $f \in L^1$ and $Ff \in L^2$, then $f \in L^2$.

0·10 INTEGRATION AND DIFFERENTIATION

NOTATION: We provide $\mathbb{R}^n$ with the norm $|x| \equiv \text{Max}(|x_1|,\ldots,|x_n|)$. By the letter C we denote a *CUBE containing the origin*, that is to say, a set such that $0 \in C$ and $\bar{C} = \{x : |x - x_0| \leqslant a\}$, where $x_0 \in \mathbb{R}^n$ and $a > 0$. The number $2a$ is called the *diameter of the cube* C and it will be denoted by $\delta(C)$.

THEOREM: (Lebesgue): *Let* $f \in L^1_{loc}(\mathbb{R}^n)$. *There exists a negligeable set* N *of* $\mathbb{R}^n$ *such that if* $x \notin N$ *then for every complex number* α *one has*

$$\lim_{\delta(C)\to 0} \frac{1}{\text{meas}(C)} \int_C |f(x + t) - \alpha| dt = |f(x) - \alpha|.$$

In particular, for almost all x,

$$f(x) = \lim_{\delta(C)\to 0} \frac{1}{\mathrm{meas}(C)} \int_C f(x + t)dt.$$

DEFINITION: *A function f is a FUNCTION of BOUNDED VARIATION ON AN INTERVAL $[a,b]$ if*

$$V(f;a,b) = \sup_{\Delta} \sum_{i=0}^{n-1} |f(x_{i+1}) - f(x_i)| < \infty,$$

the supremum being taken over all decompositions

$$\Delta = (a = x_0 < x_1 < \cdots < x_n = b)$$

of the interval $[a,b]$.

If f is of bounded variation on $[a,b]$ and on $[b,c]$, then it is on $[a,c]$, and

$$V(f;a,c) = V(f;a,b) + V(f;b,c).$$

THEOREM: (Jordan): *A real function is of bounded variation on $[a,b]$ if and only if it is the difference of two increasing functions on $[a,b]$.*

A complex function is of bounded variation if its real and imaginary parts are. Every function f with bounded variation is regular, and upon setting $V(x) = V(f;a,x)$, $a \leqslant x \leqslant b$, one has

$$V(x + 0) - V(x) = |f(x + 0) - f(x)|,$$

$$V(x) - V(x - 0) = |f(x) - f(x - 0)|,$$

at every point where these symbols have a meaning. Therefore, in Jordan's Theorem, if f is real, continuous, and of bounded variation on $[a,b]$, then on this interval it is the difference of two continuous increasing functions. Jordan's Theorem extends to real functions defined on an open interval I and with bounded variation on every compact interval contained in I.

DEFINITION: *A function f is said to be a* FUNCTION WITH BOUNDED VARIATION *on* $\mathbb{R}$ *if* $V(f;0,x)$ *and* $V(f;-x,0)$ *have finite limits when* $x \mapsto \infty$*; the sum of these limits is called the* VARIATION OF THE FUNCTION f *on* $\mathbb{R}$.

THEOREM: (Lebesgue): *Every function with bounded variation on* $[a,b]$ *is differentiable almost everywhere, its derivative is integrable, and*

$$\int_a^b |f'(x)|\,dx \leqslant V(f;a,b).$$

DEFINITION: *A function f is called* ABSOLUTELY CONTINUOUS *on* $[a,b]$ *if, for all* $\varepsilon > 0$*, there exists* $\delta > 0$ *such that for every finite sequence of mutually disjoint sub-intervals* $]\alpha_i,\beta_i[$ *of* $[a,b]$ *one has*

$$\sum_i |f(\beta_i) - f(\alpha_i)| < \varepsilon \quad \textit{whenever} \quad \sum_i (\beta_i - \alpha_i) < \delta.$$

THEOREM: (Lebesgue): *If f is absolutely continuous on* $[a,b]$ *f is differentiable almost everywhere, f' is integrable and*

$$f(b) - f(a) = \int_a^b f'(x)\,dx,$$

$$V(f;a,b) = \int_a^b |f'(x)|\,dx.$$

Conversely, if F is integrable on $[a,b]$*, and if*

$$f(x) = \int_a^x F(t)\,dt, \qquad a \leqslant x \leqslant b,$$

then f is absolutely continuous and $f' = F$ *almost everywhere.*

The theory of differentiation makes use of the two following Lemmas, which are interesting in their own right.

LEMMA (1): *Let K be a compact set of* $\mathbb{R}^n$ *covered by a family of open cubes. From this family there can be chosen a finite sequence* $C_1,\dots,C_p$ *of mutually disjoint cubes such that*

$$\operatorname{meas}(K) \leqslant 3^n \sum_{k=1}^{p} \operatorname{meas}(C_p).$$

LEMMA (2): (The Setting Sun Lemma): *Let f be a real continuous function on* $[a,b]$, *E the set of points x of this interval for which there exists a y such that* $x < y < b$ *and* $f(x) < f(y)$. *Then the set E is the disjoint union of a sequence of intervals with end points* $a_n < b_n$ *such that* $f(a_n) \leqslant f(b_n)$.

0·11 TRIGONOMETRIC SERIES

If $(u_n)_{n\in\mathbb{Z}}$ is a sequence of complex numbers indexed by some positive or negative integers, one sets

$$\sum_{-\infty}^{+\infty} u_n = u_0 + \sum_{n=1}^{\infty} (u_n + u_{-n}) = \lim_{N\to\infty} \sum_{n=-N}^{N} u_n,$$

when this last limit exists. When $\sum_{-\infty}^{+\infty} |u_n| < \infty$ one also has

$$\sum_{-\infty}^{+\infty} u_n = \lim_{\substack{N\to\infty \\ M\to\infty}} \sum_{n=-N}^{M} u_n.$$

A trigonometric series is a formal series of the type

$$\sum_{-\infty}^{+\infty} c_n e^{inx} = \frac{a_0}{2} + \sum_{n=1}^{\infty} (a_n \cos nx + b_n \sin nx),$$

where, upon agreeing to set $b_0 = 0$,

$$\begin{aligned} a_n &= c_n + c_{-n}, & c_n &= \tfrac{1}{2}(a_n - ib_n), \\ & & & & n \geq 0. \\ b_n &= i(c_n - c_{-n}), & c_{-n} &= \tfrac{1}{2}(a_n + ib_n), \end{aligned}$$

In what follows it is assumed that $f \in L^1(\mathbb{T})$ ($L^1(\mathbb{T})$ has been defined in Section 0·7). For every $n \in \mathbb{Z}$ we set

$$\hat{f}(n) = \frac{1}{2\pi}\int_0^{2\pi} e^{-inx} f(x)dx.$$

The $\hat{f}(n)$ are called the *FOURIER COEFFICIENTS of f*, and the formal series

$$\sum_{-\infty}^{+\infty} \hat{f}(n)e^{inx}$$

is the *FOURIER SERIES of f*. We use the notation

$$f(x) \sim \sum_{-\infty}^{+\infty} c_n e^{inx}$$

to indicate that the second member is the Fourier series of f, that is to say that $c_n = \hat{f}(n)$ for every $n \in \mathbb{Z}$. It will be noted that:

$$a_n = \frac{1}{\pi}\int_0^{2\pi} f(x)\cos nx dx, \qquad b_n = \frac{1}{\pi}\int_0^{2\pi} f(x)\sin nx dx.$$

On setting $e_n(x) = e^{inx}$ we have the following *FORMULAS*

$$\hat{f}(n) = (f|e_n), \tag{0·11.1}$$

$$f*e_n = \hat{f}(n)e_n, \tag{0·11.2}$$

$$\hat{\bar{f}}(n) = \overline{\hat{f}(n)} \quad \text{and} \quad \hat{\check{f}}(n) = \hat{f}(-n), \tag{0·11.3}$$

$$\widehat{f*g}(n) = \hat{f}(n)\hat{g}(n), \tag{0·11.4}$$

$$\widehat{f'}(n) = in\hat{f}(n) \quad \text{if } f \text{ is absolutely continuous.} \tag{0·11.5}$$

For the explicit calculation of the Fourier series of a function f the following result, which generalises (5) in the Formulas above, is useful:

PROPOSITION: *If f is a piecewise continuously differentiable function, that is to say, if there exist points*

$$-\pi \leqslant a_1 < a_2 < \cdots < a_p < \pi$$

such that f coincides on each open interval $]a_s, a_{s+1}[$, $1 \leqslant s \leqslant p$ and $a_{p+1} = a_1 + 2\pi$, with the restriction of a function f_s continuously differentiable on the closed interval $[a_s, a_{s+1}]$, upon denoting by $[f']$ a function with period 2π that coincides with f'_s on each $]a_s, a_{s+1}[$, one has:

$$in\hat{f}(n) = \widehat{[f']}(n) + \frac{1}{2\pi}\sum_{s=1}^{p}\{f(a_s+0) - f(a_s-0)\}e^{-ina_s}.$$

EXAMPLE 1. If $f(x) = x$ when $-\pi < x < \pi$,

$$f(x) \sim 2\sum_{n=1}^{\infty}(-1)^{n+1}\frac{\sin nx}{n}.$$

EXAMPLE 2. If $f(x) = -1$ when $-\pi < x < 0$ and $f(x) = 1$ when $0 < x < \pi$,

$$f(x) \sim \frac{4}{\pi} \sum_{n=0}^{\infty} \frac{\sin(2n + 1)x}{2n + 1}.$$

EXAMPLE 3. If $f(x) = \pi - |x|$ when $-\pi \leqslant x \leqslant \pi$,

$$f(x) \sim \frac{\pi}{2} + \frac{4}{\pi} \sum_{n=0}^{\infty} \frac{\cos(2n + 1)x}{(2n + 1)^2}.$$

THEOREM: (Riemann-Lebesgue): $\lim_{|n|\to\infty} \hat{f}(n) = 0$.

THEOREM: (Fourier): *If* $\hat{f}(n) = 0$ *for all* $n \in \mathbb{Z}$, *then* $f = 0$ *almost everywhere.*

If f is p-fold continuously differentiable, $\hat{f}(n) = o(|n|^{-p})$; if f has bounded variation on $[0,2\pi]$, $\hat{f}(n) = O(1/|n|)$. Conversely, if for an integer p one has $\sum |n|^p|\hat{f}(n)| < \infty$, then f coincides almost everywhere with a p-fold continuously differentiable function.

NOTATION: We set:

$$s_N(f;x) = \sum_{n=-N}^{N} \hat{f}(n)e^{inx} \qquad \text{(FOURIER SUMS)},$$

$$\sigma_N(f;x) = \frac{s_0(f;x) + \cdots + s_N(f;x)}{N + 1} \qquad \text{(FÉJER SUMS)},$$

then:

$$s_N(f;x) = (f*D_N)(x), \qquad \sigma_N(f;x) = (f*F_N)(x),$$

with

$$D_N(x) = \frac{\sin(N + \frac{1}{2})x}{\sin\frac{1}{2}x} \qquad \text{(DIRICHLET'S KERNEL)},$$

$$F_N(x) = \frac{1}{N + 1}\left(\frac{\sin(N + 1)\frac{1}{2}x}{\sin\frac{1}{2}x}\right)^2 \qquad \text{(FEJER'S KERNEL)}.$$

THEOREM: (Localisation): *If $f(x + 0)$ and $f(x - 0)$ exist at a point x, we set*

$$\varphi_x(t) = f(x + t) + f(x - t) - f(x + 0) - f(x - 0).$$

For every $0 \leqslant \delta < \pi$ one has:

$$s_N(f;x) - \tfrac{1}{2}(f(x + 0) + f(x - 0))$$

$$= \frac{1}{\pi}\int_0^\delta \varphi_x(t)\,\frac{\sin(N + \frac{1}{2})t}{t}\,dt + \varepsilon_N(x,\delta),$$

with

$$\lim_{N\to\infty} \varepsilon_N(x,\delta) = 0. \qquad (0\cdot 11.6)$$

Furthermore, if f is continuous on $I =]a,b[$, the convergence in $(0\cdot 11.6)$ is uniform for x belonging to a compact set contained in I.

THEOREM: (Jordan-Dirichlet): *If f is of bounded variation on every compact interval contained in an open interval I, its Fourier series converges at every point x of I to $\frac{1}{2}(f(x + 0) + f(x - 0))$.*

Furthermore, if f is continuous on I, its Fourier series converges uniformly towards f on every compact set of I.

If f is of bounded variation on $[0,2\pi]$ there exists a constant A such that $|s_N(f;x)| \leqslant A$ for every integer $N \geqslant 0$ and for all $x \in \mathbb{R}$.

THEOREM: *If one of the functions f or g is of bounded variation on* $[0,2\pi]$, *then:*

$$\frac{1}{2\pi}\int_0^{2\pi} f(x)g(x)dx = \sum_{n=-\infty}^{\infty} \hat{f}(n)\hat{g}(-n)$$

$$= \sum_{n=-\infty}^{\infty} \hat{f}(n)\int_0^{2\pi} e^{inx}g(x)dx.$$

In other words: *In order to integrate f with respect to* $g(x)dx$ *one can integrate its Fourier series term by term.*

THEOREM: (Féjer): *If* $f(x + 0)$ *and* $f(x - 0)$ *exist at a point x, then*

$$\lim_N \sigma_N(f;x) = \tfrac{1}{2}(f(x + 0) + f(x - 0)).$$

Whenever f is continuous on a compact interval, one has $\sigma_N(f;x) \to f(x)$ *uniformly on this interval.*

The Féjer kernels form an approximate identity, so that if $f \in L^p(\mathbb{T})$, $1 \leqslant p < \infty$, then $\sigma_N(f) \to f$ in $L^p(\mathbb{T})$.

THEOREM: (Féjer-Lebesgue):

$$\lim_{N\to\infty} \sigma_N(f;x) = f(x) \qquad (0\cdot11.7)$$

at every point x such that

$$\lim_{h\to 0} \frac{1}{h}\int_0^h |f(x + t) - f(x)|dt = 0.$$

In particular, Equation (0·11.7) *is true for almost all x.*

THEOREM: (Plancherel): *In order that* $f \in L^2(\mathbb{T})$ *it is necessary and sufficient that*

$$\sum_{-\infty}^{+\infty} |\hat{f}(n)|^2 < \infty,$$

and in this case

$$\frac{1}{2\pi}\int_0^{2\pi} |f(x)|^2 dx = \sum_{-\infty}^{+\infty} |\hat{f}(n)|^2.$$

NOTE

In the statements of the exercises in the following chapter, the functions considered are always complex valued when no further indication is given. Similarly, unless otherwise indicated, the sets considered are measurable sets of $\mathbb{R}^n$.

CHAPTER 1
Measurable Sets

EXERCISE 1·1: Let $E_1,\dots,E_n$ be a finite sequence of sets of finite measure. For every integer p $(1 \leqslant p \leqslant n)$ set

$$\sigma_p = \sum_{i_1<\cdots<i_p} \text{meas}(E_{i_1} \cap \cdots \cap E_{i_p}).$$

(a): Show that

$$\text{meas}(E_1 \cup \cdots \cup E_n) = \sum_{p=1}^{n} (-1)^{p-1}\sigma_p \qquad (\textit{POINCARE'S FORMULA}).$$

(b): For every integer s $(1 \leqslant s \leqslant n)$ we *denote* by G_s the set of points which belong to exactly s of the sets $E_1,\dots,E_n$. Show that

$$\text{meas}(G_s) = \sum_{p=s}^{n} (-1)^{p-s}\binom{p}{s}\sigma_p.$$

(c): Let H_s be the set of points which belong to at least s of the sets $E_1,\dots,E_n$. Show that

$$\text{meas}(H_s) = \sum_{p=s}^{n} (-1)^{p-s} \binom{p-1}{s-1} \sigma_p.$$

ΔVΔ = VΔV = ΔVΔ = VΔV = ΔVΔ

SOLUTION: For every non-empty subset A of $\{1,\ldots,n\}$ let us set

$$E_A = \bigcap_{i \in A} E_i, \qquad E'_A = E_A - \bigcup_{i \notin A} E_i.$$

The E'_A are mutually disjoint and it is clear that

$$E_A = \bigcup_{B \supset A} E'_B,$$

$$(-1)^p \sigma_p = \sum_{\text{Card}A=p} (-1)^{\text{Card}A} \text{meas}(E_A).$$

SOLUTION: (a): One has:

$$\sum_{p=1}^{n} (-1)^p \sigma_p = \sum_{A} (-1)^{\text{Card}A} \text{meas}(E_A) = \sum_{A} (-1)^{\text{Card}A} \sum_{B \supset A} \text{meas}(E'_B)$$

$$= \sum_{B} \text{meas}(E'_B) \sum_{A \subset B} (-1)^{\text{Card}A}.$$

Note that if $p = \text{Card}B$, then

$$\sum_{A \subset B} (-1)^{\text{Card}A} = \sum_{r=1}^{p} (-1)^r \binom{p}{r} = -1.$$

From this it follows that

$$\sum_{p=1}^{n} (-1)^p \sigma_p = -\sum_{B} \text{meas}(E'_B) = -\text{meas}(E_1 \cup \cdots \cup E_n).$$

SOLUTION: (b): One has:

$$\sum_{p=s}^{n} (-1)^{p}\binom{p}{s}\sigma_p = \sum_{\mathrm{Card}A\geqslant s} (-1)^{\mathrm{Card}A}\binom{\mathrm{Card}A}{s}\mathrm{meas}(E_A)$$

$$= \sum_{\mathrm{Card}A\geqslant s} (-1)^{\mathrm{Card}A}\binom{\mathrm{Card}A}{s} \sum_{B\supset A} \mathrm{meas}(E'_B)$$

$$= \sum_{\mathrm{Card}B\geqslant s} \mathrm{meas}(E'_B) \sum_{\substack{A\subset B\\ \mathrm{Card}A\geqslant s}} (-1)^{\mathrm{Card}A}\binom{\mathrm{Card}A}{s} .$$

On setting $\mathrm{Card}B = p \geqslant s$ again, one has:

$$\sum_{\substack{A\subset B\\ \mathrm{Card}A\geqslant s}} (-1)^{\mathrm{Card}A}\binom{\mathrm{Card}A}{s} = \sum_{r=s}^{p} (-1)^{r}\binom{r}{s}\binom{p}{r}$$

$$= \binom{p}{s} \sum_{r=s}^{p} (-1)^{r}\binom{p-s}{r-s}$$

$$= (-1)^{s}\binom{p}{s} \sum_{r=0}^{p-s} (-1)^{r}\binom{p-s}{r}$$

$$= \begin{cases} 0 & \text{if } p > s, \\ (-1)^{s} & \text{if } p = s. \end{cases}$$

From this it results that:

$$\sum_{p=s}^{n} (-1)^{p}\binom{p}{s}\sigma_p = (-1)^{s} \sum_{\mathrm{Card}B=s} \mathrm{meas}(E'_s) = (-1)^{s}\mathrm{meas}(G_s).$$

SOLUTION: (c): The formula is true for $s = 1$ (for this is none other than that obtained in (a)). Also,

$$H_{s+1} = H_s - G_s,$$

whence, proceeding by induction and taking account of (b),

$$\text{meas}(H_{s+1}) = \sum_{p=s}^{n} (-1)^{p-s-1}\left\{\binom{p}{s} - \binom{p-1}{s-1}\right\}\sigma_p$$

$$= \sum_{p=s+1}^{n} (-1)^{p-s-1}\binom{p-1}{s}\sigma_p.$$

EXERCISE 1·2: For all $\varepsilon > 0$ construct an open set U everywhere dense in $\mathbb{R}$, and such that $\text{meas}(U) < \varepsilon$.

Δ∇Δ = ∇Δ∇ = Δ∇Δ = ∇Δ∇ = Δ∇Δ

SOLUTION: Let (r_n) be the sequence of rational numbers, and for all n let us denote by I_n the open interval with centre r_n and length $\varepsilon 2^{-n}$. The union U of the I_n is an everywhere dense open set of $\mathbb{R}$ (for it contains all the rational numbers) and furthermore,

$$\text{meas}(U) \leqslant \sum_{n=1}^{\infty} \varepsilon 2^{-n} = \varepsilon.$$

EXERCISE 1·3: Let (E_n) be a sequence of measurable sets such that

$$\sum_{n} \text{meas}(E_n) < \infty.$$

Show that the set of points which belong to an infinity of E_n's has measure zero (The Borel-Cantelli Lemma).

Δ∇Δ = ∇Δ∇ = Δ∇Δ = ∇Δ∇ = Δ∇Δ

FIRST SOLUTION: The set considered is

$$A = \bigcap_{n=1}^{\infty} \left(\bigcup_{p \geqslant n} E_p \right)$$

For every integer n one has

$$\text{meas}\left(\bigcup_{p \geqslant n} E_p \right) \leqslant \sum_{p \geqslant n} \text{meas}(E_p),$$

and consequently

$$\text{meas}(A) = \lim_n \text{meas}\left(\bigcup_{p \geqslant n} E_p \right)$$

$$\leqslant \lim_n \sum_{p \geqslant n} \text{meas}(E_p) = 0.$$

SECOND SOLUTION: Let φ_n be the characteristic function of E_n. By virtue of the theorem on the term by term integration of series of positive functions, one has

$$\int \sum \varphi_n = \sum \int \varphi_n = \sum \text{meas}(E_n) < \infty.$$

From this it follows that the set of points where $\sum \varphi_n = \infty$ has measure zero. But this set if precisely A.

EXERCISE 1·4: Let (E_n) be a sequence of measurable sets such that

$$\sum_n \text{meas}(E_n) < \infty.$$

For every integer s, H_s is denoted as the set of points which belong to at least s of the sets E_n. Show that

$$\text{meas}(H_s) \leqslant \frac{1}{s} \sum_n \text{meas}(E_n).$$

FIRST SOLUTION: First, suppose we had a finite sequence $E_1, \ldots, E_n$ and for every non-empty subset A of $(1,\ldots,n)$ let us denote by E'_A the set of points belonging to the E_i for which $i \in A$ and not belonging to those for which $i \notin A$ (in Exercise 1). Then

$$s\,\mathrm{meas}(H_s) = \sum_{\mathrm{Card}A \geqslant s} s\,\mathrm{meas}(E'_A)$$

$$\leqslant \sum_{\mathrm{Card}A \geqslant 1} \mathrm{Card}A\,\mathrm{meas}(E'_A) = \sum_{i=1}^{n} \mathrm{meas}(E_i).$$

In the general case one considers, for $n \geqslant s$, the sets $H_{s,n}$ formed by the points which belong to at least s of the sets $E_1, \ldots, E_n$. It is clear that $H_{s,n} \subset H_{s,n+1}$ and that

$$H_s = \bigcup_{n=s}^{\infty} H_{s,n}.$$

Consequently

$$\mathrm{meas}(H_s) = \lim_n \mathrm{meas}(H_{s,n})$$

$$\leqslant \frac{1}{s} \lim_n \sum_{i=1}^{n} \mathrm{meas}(E_i) = \frac{1}{s} \sum_{i=1}^{\infty} \mathrm{meas}(E_i).$$

SECOND SOLUTION: Let φ_n be the characteristic function of E_n and $\varphi = \sum \varphi_n$. One has

$$\sum \mathrm{meas}(E_n) = \int \varphi \geqslant \int_{H_s} \varphi \geqslant s\,\mathrm{meas}(H_s).$$

EXERCISE 1·5: For every finite sequence $E_1, \ldots, E_n$ of measurable sets one sets

$$D(E_1,\ldots,E_n) = \left(\bigcup_{i=1}^{n} E_i\right) - \left(\bigcap_{i=1}^{n} E_i\right),$$

and if the E_i are not all negligeable,

$$\sigma(E_1,\ldots,E_n) = \frac{\text{meas}(D(E_1,\ldots,E_n)}{\text{meas}(E_1 \cup \cdots \cup E_n)}.$$

Show that if none of the E_i is negligeable one has

$$\sigma(F_1,\ldots,F_n) \leqslant \frac{1}{n-1} \sum_{i<j} \sigma(E_i,E_j).$$

Δ∇Δ = ∇Δ∇ = ΔVΔ = ∇Δ∇ = Δ∇Δ

SOLUTION: Let us show, by induction on n, that if $x \in D(E_1,\ldots,E_n)$ then $x \in D(E_i,E_j)$ for at least $n - 1$ pairs $i < j$. This is clear if $n = 2$. Let us assume that it be true up to $n - 1$ and that $x \in E_2 \cup \cdots \cup E_n$, $x \in E_1$. One then has $x \in D(E_i,E_j)$ for at least n 2 pairs such that $2 \leqslant i < j \leqslant n$ and for at least one pair $(1,i)$ with $i \geqslant 2$.

From this it follows, by Exercise 4, that

$$\text{meas}(D(E_1,\ldots,E_n) \leqslant \frac{1}{n-1} \sum_{i<j} \text{meas}(D(E_i,E_j)).$$

Since

$$\text{meas}(E_1 \cup \cdots \cup E_n) \geqslant \text{meas}(E_i \cup E_j),$$

from this one deduces that

$$\sigma(E_1,\ldots,E_n) \leqslant \frac{1}{n-1} \sum_{i<j} \frac{\text{meas}(D(E_i,E_j)}{\text{meas}(E_1 \cup \cdots \cup E_n)}$$

$$\leqslant \frac{1}{n-1} \sum_{i<j} \sigma(E_i,E_j).$$

EXERCISE 1·6: Consider a sequence (E_n) of measurable sets such that

$$\text{meas}\left(\bigcup_n E_n\right) < \infty$$

and

$$\inf_n \text{meas}(E_n) = \alpha \geqslant 0.$$

Show that the set A of points that belong to an infinity of sets E_n is measurable and that meas$(A) > \alpha$.

Δ∇Δ = ∇Δ∇ = Δ∇Δ = ∇Δ∇ = ΔΔ∇Δ

SOLUTION: We have

$$A = \bigcap_{n=1}^{\infty} \left(\bigcup_{s=n}^{\infty} E_s\right),$$

which shows that A is measurable. Now we also have

$$\text{meas}\left(\bigcup_{s=n}^{\infty} E_s\right) \geqslant \text{meas}(E_n) \geqslant \alpha.$$

As the sets $\bigcup_{s \geqslant n} E_s$ form a decreasing sequence and have finite measure, we have

$$\text{meas}(A) = \lim_n \text{meas}\left(\bigcup_{s=n} E_s\right) > \alpha.$$

REMARK: The first condition is essential; for example, consider the sequence $E_n = [n,n + 1[$.

EXERCISE 1·7: Let A be the set of real numbers x for which there exists an infinite number of pairs (p,q) of integers such that $q \geqslant 1$ and

$$\left|x - \frac{p}{q}\right| \leqslant \frac{1}{q^3} .$$

Show that A is negligeable.

Δ∇Δ = ∇Δ∇ = ΔVΔ = ∇Δ∇ = ΔVΔ

SOLUTION: Since

$$x + n - \frac{p + nq}{q} = x - \frac{p}{q} ,$$

by setting

$$B = A \cap [0,1]$$

from the former relation one deduces that

$$A = \bigcup_{n=-\infty}^{+\infty} (B + n).$$

Hence it suffices to prove that B is negligeable.

For p and q integers, and $q \geqslant 1$, let us set

$$I_{p,q} = \left[\frac{p}{q} - \frac{1}{q^3} , \frac{p}{q} + \frac{1}{q^3}\right] ,$$

and let us note that $x \in I_{p,q}$ is equivalent to

(1) $$qx - \frac{1}{q^2} \leqslant p \leqslant qx + \frac{1}{q^2} .$$

Since in an interval of length $2/q^2$ there can be only at most one or three integers according as $q \geqslant 2$ or $q = 1$, from this it results that $x \in B$ if and only if x belongs to an infinite number of sets

$$B_q = [0,1[\cap (\bigcup_p I_{p,q}).$$

By Exercise 1·3 it therefore suffices to show that

$$(2) \quad \sum_q \text{meas}(B_q) < \infty.$$

Now by virtue of Equation (1) above, the integers for which $I_{p,q} \cap [0,1[\neq \emptyset$ are such that

$$-\frac{1}{q^2} \leqslant p \leqslant q + \frac{1}{q^2}.$$

When $q \geqslant 2$ this is equivalent to

$$0 \leqslant p \leqslant q.$$

Consequently

$$\text{meas}(B_q) \leqslant \frac{2(q+1)}{q^3},$$

which proves Equation (2) above.

EXERCISE 1·8: Let (D_n) be a sequence of discs in the plane, of unit diameter and mutually disjoint. The number of discs that are contained within the disc with center 0 and radius n will be denoted $\nu(n)$.

Show that if

$$\liminf_n \frac{\nu(n)}{n^2} = \alpha > 0$$

there exists a half-line issuing from 0 that meets an infinite number of these discs D_n.

Δ∇Δ = ∇Δ∇ = ΔϒΔ = ∇Δ∇ = Δ∇Δ

SOLUTION: Let us consider the cones C_n with vertex 0 generated by the D_n, and let A_n be the intersection of C_n with the circle with center 0 and unit radius. For some integer p and q, $1 \leqslant p < q$, let $A_{p,q}$ be the union of the A_n corresponding to the discs D_n contained in the disc with center 0 and radius q and not contained in that of radius p. Furthermore, let us set

$$B_p = \bigsqcup_{q>p} A_{p,q}.$$

Everything reduces to proving that the B_p, which form a decreasing sequence, have a non-empty intersection. The union of the radii of the disc with center 0 and radius q that meet $A_{p,q}$ covers $(\nu(q) - \nu(p))$ discs of unit diameter and mutually disjoint. From this it follows that if $\text{meas}(A_{p,q})$ denotes the Lebesgue measure (evaluate in radians) of $A_{p,q}$ on the unit circle, one has

$$\tfrac{1}{2}q^2 \text{meas}(A_{p,q}) \geqslant \frac{\pi}{4} (\nu(q) - \nu(p)).$$

Since $A_{p,q} \subset A_{p,q+1}$, one therefore has

$$\text{meas}(B_p) = \lim_q \text{meas}(A_{p,q})$$

$$\geqslant \frac{\pi}{2} \liminf_q \frac{\nu(q) - \nu(p)}{q^2} = \frac{\pi\alpha}{2} .$$

From this it follows that

$$\text{meas}\left(\bigcap_p B_p\right) = \lim_p \text{meas}(B_p) \geqslant \frac{\pi\alpha}{2} ,$$

which proves that the intersection of the B_p is non-empty.

EXERCISE 1·9: Let A be the set of real numbers of the form

$$n + \sum_{p=1}^{\infty} a_p 2^{-2p},$$

where $n \in \mathbb{Z}$ and $a_p = 0$ or 1.

(a): Show that A is negligeable.

(b): From this deduce that there exist two negligeable sets, the sum of which is $\mathbb{R}$.

Δ∇Δ = ∇Δ∇ = Δ∇Δ = ∇Δ∇ = Δ∇Δ

SOLUTION: (a): Let $B = A \cap [0,1[$. Then

$$A = \bigcap_{n=-\infty}^{\infty} (B + n).$$

Hence it suffices to show that B is negligeable. Let u be the function, with unit period, equal to zero on $[0,\frac{1}{2}[$ and to unity on $[\frac{1}{2},1[$. For every integer $p \geqslant 1$ the number

$$a_p(x) = u(2^{p-1}x)$$

is the p-th term in the binary development of $x \in [0,1[$.

Whatever may be the numbers $\varepsilon_1,\ldots,\varepsilon_p$ ($\varepsilon_i = 0$ or 1), the set of x's such that $0 \leqslant x < 1$ and

$$a_1(x) = \varepsilon_1, \quad \ldots, \quad a_p(x) = \varepsilon_p,$$

has measure 2^{-p}. From this it follows that the set B_p of the x's such that $0 \leqslant x < 1$ and

$$a_1(x) = 0, \quad a_3(x) = 0, \quad \ldots, \quad a_{2p-1}(x) = 0,$$

has as its measure

$$2^{-2p+1} \times 2^{p-1} = 2^{-p}.$$

Since B is the intersection of the B_p one certainly has meas$(B) = 0$.

SOLUTION: (b): The set $A' = 2B$ is also negligeable. Now A' is the set of numbers of the type

$$\sum_{p=0}^{\infty} a_p 2^{-(2p+1)}, \qquad a_p = 0 \text{ or } 1.$$

As every real number may be written

$$n + \sum_{p=1}^{\infty} a_p 2^{-p}, \qquad a_p = 0 \text{ or } 1,$$

it follows that $\mathbb{R} = A + A'$.

EXERCISE 1·10: Let f be a complex measurable function on $\mathbb{R}$ such that $f(x + 1) = f(x)$ for almost all x. Show that there exists a function g such that $f = g$ almost everywhere and $g(x + 1) = g(x)$ for all x.

ΔVΔ = VΔV = ΔVΔ = VΔV = ΔVΔ

SOLUTION: Let $E = \{x, f(x) \neq f(x + 1)\}$ and let us set

$$F = \sum_{n=-\infty}^{+\infty} (E + n),$$

as well as

$$g(x) = \begin{cases} f(x) & \text{if } x \notin F, \\ 0 & \text{if } x \in F. \end{cases}$$

It is clear that g has all the properties desired.

EXERCISE 1·11: Let A be a bounded set of $\mathbb{R}^p$ and let

$$\Omega_p = \text{meas}\{x : \|x\| \leqslant 1\},$$

where $\|x\|$ denotes the Euclidian norm of x.

Let $(x_i)_{i\geqslant 1}$ be a sequence of points of A; we set

$$d_n = \inf\{\|x_i - x_j\|, 1 \leqslant i < j \leqslant n\}.$$

Prove that

$$\liminf_n n d_n^p \leqslant \frac{1}{c_p} \frac{\text{meas}(\bar{A})}{\Omega_p},$$

where

$$c_p = 2^{-p}\left\{1 + p\int_0^1 \frac{(1-t)^p}{1+t}\, dt\right\}.$$

For this one will consider a number γ such that $nd_n^p > \gamma$ for n sufficiently large, as well as the balls B_i with center x_i and radius r_i, where

$$r_i = \begin{cases} \frac{1}{2}\gamma^{1/p} n^{-1/p} & \text{if } 1 \leqslant i \leqslant n, \\ \frac{1}{2}\gamma^{1/p}(2i^{-1/p} - n^{-1/p}) & \text{if } n \leqslant i \leqslant 2^p n. \end{cases}$$

ΔVΔ = VΔV = ΔVΔ = VΔV = ΔVΔ

SOLUTION: It will be noted that the r_i form a decreasing sequence.

If $1 \leqslant i < j \leqslant n$ one has

$$\|x_i - x_j\| \geqslant d_n > \gamma^{1/p} n^{-1/p}$$

whenever n is large enough, and

$$r_i + r_j = \gamma^{1/p} n^{-1/p}.$$

When $1 \leqslant i < j \leqslant 2^p n$ and $j \geqslant n$, one has

$$\|x_i - x_j\| \geqslant d_j > \gamma^{1/p} j^{-1/p},$$

and we also have

$$r_i + r_j \leqslant \tfrac{1}{2}\gamma^{1/p} n^{-1/p} + \tfrac{1}{2}\gamma^{1/p}(2j^{-1/p} - n^{-1/p}) = \gamma^{1/p} j^{-1/p}.$$

From this it follows that the balls B_i are mutually disjoint.

If one sets $A(\varepsilon) = \{x : d(x,A) \leqslant \varepsilon\}$, one will then have

$$\frac{\gamma\Omega p}{2^p}\left[\frac{n-1}{n} + \sum_{i=n}^{2^p n} (2i^{-1/p} - n^{-1/p})^p\right] \leqslant \operatorname{meas}(A(\varepsilon_n)),$$

where

$$\varepsilon_n = \tfrac{1}{2}\gamma^{1/p} n^{-1/p}$$

We have

$$\lim_n \sum_{i=n}^{2^p n} (2i^{-1/p} - n^{-1/p})^p = \lim_n \frac{1}{n} \sum_{i=n}^{2^p n} \left[2\left(\frac{i}{n}\right)^{-1/p} - 1\right]^p =$$

(Contd)

(Contd)
$$= \int_1^{2^p} (2x^{-1/p} - 1)^p dx$$

$$= p\int_0^1 \frac{(1-t)^p}{1+t}\, dt,$$

(the last integral is obtained by setting $x = (1+t)^p$). On the other hand, the sets $A(\varepsilon_n)$ are decreasing and have $\bar{A}$ as intersection; furthermore, as A is bounded they are of finite measure. From this it follows that

$$\gamma \leqslant \frac{\text{meas}(\bar{A})}{\Omega_p}\frac{1}{c_p},$$

and consequently that

$$\liminf_n n d_n^p \leqslant \frac{1}{c_p}\frac{\text{meas}(\bar{A})}{\Omega_p}.$$

EXERCISE 1·12: Let X be a measurable set of $\mathbb{R}^s$ such that meas(X) = 1. Let u be a bijection of X onto itself such that for every subset E of X, E is measurable if and only if $u(E)$ is measurable, and then meas(E) = meas$(u(E))$. Furthermore, assume that if N is a measurable subset of X and if $u(x) \in N$ for almost all the points of N, then N or $X - N$ is negligeable.

Let E be a measurable subset of X such that meas$(E) > 0$, and if $x \in X$,

$$n(x) = \begin{cases} \infty & \text{if } u^p(x) \notin E \text{ for all } p \geqslant 1, \\ \inf\{p : p \geqslant 1, u^p(x) \in E\} & \text{otherwise} \end{cases}$$

(one sets $u^1 = u$, $u^{p+1} = u_0 u^p$).

Show that

$$\int_E n(x)\mathrm{d}x = 1.$$

$$\Delta\nabla\Delta = \nabla\Delta\nabla = \Delta\nabla\Delta = \nabla\Delta\nabla = \Delta\nabla\Delta$$

SOLUTION: For $1 \leqslant n \leqslant \infty$ let us set

$$E_n = \{x : x \in E \text{ and } n(x) = n\}.$$

Let us also set

$$G_0 = E,$$

$$G_n = E - \bigcup_{p=1}^{n} u^{-p}(E), \qquad n \geqslant 1,$$

where $u^{-p} = (u^{-1})^p$. It is clear that

$$E_n = G_{n-1} - G_n, \qquad 1 \leqslant n < \infty,$$

$$E_\infty = \bigcap_{n \geqslant 0} G_n.$$

For $n \geqslant 1$ one has

$$u^n(G_n) = u^n(E) - \bigcup_{p=0}^{n-1} u^p(E).$$

From this it results that the sets $u^n(G_n)$, $n \geqslant 0$, are mutually disjoint, and that

$$\gamma = \bigcup_{n=0}^{\infty} u^n(G_n) = \bigcup_{n=0}^{\infty} u^n(E).$$

Since $u(\gamma) \subset \gamma$ and $\mathrm{meas}(\gamma) \geqslant \mathrm{meas}(E) > 0$, the second hypothesis made on u implies that meas(γ) = 1; in other words, since u preserves the measure,

$$1 = \sum_{n=0}^{\infty} \text{meas}\, u^n(G_n) = \sum_{n=0}^{\infty} \text{meas}(G_n).$$

Let us note that the G_n are decreasing; consequently

$$\text{meas}(E_\infty) = \lim_n \text{meas}(G_n) = 0.$$

Thus one has

$$\int_E n(x)dx = \sum_{n=1}^{\infty} n\text{meas}(E_n) = \lim_N \sum_{n=1}^{N} n\{\text{meas}(G_{n-1}) - \text{meas}(G_n)\}$$

$$= \lim_{N\to\infty}\left(\sum_{n=0}^{N-1} (n + 1)\text{meas}(G_n) - \sum_{n=0}^{N} n\text{meas}(G_n)\right)$$

$$= \lim_N \left(\sum_{n=0}^{N-1} \text{meas}(G_n) - N\text{meas}(G_N)\right).$$

It remains to be observed that

$$\sum \text{meas}(G_n) = 1 \quad \text{and} \quad \text{meas}(G_n) \geqslant \text{meas}(G_{n+1})$$

implies

$$N\text{meas}(G_N) \to 0.$$

Thus

$$\int_E n(x)dx = 1.$$

REMARK: We have used the following classical result:

If $u_n \geqslant u_{n+1} \geqslant 0$ and $\sum u_n < \infty$, then $nu_n \to 0$.

We shall briefly recall the proof: let $\varepsilon > 0$ and let p be such that

$$(n - p)u_n \leqslant u_{p+1} + \cdots + u_n \leqslant \varepsilon$$

for all $n \geqslant p$; then

$$\liminf_n nu_n \leqslant \varepsilon. \text{ QED}$$

EXERCISE 1·13: One says that a set A of $\mathbb{R}^p$ is *ALMOST OPEN* if almost all the points of A are interior points of A.

Let f be a real function defined on an open set U of $\mathbb{R}^p$.

Prove that the following conditions are equivalent:

(a): f is continuous at almost all the points of U;

(b): For all $\alpha \in \mathbb{R}$ the sets $(f > \alpha)$ and $(f < \alpha)$ are almost open sets.

Δ∇Δ = ∇Δ∇ = Δ∇Δ = ∇Δ∇ = Δ∇Δ

SOLUTION: (a): Let E be a negligeable set in U such that f is continuous at every point of $U - E$. If $x \in (f > \alpha) - E$ one will have $f(y) > \alpha$ for all the points y of a neighbourhood of x, therefore x is interior to $(f > \alpha)$. This set is therefore almost open. One argues similarly for $(f < \alpha)$.

(b): f is continuous at x if for every rational number $r < f(x)$, x is interior to $(f > r)$, and if, for every rational number $s > f(x)$ x is interior to $(f > s)$. If r is rational, let us denote by A_r (*resp.* B_r) the set of points of $(f > r)$ (*resp.* $(f < r)$) not interior to this set. If these sets are negligeable then so is their union, which, by the preceding, contains the set of points of discontinuity of f.

EXERCISE 1·14: One says that a bounded real function f defined on $\mathbb{R}$ is *ALMOST EVERYWHERE CONTINUOUS* if the set of its points of discontinuity is negligeable.

(a): Give an example of a function that is almost everywhere continuous and such that there exists no continuous function coinciding with it almost everywhere.

(b): Show that in order for a bounded real function f to be almost everywhere equal to an almost everywhere continuous function, it is necessary and sufficient that there exists a set A of $\mathbb{R}$ such that $\mathbb{R} - A$ is negligeable, and that the restriction of f to A is continuous.

(c): Deduce from (b) that f is measurable and that there exists a sequence of continuous functions f_n which is convergent at every point of $\mathbb{R}$ and whose limit is almost everywhere equal to f.

(d): Show that a right-continuous function is continuous except at the points of a set that is at most denumerable, and therefore is almost everywhere continuous.

Δ∇Δ = ∇Δ∇ = Δ∇Δ = ∇Δ∇ = Δ∇Δ

SOLUTION: (a): If $f(x) = 0$ for $x < 0$ and $f(x) = 1$ for $x \geqslant 0$, f cannot coincide almost everywhere with a continuous function g, for with the complement of a negligeable set being everywhere dense in $\mathbb{R}$, one would be able to find two sequences $x_i < 0 < y_i$ tending to zero, and such that

$$g(x_i) = f(x_i) = 0, \qquad g(y_i) = f(y_i) = 1.$$

On passing to the limit one would have $g(0) = 0$ and $g(0) = 1$ at the same time, which is absurd.

SOLUTION: (b): The condition is evidently necessary. Let us show that it is sufficient. To do that let us set:

$$g(x) = \lim_{\substack{\alpha\to 0\\ \alpha>0}}\{\sup(f(y):y\in A, |y - x| < \alpha)\}.$$

This definition has a meaning, for A is everywhere dense. On A one has $f = g$. Furthermore, if $x \in A$ and if $\varepsilon > 0$ there exists $\alpha > 0$ such that $f(x) - \varepsilon \leqslant f(y) \leqslant f(x) + \varepsilon$ if $y \in A$ and $|y - x| < \alpha$. Then if $|x - x'| < \alpha$ one has $|y - x| < \alpha$ for all $y \in A$ such that

$$|y - x'| < \alpha' = \alpha - |x - x'|.$$

From this it follows that

$$g(x) - \varepsilon \leqslant g(x') \leqslant g(x) + \varepsilon,$$

which proves that g is continuous at each point of A.

SOLUTION: (c): For every x and all $\varepsilon > 0$ there exists $\alpha > 0$ such that

$$f(y) \leqslant g(x) + \varepsilon$$

if $y \in A$, $|y - x| < \alpha$. As above, from this one deduces that

$$g(x') \leqslant g(x) + \varepsilon$$

if $|x' - x| < \alpha$. The function g is therefore bounded and upper semi-continuous. There then exists (cf., a course on Topology) a sequence f_n of continuous functions that converges everywhere towards g.

SOLUTION: (d): Let us assume that f is right-continuous. If A is a non-empty set of $\mathbb{R}$ we shall denote by $\partial(A)$ the diameter of A, that is to say the upper bound of the numbers $|a - b|$ for $a \in A$, $b \in A$. For all $x \in \mathbb{R}$ let us then set

$$\omega(x) = \inf\{\partial(f(V)):V \text{ a neighbourhood of } x\},$$

(one says that $\omega(x)$ is the *OSCILLATION OF f AT x*). The set of points of discontinuity of f is then:

$$\{\omega > 0\} = \bigcup_{n=1}^{\infty} \left\{\omega > \frac{1}{n}\right\} .$$

It therefore suffices to prove that for all $\alpha > 0$ the set $A = \{\omega > \alpha\}$ is denumerable. By reason of the right-continuity of f, for all $x \in A$ there exists $\partial_x > 0$ such that

$$\partial(f(]x,x + \partial_x[)) < \alpha.$$

But then:

$$]x,x + \partial_x[\cap A = \emptyset.$$

Let us choose a rational number r_x in $]x,x + \partial_x[$. If $y \in A$ and $x < y$, one therefore has

$$r_x < y < r_y,$$

which proves that $x \in A \mapsto r_x$ is an injection of A into Q, and consequently that A is denumerable.

EXERCISE 1·15: Let A be a measurable set of $\mathbb{R}$ such that $\text{meas}(A) < \infty$. Show that the function $x \mapsto \text{meas}(A \cap]-\infty,X])$ is continuous.

$$\Delta\nabla\Delta = \nabla\Delta\nabla = \Delta\nabla\Delta = \nabla\Delta\nabla = \Delta\nabla\Delta$$

SOLUTION: If x_n is a decreasing sequence, and tends to x, one has

$$A \cap]-\infty,x] = \bigcap_n \{A \cap]-\infty,x_n]\},$$

and consequently

$$\text{meas}(A \cap]-\infty,x]) = \lim_n \text{meas}(A \cap]-\infty,x_n]).$$

If x_n is a strictly increasing sequence that converges towards x then

$$A \cap]-\infty,x[= \bigcup_n \{A \cap]-\infty,x_n]\},$$

and consequently, because $\text{meas}(\{x\}) = 0$,

$$\text{meas}(A \cap]-\infty,x]) = \text{meas}(A \cap]-\infty,x[)$$

$$= \lim_n \text{meas}(A \cap]-\infty,x_n]),$$

which proves that the function $x \mapsto \text{meas}(A \cap]-\infty,x])$ is continuous.

EXERCISE 1·16: Let $0 < \lambda < 1$. For any measurable sets $A,B \subset [0,1]$ of positive measure, do there exist $0 \leqslant x < y \leqslant 1$ such that

$$\text{meas}(A \cap [x,y]) = \lambda\text{meas}(A),$$

$$\text{meas}(B \cap [x,y]) = \lambda\text{meas}(B)?$$

$\Delta\nabla\Delta = \nabla\Delta\nabla = \Delta\nabla\Delta = \nabla\Delta\nabla = \Delta\nabla\Delta$

SOLUTION: The answer is affirmative if and only if $\lambda = 1/n$, where $n = 2,3,\ldots$. First of all let us assume that $\lambda = 1/n$. Since the function

$$f(x) = \text{meas}(A \cap [0,x])$$

is continuous and increasing on $[0,1]$, there exist points $0 = t_0 < t_1 < \cdots < t_n = 1$ such that

$$f(t_s) = \frac{s}{n}\,\text{meas}(A).$$

Under these conditions one has:

$$\text{meas}(A\cap[t_s,t_{s+1}]) = \frac{1}{n}\,\text{meas}(A), \qquad 0 \leqslant s \leqslant n-1.$$

Furthermore, the n numbers $\text{meas}(B\cap[t_s,t_{s+1}])$ have $\text{meas}(B)$ as sum. Therefore they cannot all be strictly less than or all strictly greater than $(1/n)\text{meas}(B)$. *Therefore there exists an* s $(1 \leqslant s \leqslant n-1)$ *such that one of the numbers* $\text{meas}(B\cap[t_{s-1},t_s])$ *and* $\text{meas}(B\cap[t_s,t_{s+1}])$ *is less than* $(1/n)\text{meas}(B)$ *and the other is greater.*

The idea of the proof is to 'vary x continuously' from t_{s-1} to t_s and y from t_s to t_{s+1} in such a way that

> $f(y) - f(x) = \text{meas}(A\cap[x,y])$ always remains equal to $(1/n)\text{meas}(A)$.

$\text{meas}(B\cap[x,y])$ will then vary continuously between two values on opposite sides of $(1/n)\text{meas}(B)$, and must therefore take this value at least once.

The rigorous proof reduces to proving that in the rectangle $t_{s-1} \leqslant x \leqslant t_s$, $t_s \leqslant y \leqslant t_{s+1}$ the set E of points such that

$$f(y) = f(x) + \frac{1}{n}\,\text{meas}(A)$$

is connected. Let us note that if $t_{s-1} \leqslant x \leqslant t_s$ one has

$$\begin{aligned} f(t_s) = f(t_{s-1}) + \frac{1}{n}\,\text{meas}(A) &\leqslant f(x) + \frac{1}{n}\,\text{meas}(A) \\ &\leqslant f(x) + f(t_{s+1}) - f(t_s) \\ &\leqslant f(t_{s+1}). \end{aligned}$$

From this it follows that

$$E_x = \{y : t_s \leqslant y \leqslant t_{s+1}, f(y) = f(x) + \frac{1}{n}\,\text{meas}(A)\}$$

is a non-empty compact interval, and that E is the union of the E_x's (if E_x is identified with $\{x\} \times E_x$). Note that E is compact; let us assume that it is the disjoint union of two non-empty compact sets E_1, E_2. Having seen that E_x is compact and connected, one has either $E_x \subset E_1$ or $E_x \subset E_2$. In other words, the projections of E_1 and E_2 onto the x-axis are two non-empty compacts sets that are disjoint and have the union $[t_{s-1}, t_s]$ which is absurd.

Let us now assume that λ is not of the above form; then there exists an integer n such that

$$\frac{1}{n+1} < \lambda < \frac{1}{n} .$$

Let $\alpha, \beta > 0$ be two numbers such that

$$(n + 1)\alpha + n\beta = 1,$$

and let us decompose $[0,1]$ into $2n + 1$ contiguous intervals, alternatively of length α and β, the two extreme intervals having length α. Let A be the union of intervals of length α, and B that of intervals of length β. If

$$\text{meas}(A \cap [x,y]) = \lambda \text{meas}(A),$$

then $[x,y]$ contains at least one of the intervals of length β, as otherwise one would have

$$\text{meas}(A \cap [x,y]) \leqslant \alpha = \frac{1}{n+1} \text{meas}(A) < \lambda \text{meas}(A).$$

But then,

$$\text{meas}(B \cap [x,y]) \geqslant \beta = \frac{1}{n} \text{meas}(B) > \lambda \text{meas}(B).$$

EXERCISE 1·17: Let I be a compact interval of $\mathbb{R}$ such that meas$(I) > 0$ and $0 < \beta < 1$. We shall say that the operation $T(\beta)$ is carried out on I if one subtracts from I the open interval having the same centre as I and of length $\beta\text{meas}(I)$. More generally, if I is a disjoint union of a finite number of compact intervals of non-zero lengths, to apply $T(\beta)$ to I consists in carrying out this operation on each of the intervals forming I.

Now let (β_n) be a sequence of real numbers $0 < \beta_n < 1$; we shall denote by I_n the compact set obtained by successively carrying out the operations $T(\beta_1), T(\beta_2), \ldots, T(\beta_n)$ starting from the interval $[0,1]$.

(a): Show that

$$I_{n+1} \subset I_n, \qquad \text{meas}(I_n) = (1 - \beta_1)(1 - \beta_2)\cdots(1 - \beta_n)$$

and that every interval contained in I_n has a length less than 2^{-n}.

(b): From this deduce that

$$K = \bigcap_n I_n$$

is compact, non-empty, nowhere dense, has no isolated point, and that

$$\text{meas}(K) = \lim_n (1 - \beta_1)(1 - \beta_2)\cdots(1 - \beta_n)$$

(c): Assume that $\beta_n = 1/3$ for all n. Show that $\text{meas}(K) = 0$ and that $X \in [0,1]$ belongs to K if and only if it can be written in base three uniquely using only the digits 0 and 2. From this deduce that K is not countable.

(d): If $\beta_n = 1 - \alpha^{1/n(n+1)}$, $0 < \alpha < 1$, show that $\text{meas}(K) = \alpha$ (which proves the existence in $[0,1]$ of nowhere dense compact sets whose measure is arbitrarily close to 1).

(e): Deduce from part (d) the existence in [0,1] of a sequence A_n of sets of the first category that are mutually disjoint and such that:

(i): $\mathrm{meas}(A_n) = 2^{-n}$;

(ii): $K_n = \bigcup_{i=1}^{n} A$ is a nowhere dense compact set;

(iii): Every interval contiguous with K_n (that is to say, every connected component of the complement of K_n in [0,1]) contains a set in A_{n+1} of measure greater than zero.

From this deduce that

$$A = \bigcup_{n=1} A_n$$

is of first category, has measure one, and that its complement in [0,1] is a set of second category of measure zero.

(f): Let

$$E = \bigcup_{n=0}^{\infty} A_{2n+1}.$$

Show that for every interval I contained in [0,1] and of non-zero length there holds

$$0 < \mathrm{meas}(E \cap I) < \mathrm{meas}(I).$$

(g): From part (f) above deduce the existence of Borel sets $E \subset \mathbb{R}$ such that for every interval I of non-zero length one has $0 < \mathrm{meas}(E \cap I) < \mathrm{meas}(I)$. Can one have $\mathrm{meas}(E) < \infty$ for such sets?

(h): Deduce from the preceding that there exist positive functions that are Lebesgue integrable, but that are not limits almost everywhere of increasing sequences of positive step functions.

SOLUTION: (a): It is clear that

$$I_{n+1} \subset I_n \quad \text{and} \quad \text{meas}(I_n) = (1 - \beta_1)\cdots(1 - \beta_n).$$

Moreover, I_n is formed by 2^n mutually disjoint intervals of equal lengths; when the latter property.

SOLUTION: (b): The set K is compact and non-empty by virtue of a well known theorem in Topology. If I is an interval contained in K one has $I \subset I_n$ for all n; by Question (a) one thus has meas(I) $< 2^{-n}$ and consequently meas(I) = 0. In other words the interior of K is empty, which is the definition of a nowhere dense compact set. If $x \in K$ and $\varepsilon > 0$, for large enough n one of the intervals forming I_n will be contained in $]x - \varepsilon, x + \varepsilon[$; for this it is sufficient that $2^{-n} < \varepsilon$. Now, the two endpoints of this interval belong to K, which shows that x is not an isolated point of K. Lastly,

$$\text{meas}(K) = \lim_n \text{meas}(I_n) = \lim (1 - \beta_1)\cdots(1 - \beta_n).$$

SOLUTION: (c): In this case, one has:

$$\text{meas}(K) = \lim_n \left(\frac{2}{3}\right)^n = 0.$$

Furthermore, I_1 is equal to the set of x's which are written in base three as

$$x = 0.a_1a_2\cdots a_n\cdots,$$

with $a_1 = 0$ or 2 (it will be noted that $1/3 = 0.0222\cdots$ and that $1 = 0.222\cdots$). Similarly it is seen that $x \in I_n$ if and only if $a_i = 0$ or 2 for $1 \leqslant i \leqslant n$. From this one deduces that $x \in K$ if for all i $a_i = 0$ or 2; the expansion of x in this form is then unique. If with every set $A \subset \mathbb{N}$ one associates $x_A = 0.a_1a_2\cdots$ with $a_i = 0$

if $i \in A$ and $a_i = 2$ if $i \notin A$, a bijection between $P(\mathbb{N})$ and K is realised; K is therefore not denumerable.

SOLUTION: (d): If $\beta_n = 1 - \alpha^{1/n(n+1)}$, $0 < \alpha < 1$, one has

$$\text{meas}(K) = \lim_n \alpha^{a_n},$$

with

$$a_n = \frac{1}{1.2} + \frac{1}{2.3} + \cdots + \frac{1}{n(n+1)}$$

$$= \left(1 - \frac{1}{2}\right) + \left(\frac{1}{2} - \frac{1}{3}\right) + \cdots + \left(\frac{1}{n} - \frac{1}{n+1}\right)$$

$$= 1 - \frac{1}{n+1},$$

whence $\text{meas}(K) = \alpha$.

SOLUTION: (e): By the preceeding it is seen that in every non-empty open interval I there exists a nowhere dense compact set whose measure is $\frac{1}{2}\text{meas}(I)$. The let A_1 be a nowhere dense compact set of $[0,1]$ such that $\text{meas}(A_1) = \frac{1}{2}$. In each interval contiguous to A_1 let us choose a nowhere dense compact set the measure of which is half that of this interval, and let us denote by A_2 the union of these compact sets; A_2 is of first category (for the set of intervals contiguous to a compact set is denumerable) and its measure is $\frac{1}{4}$. Furthermore, $K_2 = A_1 \cup A_2$ is closed; in fact, if x is a limit point of K_2 and does not belong to A_1 it belongs to an interval I contiguous to A_1 and is therefore a limit point of $I \cap A_2$, which is compact, whence $x \in A_2$. On the other hand, it is clear that $[0,1] - K_2$ is dense in $[0,1] - K_1$, which itself is dense in $[0,1]$, which proves that K_2 is nowhere dense. Quite generally, A_{n+1} will be constructed by choosing in every interval contiguous to the compact set K_n a nowhere dense compact set of measure equal to half the length of this interval, and by taking the union of these compact sets. As above, it is seen that A_{n+1} is of first category and that $K_{n+1} = K_n \cup A_{n+1}$ is closed. Furthermore,

$$\operatorname{meas}(K_n) = \sum_{i=1}^{n} 2^{-i} = 1 - 2^{-n},$$

whence:

$$\operatorname{meas}(A_{n+1}) = \tfrac{1}{2}[1 - (1 - 2^{-n})] = 2^{-(n+1)}.$$

Finally, Condition (iii) is satisfied by construction. If

$$A = \bigcup_{n=1} A_n$$

this set is of first category, for it is the countable union of sets of first category, and

$$\operatorname{meas}(A) = \sum_{n=1}^{\infty} 2^{-n} = 1.$$

Its complement is not of first category, by a theor m of Baire, and its measure is zero.

SOLUTION: (f): Let

$$E = \bigcup_{n=0}^{\infty} A_{2n+1}, \qquad F = \bigcup_{n=1}^{\infty} A_{2n}.$$

If I is an interval of length greater than zero contained in $[0,1]$, its intersection with E or F is of measure greater than zero, since $\operatorname{meas}(E \cup F) = 1$. Let us assume, for example, that $\operatorname{meas}(I \cap E) > 0$. Then $I \cap E$ contains at least two points $x < y$. Let n be such that these two points belong to $A_1 \cup A_2 \cup \cdots \cup A_{2n+1}$.

They therefore belong to K_{2n+1}, and as this set is compact and nowhere dense there exists an interval J contained in $[x,y]$ which is contiguous to it; one then has

$$\text{meas}(I \cap F) \geqslant \text{meas}(J \cap A_{2n+2}) > 0.$$

Since

$$\text{meas}(I) = \text{meas}(I \cap E) + \text{meas}(I \cap F),$$

it follows from this that

$$\text{meas}(E \cap I) < \text{meas}(I).$$

SOLUTION: (g): If E is the set studied in Question (f) above, then

$$\tilde{E} = \bigcup_{n \in \mathbb{Z}} (E + n)$$

answers the question. By going back to the proof of Question (f) again one sees that for E one can take the set

$$E_N = \bigcup_{n \geqslant N} A_{2n+1}.$$

If

$$\tilde{E} = \bigcup_{n \in \mathbb{Z}} (E_{|N|} + N),$$

then for every interval I of positive length one has:

$$0 < \text{meas}(I \cap \tilde{E}) < \text{meas}(I),$$

and furthermore,

$$\text{meas}(\tilde{E}) = \sum_{n=0}^{\infty} 2^{-(2n+1)} + 2 \sum_{N=1}^{\infty} \sum_{n=N}^{\infty} 2^{-(2n+1)} = \frac{10}{9}.$$

SOLUTION: (h): Let f be the characteristic function of the set E defined above, and let φ be a positive step function such that $\varphi \leq f$ almost everywhere. If I is an interval of length greater than zero on which φ is equal to a constant, since meas$(I - \tilde{E}) > 0$ this constant is zero. Thus $\varphi = 0$ almost everywhere. From this it follows in particular that f cannot be the limit almost everywhere of an increasing sequence of step functions.

REMARK: To prove the existence of a nowhere dense compact set of $[0,1]$, the measure of which is arbitrarily close to unity, one may also consider the set $E = [0,1] - Q$. There holds

$$1 = \text{meas}(E) = \sup\{\text{meas}(K) : K \text{ compact}, K \subset E\},$$

and every compact set contained in E is evidently nowhere dense.

EXERCISE 1·18: Consider a double sequence $(f_{m,n})$ of measurable complex functions on $X = [0,1]$ such that for all m the sequence $(f_{m,n})_{n\geq 1}$ converges almost everywhere towards a function g_m and that the sequence (g_m) converges almost everywhere towards a function h.

Show that there exist two sequences of strictly increasing integers (m_s) and (n_s) such that the sequence $(f_{m_s,n_s})_{s\geq 1}$ converges almost everywhere to h. Generalise this result to the case where $X = \mathbb{R}$ (or $\mathbb{R}^p$).

Δ∇Δ = ∇Δ∇ = ΔVΔ = ∇Δ∇ = Δ∇Δ

SOLUTION: First assume that every convergence is uniform on a set $E \subset X$. Then there exists $m_1 < m_2 < \cdots$ and $n_1 < n_2 < \cdots$ and such that on E there holds

$$\left|h - g_{m_s}\right| < \frac{1}{s}, \qquad \left|g_{m_s} - f_{m_s,n_s}\right| < \frac{1}{s}.$$

Thus

$$\left|h - f_{m_s,n_s}\right| < \frac{2}{s} \quad \text{on } E,$$

which proves that $f_{m_s,n_s} \to h$ on E.

We are going to show that for all $\varepsilon > 0$ there exists such a set E, which, moreover, is measurable and satisfies $\text{meas}(X - E) < \varepsilon$. In fact, by Egoroff's Theorem there exist measurable sets $E_0, E_1, \ldots$ such that

$$\text{meas}(X - E_m) < \varepsilon 2^{-(m+1)}, \qquad m \geqslant 0,$$

and, on the other hand, one has

$$\lim_m g_m = h \qquad \text{uniformly on } E_0,$$

$$\lim_n f_{m,n} = g_m \quad \text{uniformly on } E_m \text{ if } m \geqslant 1.$$

On setting

$$E = \bigcap_{m=0} E_m$$

the desired result is obtained.

Let us now use an argument known as the "diagonal process". By applying the preceding result to $\varepsilon = 1$ one first determines a measurable set E_1 and two strictly increasing mappings φ_1, ϑ_1 of $\mathbb{N}^*$ into itself, such that $\text{meas}(X - E_1) < 1$ and

$$f_{\varphi_1(n),\vartheta_1(n)} \to h \quad \text{on } E_1.$$

Now applying the same result to $\varepsilon = \frac{1}{2}$ and the double sequence $f_{\varphi_1(m),\vartheta_1(n)}$ one obtains a measurable set E_2 and two strictly increasing mappings φ_2, ϑ_2 of $\mathbb{N}^*$ into itself such that $\text{meas}(X - E_2) < \frac{1}{2}$ and

$$f_{\varphi_1 \circ \varphi_2(n), \vartheta_1 \circ \vartheta_2(n)} \to h \quad \text{on } E_2.$$

Proceeding thus repeatedly, one obtains finally a sequence (E_r) of measurable sets and two sequences (φ_r) and (ϑ_r) of strictly increasing mappings of $\mathbb{N}^*$ into itself, such that $\text{meas}(X - E_r) < 1/r$ and

$$f_{\varphi_1 \circ \cdots \circ \varphi_r(n), \vartheta_1 \circ \cdots \circ \vartheta_r(n)} \to h \quad \text{on } E_r.$$

Let us then set

$$E = \bigcup_r E_r,$$

$$m_s = \varphi_1 \circ \cdots \circ \varphi_s(s), \qquad n_s = \vartheta_1 \circ \cdots \circ \vartheta_s(s).$$

The set E is measurable and $\text{meas}(X - E) = 0$. Furthermore, for every $r \geqslant 1$ the sequence $(f_{m_s, n_s})_{s \geqslant r}$ is a subsequence of the sequence $(f_{\varphi_1 \circ \cdots \circ \varphi_r(n), \vartheta_1 \circ \cdots \circ \vartheta_r(n)})_{n \geqslant 1}$. From this it follows that

$$f_{m_s, n_s} \to h \quad \text{on } E.$$

It will be noted that in order to use Egoroff's Theorem it suffices to assume that $X \subset \mathbb{R}^p$ and $\text{meas}(X) < \infty$. By writing $\mathbb{R}^p$ as the union of a sequence of such sets, a new application of the diagonal process allows this result to be generalised to the case where $X = \mathbb{R}^p$.

EXERCISE 1·19: A mapping $t \to F(t)$ of $\mathbb{R}$ into the set $\mathcal{F}_p$ of closed sets of $\mathbb{R}^p$ is called a *MEASURABLE MAPPING* if for every compact set K of $\mathbb{R}^p$ the set $\{t : F(t) \cap K \neq \emptyset\}$ is measurable.

(a): Show that $\{t:F(t) = \emptyset\}$ is measurable.

(b): Show that $t \to F(t)$ is measurable if and only if $\{t:F(t)\cap B \neq \emptyset\}$ is measurable for every open ball B of $\mathbb{R}^p$, or again if $t \to F(t)\cap K$ is measurable for every compact set K of $\mathbb{R}^p$.

(c): Show that if $t \to F_1(t)$ and $t \to F_2(t)$ are measurable mappings of $\mathbb{R}$ into the closed sets of $\mathbb{R}^p$ and $\mathbb{R}^q$ respectively, then $t \to F_1(t)\times F_2(t)$ is measurable.

(d): Show that if $t \to K(t)$ is a measurable mapping of $\mathbb{R}$ into the compact sets of $\mathbb{R}^p$, and if f is a continuous mapping of $\mathbb{R}^p$ into $\mathbb{R}^q$, then $t \to f(K(t))$ is measurable.

(e): Show that if $t \to F_n(t)$ are measurable, then

$$t \to \bigcap_n F_n(t)$$

is also measurable.

(f): Show that $t \to F(t)$ and $t \to K(t)$ are measurable mappings of $\mathbb{R}$ into the sets of $\mathbb{R}^p$ which are respectively closed and compact, then $t \to F(t) + K(t)$ is measurable.

$\Delta\nabla\Delta = \nabla\Delta\nabla = \Delta\nabla\Delta = \nabla\Delta\nabla = \Delta\nabla\Delta$

SOLUTION: (a): Let B_n be the closed ball with center 0 and radius n. We have:

$$\{t:F(t) = \emptyset\} = \mathbb{R} - \bigcup_{n=1}^{\infty} \{t:F(t)\cap B_n \neq \emptyset\}.$$

SOLUTION: (b): If $t \to F(t)$ is measurable and if V is an open set of $\mathbb{R}^p$, there exists a sequence of compact sets K_n of which V is the union, and:

$$\{t:F(t)\cap V \neq \emptyset\} = \bigcup_{n-1}^{\infty} \{t:F(t)\cap K_n \neq \emptyset\}.$$

If now $\{t:F(t)\cap B \neq \emptyset\}$ is measurable for every open ball B, as every open set V is the countable union of such balls, the set $\{t:F(t)\cap V \neq \emptyset\}$ is measurable. If K is compact, let us consider the open sets $V_n = \{x:d(x,K) < 1/n\}$ where $d(x,K)$ denotes the distance from x to K for a norm on $\mathbb{R}^p$. One has:

$$\{t:F(t)\cap K \neq \emptyset\} = \bigcup_{n=1} \{t:F(t)\cap V_n \neq \emptyset\}.$$

In fact it is clear that the first set is contained in the second; moreover, if for all n there exists $x_n \in F(t)\cap V_n$ one can find a subsequence x_{n_i} such that $x_{n_i} \to x$. It is clear that $x \in K$ (as $d(x,K) = \lim_i d(x_{n_i},K) = 0$), and that $x \in F(t)$ also, for $F(t)$ is closed.

If $t \to F(t)$ is measurable and K is compact, $t \to F(t)\cap K$ is evidently measurable (this is true, moreover, if K is closed). If $t \to F(t)\cap K$ is now measurable for every compact set K, by considering afresh the balls B_n (cf., Question (a) above), we have

$$\{t:F(t)\cap K \neq \emptyset\} = \bigcup_{n=1} \{t:(F(t)\cap K)\cap B_n \neq \emptyset\},$$

which proves that $t \to F(t)$ is measurable.

SOLUTION: (c): Note that we have not specified the norm on $\mathbb{R}^p$. If we choose the norm $\|(x_1,\ldots,x_p)\| = \mathrm{Max}|x_i|$, every open ball of $\mathbb{R}^{p+q}$ is of the form $B_1 \times B_2$, where B_1 and B_2 are open balls of $\mathbb{R}^p$ and $\mathbb{R}^q$ respectively. Then:

$$\{t:(F_1(t)\times F_2(t))\cap(B_1\times B_2) \neq \emptyset\} = \bigcap_{i=1,2} \{t:F_i(t)\cap B_i \neq \emptyset\}.$$

SOLUTION: (d): For every open set V of $\mathbb{R}^p$ we have:

$$\{t:f(K(t))\cap V \neq \emptyset\} = \{t:K(t)\cap f^{-1}(V) \neq \emptyset\}.$$

SOLUTION: (e): Consider first the case of two measurable mappings $t \to F_1(t)$ and $t \to F_2(t)$. Let us denote the 'diagonal' of $\mathbb{R}^p \times \mathbb{R}^q$ by Δ. For every compact set K of $\mathbb{R}^p$,

$$\{t:F_1(t)\cap F_2(t)\cap K = \emptyset\} = \{t:(F_1(t)\times(F_2(t)\cap K))\cap\Delta \neq \emptyset\}.$$

By what has gone before, $t \to F_1(t)\times(F_2(t)\cap K)$ is measurable, and on the other hand Δ is a countable union of compact sets. This shows that $t \to F_1(t)\cap F_2(t)$ is measurable. By recurrence one obtains the result for a finite intersection. In the general case one writes

$$\{t: (\bigcap_{n=1}^{\infty} F\ (t))\cap K \neq \emptyset\} = \bigcap_{n=1}^{\infty} \{t:(\bigcap_{r=1}^{n} F_r(t))\cap K \neq \emptyset\},$$

an equality that results because the decreasing compact sets

$$\bigcap_{r=1}^{n} (F_r(t)\cap K)$$

have a non-empty intersection if and only if each of them is non-empty.

SOLUTION: (f): It is known (see a course on Topology) that $F(t) + K(t)$ is closed if $F(t)$ is closed and $K(t)$ is compact. Moreover, if u denotes the mapping $(x,y) \to x + y$ of $\mathbb{R}^p \times \mathbb{R}^q$ into $\mathbb{R}^p$, we have

$$u(F(t)\times K(t)) = F(t) + K(t)$$

By the preceding, $t \to F(t)\times K(t)$ is measurable, and it is seen, as in Question (d) above, that $u(F(t)\times K(t))$ is also.

EXERCISE 1·20: Let $\mathcal{F}_p$ be the set of closed sets of $\mathbb{R}^p$, $\mathcal{K}_p^0$ the set of non-empty compact sets of $\mathbb{R}^p$, $x \to \|x\|$ a norm on $\mathbb{R}^p$, and for every non-empty set F of $\mathcal{F}_p$ $d(F) = \text{Min}\{\|x\| : x \in F\}$.

(a): For $F \in \mathcal{F}_p$ write:

$$\varphi_0(F) = \begin{cases} \{x : x \in F, \|x\| = d(F)\} & \text{if } F \neq \emptyset, \\ \{0\} & \text{if } F = \emptyset. \end{cases}$$

Show that $t \mapsto \varphi_0(F(t))$ is measurable if $t \mapsto F(t)$ is measurable (for the definition of the measurability of a one-parameter family of closed sets see the preceding Exercise).

(b): Let $e_i(x) = x_i$ if $x = (x_1, \ldots, x_p) \in \mathbb{R}^p$. For $K \in \mathcal{K}_p^0$ set $e_i(K) = \text{Min}\{e_i(x) : x \in K\}$ and

$$\varphi_i(K) = \{x : x \in K, e_i(x) = e_i(K)\}.$$

Show that $t \mapsto \varphi_i(K(t))$ is measurable if $t \mapsto K(t) \in \mathcal{K}_p^0$ is measurable.

(c): From this deduce that there exists a mapping $\varphi : \mathcal{F}_p \to \mathbb{R}^p$ such that:

(i): $\varphi(F) \in F$ if $F \in \mathcal{F}_p$ and $F \neq \emptyset$;

(ii): $t \mapsto \varphi(F(t))$ is measurable if $t \mapsto F(t)$ is.

(d): Let $f : \mathbb{R}^p \to \mathbb{R}^q$ be continuous. Show that there exists $\sigma : \mathbb{R}^q \times \mathcal{F}_p \to \mathbb{R}^p$ such that:

(i): If $F \in \mathcal{F}_p$ and $x \in f(F)$, then $\sigma(x,F) \in F$ and $f(\sigma(x,F)) = x$;

(ii): If $t \mapsto x(t) \in \mathbb{R}^q$, $t \mapsto F(t) \in \mathcal{F}_p$ are measurable, then $t \mapsto \sigma(x(t),F(t))$ is.

(e): Assume that the mapping $t \mapsto F_i(t) \in \mathcal{F}_p$ $(i = 1,2)$ and

$t \to x(t) \in F_1(t) + F_2(t)$ are measurable.

Show that there exist $t \to x_i(t) \in F_i(t)$ that are measurable and such that $x(t) = x_1(t) + x_2(t)$ for all t.

Δ∇Δ = ∇Δ∇ = Δ∇Δ = ∇Δ∇ = Δ∇Δ

SOLUTION: (a): Let K be a non-empty compact set of $\mathbb{R}^p$. The set

$$\{t:\varphi_0(F(t))\cap K \neq \emptyset\}$$

is equal to

$$\{t:F(t)\cap K \neq \emptyset, d(F(t)) = d(F(t)\cap K)\} \qquad (*)$$

when $0 \notin K$, otherwise it would be necessary to add $\{t:F(t) = \emptyset\}$. The latter set is measurable (cf., Exercise 1·19(a)). Since the set $\{t:F(t)\cap K \neq \emptyset\}$ is measurable by definition, and as $t \to F(t)\cap K$ is measurable it suffices to prove that $t \to d(F(t))$ is measurable on its defining set. Now, for all $\alpha \geqslant 0$

$$\{t:d(F(t)) \leqslant \alpha\} - \{t:F(t)\cap B_\alpha \neq \emptyset\},$$

where B_α denotes the closed ball with centre 0 and radius α.

SOLUTION: (b): Similarly, on the measurable set $\{t:K(t)\cap K \neq \emptyset\}$ one has:

$$\{t:e_i(K(t))\cap K \neq \emptyset\} = \{t:e_i(K(t)) = e_i(K(t)\cap K)\},$$

and one sees, as above, that $t \to e_i(K(t))$ is measurable (by noting that $\{t:K(t)\cap F \neq \emptyset\}$ is measurable for F closed).

SOLUTION: (c): For every $F \in \mathcal{F}_p$ the set $(\varphi_p \circ \cdots \circ \varphi_1 \circ \varphi_0)(F)$ is reduced to a point $\varphi(F)$. It is clear that $F \to \varphi(F)$ has all the required properties.

SOLUTION: (d): Let us, for $x \in \mathbb{R}^q$ and $F \in \mathcal{F}_p$, write:

$$\sigma(x,F) = \varphi(f^{-1}(x) \cap F).$$

It is clear that Condition (i) is satisfied. To prove that (ii) is also satisfied it suffices to show that $t \to f^{-1}(x(t)) \cap F(t)$ is measurable. Now, for every compact set K of $\mathbb{R}^p$

$$\{t: f^{-1}(x(t)) \cap K \neq \emptyset\} = \{t: x(t) \in f(K)\},$$

and $f(K)$ is compact.

SOLUTION: (e): Let $f: \mathbb{R}^p \times \mathbb{R}^p \to \mathbb{R}^p$ be defined by $(x,y) \to x + y$, and let σ be defined as above. Let us set

$$\sigma(x(t), F_1(t) \times F_2(t)) = (x_1(t), x_2(t)).$$

Then $x_1(t)$ and $x_2(t)$ satisfy the properties required.

EXERCISE 1·21: In this Exercise it is proposed, by consideration of the Axiom of Choice, to prove the existence of non-measurable sets of $\mathbb{R}$. The Axiom of Choice appears in the following form: Given a non-empty family (B_i) of mutually disjoint non-empty sets of $\mathbb{R}$ there exists a set E of $\mathbb{R}$ which contains one and only one point of each B_i.

(a): Show that there exists a set $E \subset [0,1]$ such that for every $x \in \mathbb{R}$ there exists an unique $y \in E$ such that $x - y$ is rational.

(b): Let S be the union of the sets $E + r$, where r runs over the set of rational numbers lying between -1 and 1.

Show that

$$[0,1] \subset S \subset [-1,2],$$

and that if r,s are two distinct rational numbers, then $E + r$ and

$E + s$ are disjoint.

(c): Deduce from this that E is not measurable.

Δ∇Δ = ∇Δ∇ = ΔVΔ = ∇Δ∇ = Δ∇Δ

SOLUTION: (a): Let Q be the set of rational numbers, and if x,y are real numbers let us express that $x - y \in Q$ by writing $x \sim y$. This defines an equivalence relation on $\mathbb{R}$ which all the equivalence classes intersect [0,1]. The existence of E follows from this and the axiom of choice.

SOLUTION: (b): It is clear that

$$E + r \subset [0,1] + [-1,1] = [-1,2],$$

and therefore $S \subset [-1,2]$. On the other hand, if $0 \leqslant x \leqslant 1$ there exists $y \in E$ and $r \in Q$ such that $x = y + r$. One has $|r| = |x - y| \leqslant 1$ which proves that $[0,1] \subset S$. Lastly, if $r \neq s$ one has $(E + r) \cap (E + s) = \emptyset$, otherwise there would exist $z \in \mathbb{R}$ and $y_1 \in E, y_2 \in E$ such that

$$z = y_1 + r = y_2 + s,$$

and consequently $y_1 \neq y_2$, which contradicts z being equivalent to a single element of E.

SOLUTION: (c): If E were measurable S would also be measurable, and

$$\text{meas}(S) = \sum_{r \in Q \cap [-1,1]} \text{meas}(E + r).$$

Now, all the numbers meas$(E + r)$ are equal to meas(E), so that meas$(S) = 0$ if meas$(E) = 0$, and meas$(S) = \infty$ if meas$(E) > 0$. This is absurd, because by virtue of the inclusions proved in (b) one would have to have $1 \leqslant \text{meas}(S) \leqslant 3$.

CHAPTER 2

σ-Algebras and Positive Measures

EXERCISE 2·22: Let (X,C) be a measurable space, (x_i) a sequence of points of X, and (m_i) a sequence of real numbers $m_i > 0$. For every set $E \in C$ set:

$$\mu(E) = \sum_{x_i \in E} m_i .$$

(a): Show that μ is a measure on C.

(b): Show that if $\{x_i\} \in C$ for all i then one has $C_\mu = P(X)$, and conversely.

Δ∇Δ = ∇Δ∇ = ΔVΔ = ∇Δ∇ = ΔVΔ

SOLUTION: (a): It is clear that $\mu(\emptyset) = 0$ (as usual, the convention is adopted that $\sum_{i\in\emptyset} m_i = 0$). Let (E_n) be a sequence of mutually disjoint sets of C, and let E be their union. For every integer N one evidently has:

$$\mu(E_1 \cup \cdots \cup E_N) = \sum_{n=1}^{N} \mu(E_n) \leqslant \mu(E),$$

whence

$$\sum_{n=1}^{\infty} \mu(E_n) \leqslant \mu(E).$$

Moreover, for every finite set A of $\mathbb{N}$ one has

$$\sum_{\substack{x_i \in E \\ i \in A}} m_i = \sum_{n=1}^{\infty} \sum_{\substack{x_i \in E_n \\ i \in A}} m_i \leqslant \sum_{n=1}^{\infty} \mu(E_n),$$

whence

$$\mu(E) = \sup_A \sum_{\substack{x_i \in E \\ i \in A}} m_i \leqslant \sum_{n=1}^{\infty} \mu(E_n) \leqslant \mu(E).$$

SOLUTION: (b): A set $E \subset X$ is measurable if and only if there exist $A, B \in \mathcal{C}$ such that $A \subset E \subset B$ and $x_i \notin B - A$ for all i. If for an index i one has $\{x_i\} \notin \mathcal{C}$ then $\{x_i\}$ is not μ-measurable, because A can only be the empty set and so $B \supset \{x_i\}$ implies that $x_i \in B - A = B$. On the other hand, if $\{x_i\} \in \mathcal{C}$ for all i and if E is an arbitrary set of X, let

$$A = \{x_i : x_i \in E\}, \qquad C = \{x_i : x_i \notin E\}.$$

Then A and $B = X - C$ belong to $\mathcal{C}$, $A \subset E \subset B$ and $\{x_i\} \notin B - A$ for all i. From this it follows that E is μ-measurable.

EXERCISE 2·23: Let $\mathbb{Z}$ be the set of positive, nul or negative integers.

(a): Show that the set of subsets A of $\mathbb{Z}$ such that for every integer $n \geqslant 1$ one has $2n \in A$ if and only if $2n + 1 \in A$ is a σ-algebra.

(b): Show that the mapping f of $\mathbb{Z}$ into itself defined by $f(n) = n + 2$ is a measurable bijection, but that f^{-1} is not measurable.

$$\Delta\nabla\Delta = \nabla\Delta\nabla = \Delta\nabla\Delta = \nabla\Delta\nabla = \Delta\nabla\Delta$$

SOLUTION: (a): It is clear that $\emptyset$ possesses the property, and that the property is preserved under complementation. If the A_i have the property for $i = 1,2,\ldots$, it is impossible that for an $n \geqslant 1$ one of the two integers $2n, 2n + 1$ belongs to one of the A_i and that the other belongs to none of them. This shows that the union of the A_i's also possesses the property.

SOLUTION: (b): It is clear that f is a bijection. Furthermore, if A has the property and if $n \geqslant 1$ one has $2n \in f^{-1}(A)$ if and only if $2(n + 1) \in A$, hence if and only if $2(n + 1) + 1 \in A$, that is to say, if and only if $2n + 1 \in f^{-1}(A)$. This proves that f is measurable. Finally note that $A = \{0\}$ has the property, but that $f(A) = \{2\}$ does not, since $2 \in f(A)$ and $3 \notin f(A)$, so f^{-1} is not measurable.

EXERCISE 2·24: Let C be a family of subsets of a set X. If $M \subset X$, set:

$$C_M = \{M \cap E ; E \in C\}.$$

(a): Show that if C is a σ-algebra on X, C_M is a σ-algebra on M (C_M is called the σ-algebra *INDUCED* on M by C).

(b): If $M \in C$ give a simple characterisation of C_M.

(c): If C is generated by a family A of subsets of X, show that C_M is generated by A_M.

(d): Deduce from part (c) that if M is a subset of a topological space X, the Borel σ-algebra associated with the topology induced by X on M is equal to the σ-algebra induced on M by the Borel σ-algebra of X. Consider in particular the case where M is a Borel set.

Δ∇Δ = ∇Δ∇ = Δ∇Δ = ∇Δ∇ = Δ∇Δ

SOLUTION: (a): $\emptyset = M \cap \emptyset$, and:

$$M - (M \cap E) = M \cap (X - E), \tag{*}$$

$$\bigcup_n (M \cap E_n) = M \cap (\bigcup_n E_n), \tag{**}$$

which shows that C_M is a σ-algebra on M.

SOLUTION: (b): If $M \in C$, then

$$C_M = \{E : E \in C \text{ and } E \subset M\}.$$

SOLUTION: (c): Since $A_M \subset C_M$, $C(A_M) \subset C_M$. Let C_0 be the set of subsets $E \subset X$ such that $M \cap E \in C(A_M)$. Evidently one has $\emptyset \in C_0$, and equalities (*) and (**) show that C_0 is a σ-algebra on X. Since $A \subset C_0$, one has $C \subset C_0$, which proves that $C_M \subset C(A_M)$.

SOLUTION: (d): This follows from T_M being the topology induced on M by the topology T of X.

If M is a Borel set, the Borel sets of M can be interpreted either as the Borel sets of X contained in M or as the Borel sets of the topological subspace M of X.

EXERCISE 2·25: Let C be a σ-algebra on X and let μ be a probability on C. Let $M \subset X$ be such that $E \in C$ and $E \supset M$ implies $\mu(E) = 1$.

Show that a probability μ_M is defined on C_M by setting

$$\mu_M(M \cap E) = \mu(E)$$

for all $E \in C$ (the induced σ-algebra C_M was defined in the preceding Exercise).

Δ∇Δ = ∇Δ∇ = Δ∇Δ = ∇Δ∇ = Δ∇Δ

SOLUTION: Let us first prove that if $E \in C$, $F \in C$, and $M \cap E = M \cap F$, then $\mu(E) = \mu(F)$. Now, in this case one has

$$(X - E) \cup (E \cap F) \supset M,$$

$$(X - F) \cup (E \cap F) \supset M,$$

whence

$$1 - \mu(E) + \mu(E \cap F) = 1,$$

$$1 - \mu(F) + \mu(E \cap F) = 1,$$

and consequently $\mu(E) = \mu(F)$. The mapping μ_M of C_M into $[0,1]$ is thus defined unambiguously. It is clear that:

$$\mu_M(\emptyset) = \mu_M(M \cap \emptyset) = \mu(\emptyset) = 0,$$

$$\mu_M(M) = \mu_M(M \cap X) = \mu(X) = 1.$$

Moreover, if $E \in C$, $F \in C$, and

$$(M \cup E) \cap (M \cup F) = M \cap (E \cup F) = \emptyset,$$

one has:

$$\mu(E \cap F) = 0,$$

whence

$$\mu_M\{(M \cap E) \cup (M \cap F)\} = \mu_M\{M \cap (E \cup F)\} = \mu(E \cup F) =$$ (Contd)

(Contd)
$$= \mu(E) + \mu(F)$$
$$= \mu_M(M \cap E) + \mu_M(M \cap F).$$

Hence μ_M is additive. It remains to prove that it is continuous. To do this let us consider a sequence (E_n) of elements of C such that

$$M \cap E_n \subset M \cap E_{n+1}$$

Let us set $F_n = E_1 \cup \cdots \cup E_n$, so that

$$M \cap E_n = M \cap F_n, \qquad F_n \subset F_{n+1}.$$

If F is the union of the F_n's, then $F \in C$ and

$$\lim_n \mu_M(M \cap F_n) = \lim_n \mu(F_n) = \mu(F) = \mu_M(M \cap F)$$
$$= \mu_M\left(\bigcup_n (M \cap F_n)\right).$$

EXERCISE 2·26: Let X be a non-empty set. Show that the σ-algebra generated by the sets $\{x\}$, $x \in X$ consists of the sets $E \subset X$ such that E or $X - E$ is countable. Prove that a positive measure is defined on this σ-algebra by setting $\mu(E) = 0$ or 1 according as E is countable or not.

Δ∇Δ = ∇Δ∇ = Δ∇Δ = ∇Δ∇ = Δ∇Δ

SOLUTION: It is clear that the σ-algebra generated by the $\{x\}$, $x \in X$ contains all the sets indicated. It remains to prove that these sets form a σ-algebra. The family of them is closed under complementation and contains the empty set. If (E_n) is a sequence of such sets their union is countable if all of them are; if the complement of one of them is countable the complement of their union will be, *a fortiori*. This proves the first part of the Exercise.

Let us now note that if the E_n are mutually disjoint there can be at most one of them that is not countable. The countable additivity of μ follows from this.

EXERCISE 2·27: Let $\mathbb{N}$ be the set of natural number, $P(\mathbb{N})$ the σ-algebra of all subsets of $\mathbb{N}$. For every natural number n denote by $n\mathbb{N}$ the set of multiples of n.

Show that there cannot exist a probability μ on $P(\mathbb{N})$ such that for every integer $n \geqslant 1$ there holds:

$$\mu(n\mathbb{N}) = \frac{1}{n} .$$

Δ∇Δ ∇Δ∇ Δ∇Δ ∇Δ∇ Δ∇Δ

SOLUTION: Since $\{0\} \subset n\mathbb{N}$ for all $n \geqslant 1$ one would have $\mu(\{0\}) \leqslant 1/n$, and consequently $\mu(\{0\}) = 0$. Furthermore, since 0 is the unique integer that is divisible by an infinite number of prime numbers, one would have:

$$\{0\} = \bigcap_{r=1}^{\infty} \left(\bigcup_{i=r}^{\infty} p_i\mathbb{N} \right) ,$$

where (p_i) denotes the sequence of prime numbers. From this it would result that

$$0 = \lim_{r\to\infty} \mu\left(\bigcup_{i=r}^{\infty} p_i\mathbb{N} \right) . \qquad (*)$$

By virtue of an elementary property of arithmetic one has:

$$p_{i_1}\mathbb{N} \cap \cdots \cap p_{i_\sigma}\mathbb{N} = p_{i_1} \cdots p_{i_\sigma}\mathbb{N} \qquad (1 \leqslant i_1 < \cdots < i_\sigma).$$

Therefore one would have

$$\mu(p_{i_1}\mathbb{N}\cap\cdots\cap p_{i_\sigma}\mathbb{N}) = \frac{1}{p_{i_1}\cdots p_{i_\sigma}} .$$

If $1 \leqslant r \leqslant s$ Poincaré's Formula (cf., Exercise 1·1) would give

$$\mu\left(\bigcup_{i=r}^{s} p_i\mathbb{N}\right) = \sum_{r\leqslant i_1<\cdots<i_\sigma\leqslant s} (-1)^{\sigma-1} \frac{1}{p_{i_1}\cdots p_{i_\sigma}}$$

$$= 1 - \prod_{i=r}^{s}\left(1 - \frac{1}{p_i}\right) .$$

Now, it is known (see the end of this solution) that

$$\lim_{s\to\infty}\prod_{i=1}^{s}\left(1 - \frac{1}{p_i}\right) = 0, \tag{**}$$

and therefore that for $r \geqslant 1$

$$\lim_{s\to\infty}\prod_{i=r}^{s}\left(1 - \frac{1}{p_i}\right) = 0.$$

Therefore for all $r \geqslant 1$ one would have

$$\mu\left(\bigcup_{i=r}^{\infty} p_i\mathbb{N}\right) = \lim_{s\to\infty}\mu\left(\bigcup_{i=r}^{s} p_i\mathbb{N}\right) = 1,$$

which would contradict (*).

Now for the proof of (**). Let A_s be the set of non-zero integers that do not have prime factors greater than p_s. Then:

$$\prod_{i=1}^{s}\frac{1}{1-\frac{1}{p_i}} = \prod_{i=1}^{s}\left(\sum_{n=0}^{\infty}\frac{1}{p_i^n}\right) = \sum_{k\in A_s}\frac{1}{k} ,$$

whence:

$$\lim_{s\to\infty} \prod_{i=1}^{s} \frac{1}{1 - \frac{1}{p_i}} = \lim_{s\to\infty} \sum_{k\in A_s} \frac{1}{k} = \sum_{k=1}^{\infty} \frac{1}{k} = \infty.$$

EXERCISE 2·28: Show that there does not exist a σ-algebra having a countably infinite number of elements.

$\Delta\nabla\Delta = \nabla\Delta\nabla = \Delta\nabla\Delta = \nabla\Delta\nabla = \Delta\nabla\Delta$

SOLUTION: Let X be a set and let C be a countable family of sets of X which is a σ-algebra on X. For all $x \in X$ the set of the $E \in C$ such that $x \in E$ is countable; therefore the intersection E_x of these sets belongs to C, and this is the smallest set of C that contains x. Since for all $E \in C$ the point $x \in (E_x - E)$ or $x \in E_x \cap E$, one has either $E_x - E = E_x$ or $E_x \cap E = E_x$, that is to say $E_x \cap E = \emptyset$ or $E_x \subset E$. In particular, for two arbitrary points $x,y \in X$ one has $E_x \cap E_y = \emptyset$ or $E_x = E_y$.

Let I be a countable set such that

$$\{E_x\}_{x\in X} = \{E_i\}_{i\in I},$$

with $E_i \neq E_j$ if $i \neq j$. For every subset $A \subset I$,

$$E_A = \bigsqcup_{i\in A} E_i \in C,$$

and $E_A \neq E_B$ if $A \neq B$. It is clear that $A \mapsto E_A$ is a bijection of $P(I)$ onto C. Then if I is finite C is finite, and if I is infinite C is not countable.

CHAPTER 3

The Fundamental Theorems

EXERCISE 3·29: Calculate

$$\int_0^1 x^n \log x dx$$

for every integer $n \geqslant 0$, and deduce from it the value of

$$\int_0^1 \frac{\log x}{1-x} dx,$$

given that

$$\sum_{n=1}^{\infty} \frac{1}{n^2} = \frac{\pi^2}{6}.$$

Δ∇Δ = ∇Δ∇ = Δ∇Δ = ∇Δ∇ = Δ∇Δ

SOLUTION: By setting $x = e^{-t}$ one has:

$$\int_0^1 x^n \log x dx = -\int_0^{\infty} te^{-(n+1)t} dt =$$ (Contd)

(Contd) $$= -\frac{1}{(n+1)^2}\int_0^\infty te^{-t}dt = -\frac{1}{(n+1)^2}.$$

Moreover, if $0 < x < 1$,

$$-\frac{\log x}{1-x} = \sum_{n=0}^{\infty} -x^n\log x.$$

As the functions $x \mapsto -x^n\log x$ are positive, one can integrate term by term, which gives

$$\int_0^1 \frac{\log x}{1-x}\,dx = \sum_{n=0}^{\infty} -\frac{1}{(n+1)^2} = -\frac{\pi^2}{6}.$$

EXERCISE 3·30: (a): Let $a > 0$. For what values of $s \in \mathbb{C}$ is the function $x \mapsto x^s e^{-ax}$ integrable on $\mathbb{R}_+$? Also calculate the value of its integral with the aid of the Γ function.

(b): Show that for $\mathrm{Re}(s) > 1$,

$$\sum_{n=1}^{\infty} n^{-s} = \frac{1}{\Gamma(s)}\int_0^\infty \frac{x^{s-1}}{e^x - 1}\,dx.$$

ΔVΔ = VΔV = ΔVΔ = VΔV = ΔVΔ

SOLUTION: (a): Since

$$|x^s e^{-ax}| = x^{\mathrm{Re}(s)}e^{-ax},$$

it is clear that this function is integrable only when $\mathrm{Re}(s) > -1$, and that then:

$$\int_0^x x^s e^{-ax}dx = a^{-(s+1)}\int_0^\infty x^s e^{-x}dx = a^{-(s+1)}\Gamma(s+1).$$

SOLUTION: (b): For $x > 0$

$$\frac{x^{s-1}}{e^x - 1} = e^{-x}\frac{x^{s-1}}{1 - e^{-x}} = \sum_{n=1}^{\infty} x^{s-1}e^{-nx}. \qquad (*)$$

Since

$$\sum_{n=1}^{\infty}\int_0^{\infty}|x^{s-1}e^{-nx}|\,dx = \sum_{n=1}^{\infty}\int_0^{x} x^{\mathrm{Re}(s)-1}e^{-nx}dx$$

$$= \Gamma(\mathrm{Re}(s))\sum_{n=1}^{\infty} n^{-\mathrm{Re}(s)} < \infty$$

if $\mathrm{Re}(s) > 1$, equation (*) can be integrated term by term, giving

$$\int_0^{\infty}\frac{x^{s-1}}{e^s - 1}\,ds = \sum_{n=1}^{\infty}\int_0^{\infty}x^{s-1}e^{-nx}dx = \Gamma(s)\sum_{n=1}^{\infty} n^{-s}.$$

EXERCISE 3·31: For every integer $n \geqslant 0$ calculate:

$$\int_0^{\infty} t^{2n}e^{-t^2}dt,$$

and from this deduce that for all $z \in \mathbb{C}$ the function $t \mapsto e^{-t^2}\cos zt$ is integrable on $[0,\infty]$, and calculate its integral.

ΔVΔ VΔV ΔVΔ VΔV ΔVΔ

SOLUTION: By making the change of variable $t \to t^{\frac{1}{2}}$ one has:

$$\int_0^{\infty} t^{2n}e^{-t^2}dt = \tfrac{1}{2}\int_0^{\infty} t^{n-\frac{1}{2}}e^{-t}dt = \tfrac{1}{2}\Gamma(n + \tfrac{1}{2}) = \qquad \text{(Contd)}$$

(Contd)
$$= \tfrac{1}{2}(n - \tfrac{1}{2}) \cdots \tfrac{1}{2}\Gamma(\tfrac{1}{2})$$

$$= \frac{1.3.\ \cdots\ (2n - 1)}{2^{n+1}} \sqrt{\pi} = \frac{(2n)!}{2^{2n+1} n!} \sqrt{\pi},$$

since $\Gamma(\tfrac{1}{2}) = \sqrt{\pi}$. Therefore

$$e^{-t^2} \cos zt = \sum_{n=0}^{\infty} (-1)^n \frac{z^{2n}}{(2n)!} t^{2n} e^{-t^2},$$

and

$$\sum_{n=0}^{\infty} \int_0^{\infty} \left| (-1)^n \frac{z^{2n}}{(2n)!} t^{2n} e^{-t^2} \right| dt = \sqrt{\pi} \sum_{n=0}^{\infty} \frac{|z|^{2n}}{2^{2n+1} n!} < \infty.$$

Thus,

$$\int_0^{\infty} e^{-t^2} \cos zt dt = \sum_{n=0}^{\infty} (-1)^n \frac{z^{2n}}{(2n)!} \int_0^{\infty} t^{2n} e^{-t^2} dt$$

$$= \frac{\sqrt{\pi}}{2} \sum_{n=0}^{\infty} (-1)^n \frac{\left(\frac{z^2}{4}\right)^n}{n!}$$

$$= \frac{\sqrt{\pi}}{2} e^{-z^2/4}.$$

EXERCISE 3·32: Establish the relation:

$$\int_0^{\infty} \frac{\sin ax}{e^x - 1} dx = \sum_{n=1}^{\infty} \frac{a}{n^2 + a^2}.$$

Δ∇Δ = ∇Δ∇ = ΔΔ∇ = ∇Δ∇ = Δ∇Δ

SOLUTION: If $x > 0$,

$$\frac{\sin ax}{e^{x} - 1} = \sum_{n=1}^{\infty} e^{-nx} \sin ax. \tag{*}$$

Furthermore, by using the inequality $|\sin u| \leqslant |u|$, one has for $n \geqslant 1$

$$\int_0^{\infty} |e^{-nx} \sin ax| dx \leqslant a \int_0^{\infty} x e^{-nx} dx = \frac{a}{n^2},$$

so that:

$$\sum_{n=1}^{\infty} \int_0^{\infty} |e^{-nx} \sin ax| dx < \infty.$$

Equation (*) can therefore be integrated term by term, which gives the stated result, since

$$\int_0^{\infty} e^{-nx} \sin ax dx = \operatorname{Im} \int_0^{\infty} e^{-x(n-ia)} dx = \frac{a}{n^2 + a^2}.$$

EXERCISE 3·33: Find a method of calculating the series

$$\sum_{n=1}^{\infty} \frac{(-1)^{n+1}}{n}$$

by the integration of a series of integrable functions.

ΔVΔ = VΔV = ΔVΔ = VΔV = ΔVΔ

SOLUTION: Evidently there is a large number of possible methods. But it is first necessary to reduce it to the calculation of an absolutely convergent series. One can proceed as follows:

$$\sum_{n=1}^{\infty} \frac{(-1)^{n+1}}{n} = \sum_{n=0}^{\infty} \left(\frac{1}{2n + 1} - \frac{1}{2n + 2}\right) = \sum_{n=0}^{\infty} \frac{1}{(2n + 1)(2n + 2)} .$$

Furthermore, for $0 \leqslant x < 1$

$$\frac{1}{2} \log \frac{1 + x}{1 - x} = \sum_{n=0}^{\infty} \frac{x^{2n+1}}{2n + 1}$$

and consequently

$$\frac{1}{2}\int_0^1 \log \frac{1 + x}{1 - x}\, dx = \sum_{n=0}^{\infty} \frac{1}{(2n + 1)(2n + 2)} .$$

Now, the integral is equal to

$$[(1 + x)\log(1 + x) + (1 - x)\log(1 - x)]_{x=0}^{x=1} = 2\log 2,$$

so we have proved that:

$$\sum_{n=1}^{\infty} \frac{(-1)^{n+1}}{n} = \log 2.$$

REMARK: Recall that the most natural method is to apply Abel's Theorem to the series development of $\log(1 + x)$.

EXERCISE 3·34: Show that if $\alpha \geqslant 0$ then

$$\int_0^{\infty} \frac{\sin \alpha x}{x(x^2 + 1)}\, dx = \frac{\pi}{2} (1 - e^{-\alpha}).$$

(Consider the function of α defined by the left hand side).

Δ∇Δ = ∇Δ∇ = ΔΔ∇ = ∇Δ∇ = ΔΔ∇

SOLUTION: Let $\varphi(\alpha)$ be this left hand side. Since

$$\left|\frac{\cos \alpha x}{x^2 + 1}\right| \leqslant \frac{1}{x^2 + 1},$$

the function φ is continuously differentiable on $\mathbb{R}^+$, and

$$\varphi'(\alpha) = \int_0^\infty \frac{\cos \alpha x}{x^2 + 1}\, dx.$$

Furthermore, if $\alpha > \alpha_0 > 0$, then for $A > 1$:

$$\int_A^\infty \frac{x \sin \alpha x}{x^2 + 1}\, dx < \frac{2}{A\alpha_0}$$

(by the second mean-value formula, since $x(x^2 + 1)^{-1}$ is decreasing for $x \geqslant 1$). This proves the uniform convergence of the improper integral

$$\int_0^\infty \frac{x \sin \alpha x}{x^2 + 1}\, dx$$

on $[\alpha_0, \infty[$. Consequently φ is twice continuously differentiable for $\alpha > 0$, and

$$\varphi''(\alpha) = -\int_0^\infty \frac{x \sin \alpha x}{x^2 + 1}\, dx = -\int_0^\infty \frac{\sin \alpha x}{x}\, dx + \int_0^\infty \frac{\sin \alpha x}{x(x^2 + 1)}\, dx$$

$$= -\frac{\pi}{2} + \varphi(\alpha).$$

On $[0, \infty[$ φ is therefore a solution of the system

$$\varphi'' - \varphi = -\frac{\pi}{2},$$

$$\varphi(0) = 0, \qquad \varphi'(0) = \frac{\pi}{2}.$$

(Indeed, as φ coincides on $]0,\infty[$ with a solution of $\varphi'' - \varphi = -\pi/2$ and as, furthermore, it is continuous on $[0,\infty[$, it satisfies this differential equation on $[0,\infty[$). From this it follows that

$$\varphi(\alpha) = \frac{\pi}{2}(1 - e^{-\alpha}).$$

EXERCISE 3·35: For every integer $n \geqslant 1$ and all real x, let

$$f_n(x) = e^{-nx} - 2e^{-2nx}.$$

Show that the series with the general term $f_n(x)$ is convergent for all $x > 0$, and calculate its sum $f(x)$. Next, show that each f_n, as well as f, is integrable on $(0,\infty)$, and compare

$$\int_0^\infty f(x)dx \quad \text{and} \quad \sum_{n=1}^{\infty} \int_0^\infty f_n(x)dx.$$

Explain this result.

Δ∇Δ = ∇Δ∇ = ΔVΔ = ∇Δ∇ = ΔVΔ

SOLUTION: For $x > 0$ one has:

$$\sum_{n=1}^{\infty} f_n(x) = \frac{e^{-x}}{1 - e^{-x}} - 2\frac{e^{-2x}}{1 - e^{-2x}}$$

$$= \frac{1}{e^x - 1} - \frac{2}{e^{2x} - 1}$$

$$= \frac{1}{e^x + 1}.$$

On the other hand,

$$\int_0^\infty f_n(x)\mathrm{d}x = \frac{1}{n} - 2\,\frac{1}{2n} = 0,$$

so that:

$$\sum_{n=1}^{\infty} \int_0^\infty f_n(x)\mathrm{d}x = 0.$$

However,

$$\int_0^\infty \frac{\mathrm{d}x}{e^x + 1} = \left[\log \frac{e^x}{e^x + 1}\right]_{x=0}^{x=\infty} = \log 2.$$

Hence one must have:

$$\sum_{n=1}^{\infty} \int_0^\infty |f_n(x)|\,\mathrm{d}x = \infty,$$

which is immediately verified by observing that

$$\int_0^\infty |f_n(x)|\,\mathrm{d}x = \frac{1}{n}\int_0^1 |1 - 2u|\,\mathrm{d}u = \frac{1}{2n}\,.$$

EXERCISE 3·36: (a): Let $(f_n)_{n\geq 1}$ be the sequence of functions defined on $\mathbb{R}$ by

$$f_n(x) = \begin{cases} n^2 x & \text{if } 0 \leq x \leq \frac{1}{n}\,, \\ -\,n^2\left(x - \frac{2}{n}\right) & \text{if } \frac{1}{n} \leq x \leq \frac{2}{n}\,, \\ 0 & \text{if } x \geq \frac{2}{n}\,. \end{cases}$$

Calculate the four numbers:

$$A = \liminf_n \int f_n, \qquad A' = \int \liminf_n f_n,$$

$$B = \limsup_n \int f_n, \qquad B' = \int \limsup_n f_n.$$

(b): The same question for the sequence $(g_n)_{n\geq 0}$ defined by:

$$g_n = \mathbb{1}_{[0,\frac{1}{4}]} \text{ if } n \text{ is even}, \qquad g_n = \mathbb{1}_{[\frac{1}{4},1]} \text{ otherwise.}$$

(c): If (h_n) is a sequence of positive measurable functions, what can be said of the four numbers A, A', B, B', and more particularly about B and B'? Is the Fatou Lemma true for sequences of real measurable functions with arbitrary sign?

Δ∇Δ — ∇Δ∇ — ΔΔ∇ — ∇Δ∇ — ΔΔ∇

SOLUTION: (a): For all n one has

$$\int f_n = 1,$$

so that $A = B = 1$. On the other hand,

$$\liminf_n f_n = \limsup_n f_n = 0,$$

and $A' = B' = 0$. In particular, $A' < A$ and $B' < B$.

SOLUTION: (b): In this case $\int g_n$ is alternately equal to $\frac{1}{4}$ and $\frac{3}{4}$, so that $A = \frac{1}{4}$, $B = \frac{3}{4}$. Furthermore,

$$\liminf_n g_n = 0 \quad \text{and} \quad \limsup_n g_n = \mathbb{1}_{[0,1]},$$

whence $A' = 0$, $B' = 1$. Thus $A' < A < B < B'$.

SOLUTION: (c): There always holds:

$$A' \leq A \leq B \quad \text{and} \quad A' \leq B'.$$

(The inequality $A' \leq A$ is Fatou's Lemma, the two others are clear).

The two preceding examples prove that in general nothing can be said about the relative values of B and B'.

However, one can extend Fatou's Lemma in the following way: let us assume that the f_n's are real and measurable, and that there exists a measurable real function g such that:

$$\begin{cases} g \leqslant f_n \quad \text{for all } n, \\ \int g_- < \infty. \end{cases}$$

One then has $A' \leqslant A$. In fact $0 \leqslant f_n - g$, and

$$-\infty < \int g \leqslant \int f_n,$$

therefore we can suppose that $\int g < \infty$, and in this case

$$\int (f_n - g) = \int f_n - \int g.$$

The classical Fatou Lemma shows that

$$\int \liminf_n (f_n - g) = \int \liminf_n f_n - \int g$$

$$\leqslant \liminf_n \int f_n - \int g,$$

whence the result. (Recall that if $\int g_- < \infty$ one sets

$$\int g = \int g_+ - \int g_- ,$$

which is an element of $]-\infty,+\infty]$, and that

$$g \leqslant h, \quad \int g_- < \infty \quad \text{imply} \quad \int h_- < \infty \quad \text{and} \quad \int g \leqslant \int h).$$

Similarly one shows that if:

$$\begin{cases} f_n \leqslant g & \text{for all } n, \\ \int g_+ < \infty, \end{cases}$$

then

$$\limsup_n \int f_n \leqslant \int \limsup_n f_n.$$

(Consider the functions $g - f_n$ and recall that

$$\limsup_n (-f_n) = -\liminf_n f_n.$$

Here one again sets

$$\int g = \int g_+ - \int g_- \quad \text{if} \quad \int g_+ < \infty,$$

and

$$g \leqslant h, \quad \int h_+ < \infty \quad \text{imply} \quad \int g_+ < \infty \quad \text{and} \quad \int g \leqslant \int h).$$

EXERCISE 3·37: (a): Show that if f is integrable on $\mathbb{R}^m$, and if K is a compact set of this space, then

$$\lim_{\|z\|\to\infty} \int_{K+z} |f(x)|dx = 0.$$

(b): Show that if f is uniformly continuous on $\mathbb{R}^m$, and that if there exists $p > 0$ such that $|f|^p$ is integrable, then

$$\lim_{\|x\|\to\infty} f(x) = 0.$$

ΔVΔ = VΔV = ΔVΔ = VΔV = ΔVΔ

SOLUTION: (a):

$$\lim_{r\to\infty}\int_{\|x\|\geqslant r}|f(x)|\,dx = 0,$$

since

$$f_r = 1\!\!1_{(\|x\|\geqslant r)}f \to 0 \quad \text{and} \quad |f_r| \leqslant |f|$$

(Lebesgue's Theorem). If $\delta = \sup\{\|y\| : y \in K\}$,

$$\int_{K+z}|f(x)|\,dx \leqslant \int_{\|x\|>\|z\|-\delta}|f(x)|\,dx,$$

whence the result.

SOLUTION: (b): Assume that $f_n \not\to 0$ as $\|x\| \to \infty$. There there exist $\varepsilon > 0$ and a sequence x_n such that

$$\|x_n\| \to \infty, \qquad |f(x_n)| \geqslant \varepsilon$$

As the function f is uniformly continuous there exists a closed ball B, with centre 0 and radius greater than zero, such that

$$|f(y) - f(x)| < \frac{\varepsilon}{2} \quad \text{if} \quad y \in B + x.$$

In particular, $|f| \geqslant \varepsilon/2$ on the balls $B + x_n$, and consequently

$$\int_{B+x_n}|f|^p \geqslant \left(\frac{\varepsilon}{2}\right)^p \operatorname{meas}(B) > 0,$$

which contradicts the first part of the Exercise.

EXERCISE 3·38: Let G be a continuous function on $\mathbb{R}$ such that $G(0) = 0$ and $G(x) > 0$ if $x \neq 0$.

Show that if f is an uniformly continuous bounded real function on $\mathbb{R}^m$ and

$$\int_{\mathbb{R}^m} G(f(x))dx < \infty,$$

then

$$\lim_{\|x\|\to\infty} f(x) = 0.$$

$$\Delta\nabla\Delta = \nabla\Delta\nabla = \Delta\nabla\Delta = \nabla\Delta\nabla = \Delta\nabla\Delta$$

SOLUTION: If $f(x) \not\to 0$ as $x \to \infty$, there would exist $\varepsilon > 0$ and a sequence x_n such that

$$\|x_n\| \to \infty, \qquad |f(x_n)| \geqslant \varepsilon.$$

As the function f is uniformly continuous there exists a closed ball B, with centre 0 and radius greater than zero, such that

$$|f(y) - f(x)| < \frac{\varepsilon}{2} \text{ if } y \in B + x.$$

In particular, if M is the upper bound of $|f|$ on $\mathbb{R}^m$, one would have

$$\frac{\varepsilon}{2} \leqslant |f| \leqslant M$$

on the balls $B + x_n$. By virtue of the assumptions made about G, one has

$$\inf\{G(u): \frac{\varepsilon}{2} \leqslant |u| \leqslant M\} = \mu > 0,$$

so that

$$\int_{B+x_n} G(f(x))dx \geqslant \mu\text{meas}(B),$$

which would contradict that $G(f)$ is integrable (cf., the preceding Exercise).

EXERCISE 3·39: (a): Let f be an integrable function on $\mathbb{R}^p$. Show that

$$\text{meas}(|f| > a) = o\left(\frac{1}{a}\right) \quad \text{as } a \to \infty.$$

(b): Show that if f is measurable (and almost everywhere finite) on $\mathbb{R}^p$ then it is integrable if and only if

$$\sum_{n=-\infty}^{+\infty} 2^n \text{meas}(2^{n-1} < |f| \leqslant 2^n) < \infty.$$

ΔVΔ = VΔV = ΔVΔ = VΔV = ΔVΔ

SOLUTION: (a): This results from

$$\lim_{a\to\infty} \int_{|f|>a} |f(x)|\,dx = 0,$$

and the inequality

$$a\,\text{meas}(|f| > a) \leqslant \int_{|f|>a} |f(x)|\,dx.$$

SOLUTION: (b): Let us set

$$A_n = (2^{n-1} < |f| \leqslant 2^n), \quad B = (f = 0), \quad C = (|f| = \infty).$$

These sets are mutually disjoint, with union $\mathbb{R}^p$, and C has measure zero. Then

$$\int_{\mathbb{R}^p} |f| = \sum_{n=-\infty}^{+\infty} \int_{A_n} |f| ,$$

whence

$$\frac{1}{2}\sum_{n=-\infty}^{+\infty} 2^n \text{meas}(A_n) \leqslant \int_{\mathbb{R}^p} |f| \leqslant \sum_{n=-\infty}^{+\infty} 2^n \text{meas}(A_n),$$

which proves that the condition is necessary and sufficient.

EXERCISE 3·40: Let f be an integrable positive real function on $\mathbb{R}^p$.

Show that there exist measurable sets $A_n \subset \mathbb{R}^p$ $(n \in \mathbb{Z})$ with finite measure such that

$$f(x) = \sum_{n=-\infty}^{+\infty} 2^n \mathbb{1}_{A_n}(x)$$

for all $x \in \mathbb{R}^p$.

ΔVΔ = VΔV = ΔVΔ = VΔV = VΔV

SOLUTION: Observe that if for all $t \in \mathbb{R}_+$ and $n \in \mathbb{Z}$ one sets:

$$a_n(t) = \mathbb{1}_{[\frac{1}{2},1[}(2^{-(n+1)}t - [2^{-(n+1)}t]),$$

where $[x]$ denotes the integral part of x, then

$$t = \sum_{n=-\infty}^{+\infty} 2^n a_n(t).$$

(Note that $a_n(t) = 0$ whenever $2^n > t$; the $a_n(t)$ are simply the digits of the binary expansion of t). In particular,

$$f(x) = \sum_{n=-\infty}^{+\infty} 2^n a_n(f(x)).$$

If one sets $A_n = \{x : x \in \mathbb{R}^p \text{ and } a_n(f(x)) = 1\}$, then A_n is measurable, and

$$f(x) = \sum_{n=-\infty}^{+\infty} 2^n \mathbb{1}_{A_n}(x).$$

Also,

$$2^n \text{meas}(A_n) \leqslant \int_{\mathbb{R}^p} f(x)\,dx,$$

so A_n has finite measure.

EXERCISE 3·41: Let X be a measurable set of $\mathbb{R}^p$ such that meas(X) = 1.

Show that if f is integrable on X then

$$\int_X f(x)\,dx = 0$$

if and only if

$$\int_X |1 + zf(x)|\,dx \geqslant 1$$

for every complex number z.

ΔVΔ = VΔV = ΔVΔ = VΔV = ΔVΔ

SOLUTION: The condition is necessary, because

$$\int_X |1 + zf(x)|\,dx \geqslant \left|\int_X (1 + zf(x))\,dx\right| = \left|1 + z\int_X f(x)\,dx\right| = 1.$$

Let us now show that it is sufficient. The inequality can be written in the form:

$$\int_X \frac{|1 + \rho e^{i\vartheta} f(x)| - 1}{\rho}\,dx \geqslant 0,$$

where $\rho > 0$, $\vartheta \in \mathbb{R}$. An easy calculation shows that:

$$\frac{|1 + \rho z| - 1}{\rho} = \frac{2\mathrm{Re}(z) + \rho|z|^2}{|1 + \rho z| + 1}$$

if $\rho > 0$, $z \in \mathbb{C}$, so that

$$\lim_{\rho \to 0} \frac{|1 + \rho e^{i\vartheta} f(x)| - 1}{\rho} = \mathrm{Re}(e^{i\vartheta} f(x)).$$

Moreover,

$$\left| \frac{|1 + \rho e^{i\vartheta} f(x)| - 1}{\rho} \right| \leqslant |f(x)|.$$

Therefore, by Lebesgue's Theorem,

$$\begin{aligned} \mathrm{Re}(e^{i\vartheta} \int_X f(x)\mathrm{d}x) &= \int_X \mathrm{Re}(e^{i\vartheta} f(x))\mathrm{d}x \\ &= \lim_{\rho \to 0} \int_X \frac{|1 + \rho e^{i\vartheta} f(x)| - 1}{\rho} \mathrm{d}x \\ &\geqslant 0, \end{aligned}$$

and this holds for all $\vartheta \in \mathbb{R}$. By choosing ϑ so that

$$e^{i\vartheta} \int_X f(x)\mathrm{d}x = - \left| \int_X f(x)\mathrm{d}x \right|$$

one obtains

$$\int_X f(x)\mathrm{d}x = 0.$$

EXERCISE 3·42: Let f be an integrable function on $\mathbb{R}$ and let $a > 0$. Show that for almost all $x \in \mathbb{R}$ the series

$$\sum_{n=-\infty}^{+\infty} f\left(\frac{x}{a} + n\right) \tag{*}$$

is absolutely convergent, and that its sum $F(x)$ is periodic with period a and is integrable on $(0,a)$.

ΔVΔ = VΔV = ΔVΔ = VΔV = ΔVΔ

SOLUTION: Setting $u = \dfrac{x}{a} + n$ on obtains

$$\sum_{n=-\infty}^{+\infty} \int_0^a \left|f\left(\frac{x}{a} + n\right)\right| dx = a \sum_{n=-\infty}^{+\infty} \int_n^{n+1} |f(x)|\,dx = a\int_{-\infty}^{+\infty} |f(x)|\,dx$$

$$< \infty,$$

so that the series in (*) converges absolutely at almost all points of $(0,a)$. As this series does not change when x is replaced by $x + a$, it follows from this that it is absolutely convergent at almost all points of $\mathbb{R}$, and that its sum coincides with a function F of period a (setting, for example, $F(x) = 0$ at the points x where the series is not absolutely convergent) that is integrable on $(0,a)$.

Since one can integrate term by term, one obtains by proceeding as above:

$$\frac{1}{a}\int_0^a F(x)\,dx = \int_{-\infty}^{+\infty} f(x)\,dx.$$

EXERCISE 3·43: Let f be an integrable function on $\mathbb{R}$, and let $\alpha > 0$. Show that for almost all $x \in \mathbb{R}$:

$$\lim_{n\to\infty} n^{-\alpha} f(nx) = 0.$$

$$\Delta\nabla\Delta = \nabla\Delta\nabla = \Delta\nabla\Delta = \nabla\Delta\nabla = \Delta\nabla\Delta$$

SOLUTION:

$$\int_{-\infty}^{+\infty} |n^{-\alpha} f(nx)|\, dx = n^{-(\alpha+1)} \int_{-\infty}^{+\infty} |f(x)|\, dx,$$

so

$$\sum_{n=1}^{\infty} \int_{-\infty}^{+\infty} |n^{-\alpha} f(nx)|\, dx < \infty.$$

From this it follows that for almost all $x \in \mathbb{R}$

$$\sum_{n=1}^{\infty} n^{-\alpha} |f(nx)| < \infty,$$

and in particular that

$$n^{-\alpha} f(nx) \to 0.$$

EXERCISE 3·44: Let f be a measurable complex function on $\mathbb{R}$ with period $T > 0$ and such that:

$$A = \int_0^T |f(x)|\, dx < \infty.$$

(a): Show that for almost all x,

$$\lim_n n^{-2} f(nx) = 0.$$

(b): From this deduce that for almost all x,

$$\lim_n |\cos nx|^{1/n} = 1.$$

$$\Delta\nabla\Delta = \nabla\Delta\nabla = \Delta\nabla\Delta = \nabla\Delta\nabla = \Delta\nabla\Delta$$

SOLUTION: (a): Set

$$\varphi_n(x) = \frac{1}{n^2} f(nx).$$

Then:

$$\int_0^T |\varphi_n(x)|\,dx = \frac{1}{n^2}\int_0^T |f(nx)|\,dx = \frac{1}{n^3}\int_0^{nT} |f(x)|\,dx$$

$$= \frac{A}{n^2} .$$

Hence

$$\sum_n \int_0^T |\varphi_n(x)|\,dx < \infty.$$

The series $\sum \varphi_n(x)$ therefore converges for almost all x, and in particular

$$\varphi_n(x) = n^{-2} f(nx) \to 0$$

for almost all x.

SOLUTION: (b): Consider the function:

$$f(x) = (\log|\cos x|)^2.$$

It has period π and is integrable on $[0,\pi]$, for in the neighbourhood of $\pi/2$ it is equivalent to $(\log|x - \frac{1}{2}\pi|)^2$. By the preceeding,

$$\lim_n (n^{-1}\log|\cos nx|)^2 = 0,$$

i.e.,

$$\lim_n |\cos nx|^{1/n} = 1$$

for almost all x.

EXERCISE 3·45: Let $(x_n)_{n\geq 1}$ be a sequence of points of $[0,1]$. If $0 \leqslant a \leqslant b \leqslant 1$ and $N \geqslant 1$ is an integer, one denotes by $\nu(N;a,b)$ the number of integers n such that $1 \leqslant n \leqslant N$ and $a \leqslant x_n \leqslant b$. The sequence $(x_n)_{n\geq 1}$ is called an *EQUIDISTRIBUTED SEQUENCE* if, for any $0 \leqslant a \leqslant b \leqslant 1$,

$$\lim_{N\to\infty} \frac{\nu(N;a,b)}{N} = b - a.$$

(a): Show that the following conditions are equivalent:

(i): The sequence $(x_n)_{n\geq 1}$ is equidistributed;

(ii): For every continuous function f on $[0,1]$

$$\int_0^1 f(x)dx = \lim_{N\to\infty} N^{-1} \sum_{n=1}^{N} f(x_n); \tag{*}$$

(iii): For every integer $p \geqslant 1$,

$$\lim_{N\to\infty} N^{-1} \sum_{n=1}^{N} e^{2\pi i p x_n} = 0.$$

APPLICATIONS: Investigate whether the sequences

$$x_n = n\alpha - [n\alpha] \qquad (\alpha \text{ irrational})$$

and

$$x_n = \log n - [\log n]$$

are equidistributed (here $[a]$ denotes the integral part of a).

(b): Show that if $(x_n)_{n\geq 1}$ is equidistributed and if f is Riemann integrable, then (*) still holds.

Δ∇Δ = ∇Δ∇ = Δ∇Δ = ∇Δ∇ = Δ∇Δ

SOLUTION: We shall prove part (b) first of all, which will also prove the step (i) => (ii) of part (a). We may assume that f is real. Let us note that if (x_n) is equidistributed then (*) holds for every characteristic function of an interval $[a,b]$, $0 \leqslant a \leqslant b \leqslant 1$. This formula therefore also holds, by linearity, for every step function. Now for every $\varepsilon > 0$ there exists a step function φ such that:

$$f \leqslant \varphi, \qquad \int_0^1 \varphi \leqslant \varepsilon + \int_0^1 f$$

From this one deduces that:

$$\limsup_N N^{-1} \sum_{n=1}^{N} f(x_n) \leqslant \lim_N N^{-1} \sum_{n=1}^{N} \varphi(x_n) \leqslant \varepsilon + \int_0^1 f,$$

and consequently

$$\limsup_{N\to\infty} N^{-1} \sum_{n=1}^{N} f(x_n) \leqslant \int_0^1 f.$$

Replacing f by $-f$, this yields:

$$\liminf_{N\to\infty} N^{-1} \sum_{n=1}^{N} f(x_n) \geqslant \int_0^1 f,$$

which shows that (*) holds for f.

That (ii) => (iii) is trivial upon setting $f(x) = e^{2\pi ipx}$ in (*).

Finally, let us show that (iii) => (i). We first show that

(iii) => (ii)[†]. Now, if (iii) holds, then (*) is true for every trigonometric polynomial. If f is continuous on [0,1], and if $\varepsilon > 0$, there exists such a trigonometric polynomial φ for which $|f - \varphi| \leqslant \varepsilon$ (the Stone-Weierstrass Theorem). To simplify the ensuing calculation, set

$$u_N(f) = N^{-1} \sum_{n=1}^{N} f(x_n) - \int_0^1 f$$

Then

$$|u_N(f)| \leqslant |u_N(\varphi)| + |u_N(f - \varphi)| \leqslant |u_N(\varphi)| + 2\varepsilon,$$

and consequently, since $u_N(\varphi) \to 0$,

$$\lim_{N\to\infty} \sup |u_N(f)| \leqslant 2\varepsilon,$$

which proves that (ii) is satisfied. Now let $0 \leqslant a \leqslant b \leqslant 1$ and let ψ be the characteristic function of $[a,b]$. For every $\varepsilon > 0$ there exist two continuous functions g_ε and f_ε such that

$$0 \leqslant f_\varepsilon \leqslant \psi \leqslant g_\varepsilon \leqslant 1,$$

$$f_\varepsilon(x) = 1 \quad \text{if } a + \varepsilon \leqslant x \leqslant b - \varepsilon,$$

$$g_\varepsilon(x) = 0 \quad \text{if } x \leqslant a - \varepsilon \quad \text{or} \quad x \geqslant b + \varepsilon.$$

Then

$$b - a - 2\varepsilon \leqslant \int f_\varepsilon = \lim_{N\to\infty} N^{-1} \sum_{n=1}^{N} f_\varepsilon(x_n) \leqslant \lim_{N\to\infty} \inf N^{-1} \sum_{n=1}^{N} \psi(x_n),$$

$$b - a + 2\varepsilon \geqslant \int g_\varepsilon = \lim_{N\to\infty} N^{-1} \sum_{n=1}^{N} g_\varepsilon(x_n) \geqslant \lim_{N\to\infty} \sup N^{-1} \sum_{n=1}^{N} \psi(x_n).$$

---

† See *Erratum*, p. 545.

Letting $\varepsilon \to 0$ yields:

$$\lim_{N\to\infty} N^{-1} \sum_{n=1}^{N} \psi(x_n) = \lim_{N\to\infty} N^{-1} \nu(N;a,b) = b - a.$$

APPLICATIONS: If $x_n = \alpha n - [\alpha n]$, and if $p \geqslant 1$ is an integer,

$$N^{-1} \sum_{n=1}^{N} e^{2\pi i p x_n} = N^{-1} \sum_{n=1}^{N} e^{2\pi i p \alpha n} = N^{-1} e^{2\pi i p \alpha} \frac{1 - e^{2\pi i p \alpha N}}{1 - e^{2\pi i p \alpha}},$$

since because α is irrational, $e^{2\pi i p \alpha} \neq 1$. Hence one clearly has

$$\lim_{N} N^{-1} \sum_{n=1}^{N} e^{2\pi i p x_n} = 0$$

in this case. When $x_n = \log n - [\log n]$ it is necessary to study the sums:

$$N^{-1} \sum_{n=1}^{N} e^{2\pi i p \log n} = N^{-1} \sum_{n=1}^{N} n^{2\pi i p}.$$

Now

$$\frac{N^{2\pi i p}}{1 + 2\pi i p} = N^{-1} \int_0^N t^{2\pi i p} dt = N^{-1} \sum_{n=1}^{N} \int_{n-1}^{n} t^{2\pi i p} d(t - n + 1),$$

whence, integrating by parts,

$$N^{-1} \sum_{n=1}^{N} n^{2\pi i p} = \frac{N^{2\pi i p}}{1 + 2\pi i p} + 2\pi i p N^{-1} \sum_{n=1}^{N} \int_{n-1}^{n} (t - n + 1) t^{2\pi i p - 1} dt.$$

When $n \geqslant 2$

$$\left| \int_{n-1}^{n} (t - n + 1) t^{2\pi i p - 1} dt \right| \leqslant \int_{n-1}^{n} t^{-1} dt = \log\left(\frac{n}{n - 1}\right).$$

From this one deduces that:

$$N^{-1}\sum_{n=1}^{N} n^{2\pi ip} = \frac{N^{2\pi ip}}{1+2\pi ip} + O\left(\frac{\log N}{N}\right),$$

which shows that in this case *the sequence* x_n *is not equidistributed.*

REMARKS: Formula (*) is not satisfied by all the Lebesgue-integrable functions. In fact, if f is the characteristic function of the set $\{x_n\}$ one has

$$\int_0^1 f(x)dx = 0,$$

but

$$N^{-1}\sum_{n=1}^{N} f(x_n) = 1 \quad \text{for all } n.$$

Let us again note that the characteristic function of a set E is Riemann integrable if and only if the set of points of discontinuity of this function has measure zero, or, in other words, if the boundary of E has measure zero. From this one deduces that if $(x_n)_{n\geq 1}$ is equidistributed and if the boundary of E has measure zero, then

$$\lim_{N\to\infty} N^{-1}\nu(N;E) = \text{meas}(E),$$

where $\nu(N;E)$ denotes the number of integers n such that $1 \leq n \leq N$ and $x_n \in E$.

EXERCISE 3·46: Let f be a continuous function on $\mathbb{R}^2$ such that:

$$f(x,y) = f(x+1,y) = f(x,y+1)$$

for any x,y.

Show that for every irrational number

$$\iint_{0\leqslant x,y\leqslant 1} f(x,y)\,dx\,dy = \lim_{T\to\infty} \frac{1}{T}\int_0^T f(t,\alpha t)\,dt.$$

Δ∇Δ = ∇Δ∇ = ΔVΔ = ∇Δ∇ = ΔVΔ

SOLUTION: For $p,q \geqslant 0$ integers, the Formula is verified without difficulty for the function:

$$\varphi_{p,q}(x) = e^{2\pi i(px+qy)}$$

by observing that $p + \alpha q \neq 0$ if $p^2 + q^2 \neq 0$ (for α is irrational). By linearity the Formula holds for the linear combinations of functions $\varphi_{p,q}$. The Stone-Weierstrass Theorem assures, for every continuous function f with period unity at x and at y, and for all $\varepsilon > 0$, the existence of such a linear combination φ satisfying $|f - \varphi| \leqslant \varepsilon$. Denoting by $u_T(f)$ the difference between the two sides of the Formula to be proved, one has:

$$|u_T(f)| \leqslant |u_T(\varphi)| + |u_T(f - \varphi)| \leqslant |u_T(\varphi)| + 2\varepsilon.$$

As we have just seen that $u_T(\varphi) \to 0$, the Formula is proved.

EXERCISE 3·47: Let g be a measurable function on $\mathbb{R}$, bounded, and of period $T > 0$.

Show that for every integrable function f, and for every sequence (a_n) of real numbers,

$$\lim_{n\to\infty}\int_{-\infty}^{+\infty} f(x)g(nx + a_n)\,dx = \left(\frac{1}{T}\int_0^T g(x)\,dx\right)\left(\int_{-\infty}^{+\infty} f(x)\,dx\right). \qquad (*)$$

(*FÉJER'S FORMULA*)

ΔVΔ = ∇Δ∇ = ΔVΔ = ∇Δ∇ = ΔVΔ

SOLUTION: Assume first that f is the characteristic function of the interval $[\alpha,\beta]$. Then

$$\int_{-\infty}^{+\infty} f(x)g(nx + a_n)dx = \int_{\alpha}^{\beta} g(nx + a_n)dx = \frac{1}{n}\int_{n\alpha+a_n}^{n\beta+a_n} g(x)dx.$$

In view of the periodicity of g,

$$\int_{n\alpha+a_n}^{n\beta+a_n} g(x)dx = \left[\frac{n(\beta - \alpha)}{T}\right]\int_0^T g(x)dx + \varepsilon_n$$

with $|\varepsilon_n| < MT$, where $M = \sup|g|$. Consequently,

$$\lim_{n\to\infty}\int_{-\infty}^{+\infty} f(x)g(nx + a_n)dx = \frac{\beta - \alpha}{T}\int_0^T g(x)dx,$$

which proves (*) in this case. By linearity the formula holds for every step function. In the general case, for all $\varepsilon > 0$ there exists a step function h such that

$$\int_{-\infty}^{+\infty} |f - h| < \varepsilon.$$

Setting

$$\gamma = \frac{1}{T}\int_0^T g(x)dx,$$

one has (taking account of $|\gamma| \leqslant M$)

$$\left|\int_{-\infty}^{+\infty} f(x)[g(nx + a_n) - \gamma]\,dx\right| < \left|\int_{-\infty}^{+\infty} h(x)[g(nx + a_n) - \gamma]dx\right| + 2M\varepsilon,$$

and consequently,

$$\limsup_{n\to\infty}\left|\int_{-\infty}^{+\infty} f(x)[g(nx + a_n) - \gamma]dx\right| \leqslant 2M\varepsilon,$$

which proves (*).

REMARK: When $g(x) = e^{ix}$ and $a_n = 0$ for all n, one obtains the Riemann-Lebesgue Lemma:

$$\lim_{n\to\infty}\int_{-\infty}^{+\infty} f(x)e^{inx}dx = 0$$

if f is integrable.

EXERCISE 3·48: Let (ρ_n) and (a_n) be two sequences of real numbers such that

$$\sum_n |\rho_n \cos(nx + a_n)| < \infty$$

for all the x's of a set A of measure greater than zero.

Show that

$$\sum_n |\rho_n| < \infty.$$

Δ∇Δ = ∇Δ∇ = ΔVΔ = ∇Δ∇ = Δ∇Δ

SOLUTION: For every integer n let A_N be the set of x's such that

$$\sum_n |\rho_n \cos(nx + a_n)| \leqslant N.$$

Since A is the union of the increasing sequence of A_N's, $\text{meas}(A_N) \to \text{meas}(A) > 0$. Hence there exists an N such that $\text{meas}(A_N) > 0$. Let $E \subset A_N$, $0 < \text{meas}(E) < \infty$. Then

(Contd)

$$\sum_n |\rho_n| \int_E |\cos(nx + a_n)| dx = \int_E \left(\sum_n |\rho_n \cos(nx + a_n)|\right) dx \leqslant$$

(Contd) $\leqslant N\text{meas}(E) < \infty.$

Furthermore, the function $x \to |\cos x|$ has period π and

$$\frac{1}{\pi}\int_0^\pi \cos x \, dx = \frac{2}{\pi},$$

so that Féjer's Formula (cf. the preceding exercise) applied to the characteristic function of E gives

$$\lim_{n\to\infty}\int_E |\cos(nx + a_n)|dx = \frac{2}{\pi}\text{meas}(E) > 0.$$

From this it follows that

$$\sum_n |\rho_n| < \infty.$$

EXERCISE 3·49: Let $n_1 < n_2 < \cdots < n_s < \cdots$ be a strictly increasing sequence of positive integers and $(\vartheta_s)_{s\geqslant 1}$ be a sequence of real numbers.

(a): Prove that for every measurable set E of $[0,2\pi]$ and for every integer $p \geqslant 1$

$$\lim_{s\to\infty}\int_E \cos^{2p}(n_s x + \vartheta_s)dx = 2^{-2p}\binom{2p}{p}\text{meas}(E).$$

(b): Deduce from this that for almost all $x \in \mathbb{R}$

$$\limsup_{s\to\infty}|\cos(n_s x + \vartheta_s)| = 1.$$

Δ∇Δ = ∇Δ∇ = ΔΔ∇Δ = ∇Δ∇ = Δ∇Δ

SOLUTION: (a): An elementary calculation shows that:

$$\cos^{2p} z = 2^{-2p}(e^{iz} + e^{-iz})^{2p}$$

$$= 2^{-2p}\binom{2p}{p} + 2^{1-2p} \sum_{k=1}^{p} \binom{2p}{p-k} \cos 2kz .$$

Furthermore,

$$\left|\int_E \cos 2k(n_s x + \vartheta_s)dx\right| < \left|\int_E \cos 2kn_s x dx\right| + \left|\int_E \sin 2kn_s x dx\right|,$$

and the two latter integrals tend towards zero by virtue of the Riemann-Lebesgue Theorem.

(b): Let $0 < \alpha < 1$ and let E_α be the set of x's $(0 \leqslant x \leqslant 2\pi)$ such that

$$\limsup_s |\cos(n_s x + \vartheta_s)| \leqslant \alpha .$$

By virtue of Fatou's Lemma and of part (a),

$$\alpha^{2p}\text{meas}(E_\alpha) \geqslant \int_{E_\alpha} \limsup_{s\to\infty} \cos^{2p}(n_s x + \vartheta_s)dx$$

$$\geqslant \limsup_{s\to\infty} \int_{E_\alpha} \cos^{2p}(n_s x + \vartheta_s)dx$$

$$= 2^{-2p}\binom{2p}{p}\text{meas}(E_\alpha).$$

Note that

$$2^{-2p}\binom{2p}{p} = 2^{-2p}\frac{(2p)!}{(p!)^2} \sim \frac{1}{\sqrt{2\pi}}\, 2^{-2p}\, \frac{(2p)^{2p+1/2}e^{-2p}}{p^{2p+1}e^{-2p}}$$

$$\sim \frac{1}{\sqrt{\pi p}} .$$

Since $\alpha^{2p}\sqrt{\pi p} \to 0$ when $p \to \infty$, it follows from the preceding that $\text{meas}(E_\alpha) = 0$. The set of x's from $[0,2\pi]$ such that

$$\limsup_{s\to\infty} |\cos(n_s x + \vartheta_s)| \neq 1$$

being the union of the $E_{1 - 1/n}$ $(n \geqslant 1)$, the proposition is proved.

EXERCISE 3·50: Let f be a positive integrable function on an open set X of $\mathbb{R}^p$ such that $\text{meas}(X) < \infty$.

Show that there exists a function g, lower semi-continuous on X, such that $g \geqslant 1/f$ and $\int_X fg < \infty$.

Δ∇Δ = ∇Δ∇ = Δ∇Δ = ∇Δ∇ = ΔΔ∇Δ

SOLUTION: Let us set:

$$A_n = \left(\frac{1}{n+1} < f \leqslant \frac{1}{n}\right) \qquad (n = 0,1,2,\ldots),$$

and

$$A_\infty = (f = 0).$$

There exist open sets V_n and W_n contained in X such that

$$V_n \supset A_\infty, \qquad \int_{V_n} f(x)\,dx < 2^{-n},$$

$$W_n \supset A_n, \qquad \int_{W_n - A_n} f(x)\,dx < (n+1)^{-1}2^{-n}.$$

The function

$$g = \sum_{n=0}^{\infty} 1\!\!1_{V_n} + \sum_{n=0}^{\infty} (n+1) 1\!\!1_{W_n}$$

is lower semi-continuous and $g \geqslant 1/f$. Furthermore,

$$\int_X fg = \sum_{n=0}^{\infty} \int_{V_n} f + \sum_{n=0}^{\infty} (n+1)\int_{W_n - A_n} f + \sum_{n=0}^{\infty} (n+1)\int_{A_n} f$$

$$\leqslant 4 + \int_{A_0} f + \sum_{n=1}^{\infty} \frac{n+1}{n} \operatorname{meas}(A_n) < \infty,$$

for $\sum \operatorname{meas}(A_n) < \infty$, since the A_n's are mutually disjoint and $\operatorname{meas}(X) < \infty$.

REMARK: If $\operatorname{meas}(X) = \infty$, the result may not hold. Indeed, if X is the disjoint union of sets X_n such that $\operatorname{meas}(X_n) = n^{-\frac{1}{2}}$ and if

$$f = \sum_{n=1}^{\infty} \frac{1}{n} \mathbb{1}_{X_n}$$

one will have, if $g \geqslant 1/f$:

$$\int_X fg = \sum_{n=1}^{\infty} \int_{X_n} fg \geqslant \sum_{n=1}^{\infty} n^{-\frac{1}{2}} = \infty.$$

EXERCISE 3·51: Let (f_n) be a sequence of measurable functions on $X = [0,1]$.

Show that the following conditions are equivalent:

(i): There exists a subsequence (f_{n_s}) of the sequence which converges to zero almost everywhere;

(ii): There exists a sequence (t_n) of real numbers such the $\limsup |t_n| > 0$ and the series $\sum t_n f_n(x)$ converges for almost all x;

(iii): There exists a sequence $|t_n|$ of real numbers such that $\sum |t_n| = +\infty$ and the series $\sum t_n f_n(x)$ is absolutely convergent for almost all x.

ΔVΔ = VΔV = ΔVΔ = VΔV = ΔVΔ

SOLUTION: (i) => (ii),(iii): Without loss of generality $f_n \to 0$ almost everywhere. By Egoroff's Theorem there exists a sequence of measurable sets $A_1 \subset A_2 \subset \cdots$ such that $\text{meas}(X - A_s) < 1/s$ and $f_n \to 0$ uniformly on each of the A_s. Hence there exists a sequence of integers $n_1 < n_2 < \cdots$ such that for all $n \geqslant n_s$ one has $|f_n| \leqslant 2^{-s}$ on A_s. Let A be the union of the A_s so that $\text{meas}(X - A) = 0$. Set $t_n = 0$ if $n \neq n_s$ for all s, and let $t_{n_s} = 1$. Then

$$\sum_n t_n f_n(x) = \sum_s f_{n_s}(x),$$

and for all $x \in A$ one will have $|f_{n_s}(x)| < 2^{-s}$ whenever $s \geqslant s_0$, where s_0 is chosen so that $x \in A_{s_0}$. This proves (ii) and (iii) hold.

(ii) => (i): If $\limsup t_n > 0$ there exists $\alpha > 0$ and a sequence $n_1 < n_2 < \cdots$ such that $|t_{n_s}| > \alpha$ for all s. If for almost all x the series $\sum t_n f_n(x)$ converges, then in particular $t_{n_s} f_{n_s}(x) \to 0$, and consequently $f_{n_s}(x) \to 0$.

(iii) => (i): Letting $g(x) = \sum |t_n f_n(x)|$, by hypothesis one has $g(x) < \infty$ almost everywhere. Then if A is the union of the $A_s = \{g \leqslant s\}$ one has $\text{meas}(X - A) = 0$. Furthermore,

$$\sum_n |t_n| \int_{A_s} |f_n| = \int_{A_s} g < \infty.$$

Since $\sum |t_n| = \infty$, one thus has:

$$\liminf_{n\to\infty} \int_{A_s} |f_n| = 0.$$

It is then possible to determine a sequence $n_1 < n_2 < \cdots$ such that:

$$\int_{A_s} |f_{n_s}| < 2^{-s}.$$

For every integer σ one has:

$$\sum_s \int_{A_\sigma} |f_{n_s}| < \infty,$$

because $A_s \supset A_\sigma$ whenever $s \geqslant \sigma$. From this it follows that at almost all points of A_σ the series $\sum f_{n_s}$ converges, and $f_{n_s} \to 0$ in particular. But then $f_{n_s} \to 0$ at almost all points of A, hence also at almost all points of X, because $\text{meas}(X - A) = 0$.

EXERCISE 3·52: Let f be an integrable function on $X = \mathbb{R}^p$.

(a): Show that for all $\varepsilon > 0$ there exists a measurable set with finite measure such that f is bounded on A and

$$\int_{X-A} |f| < \varepsilon$$

(b): From this deduce that:

$$\lim_{\text{meas}(E)\to 0} \int_E |f| = 0$$

(which means that for all $\varepsilon > 0$ there exists $\delta > 0$ such that if E is measurable and $\text{meas}(E) < \delta$ then

$$\int_E |f| < \varepsilon.$$

$\Delta\nabla\Delta = \nabla\Delta\nabla = \Delta\nabla\Delta = \nabla\Delta\nabla = \Delta\nabla\Delta$

SOLUTION: We may assume that $f \geqslant 0$ (since only $|f|$ appears in the statement of the exercise).

SOLUTION: (a): Let:

$$A_0 = (f = 0), \qquad A_n = \left(\frac{1}{n} \leqslant f \leqslant n\right) \quad (n \geqslant 1), \qquad A_\infty = (f = \infty).$$

Since $A_1 \subset A_2 \subset \cdots$, and as

$$\bigcap_{n=1}^{\infty} (X - A_n) = A_0 \cup A_\infty,$$

one has, because $\text{meas}(A_\infty) = 0$ since f is integrable,

$$\lim_{n\to\infty}\int_{X-A_n} f = \int_{A_0} f + \int_{A_\infty} f = 0.$$

Hence there exists an integer n_0 such that, letting $A = A_{n_0}$,

$$\int_{X-A} f < \varepsilon.$$

Furthermore, f is bounded on A by n_0. Finally A has finite measure, for

$$\frac{1}{n_0}\,\text{meas}(A) \leqslant \int_A f < \infty.$$

SOLUTION: (b): Let A be as in part (a) with $\varepsilon/2$ instead of ε. Set $M = \sup_A f$. Then if E is measurable and $\text{meas}(E) < \varepsilon/2M$, one has

$$\int_E f \leqslant \int_{X-A} f + \int_{E\cap A} f \leqslant \frac{\varepsilon}{2} + M\text{meas}(E) < \varepsilon.$$

EXERCISE 3·53: Let (f_n) be a sequence of integrable functions on $X = \mathbb{R}$ such that $f_n \to f$ almost everywhere.

(a): Show that if f is integrable and if

$$\int_X f_n \to \int_X f,$$

then for all $\varepsilon > 0$ there exists a measurable set A with finite measure, an integrable function $g \geqslant 0$, and an integer n_0, such that for all $n \geqslant n_0$

$$\left|\int_{X-A} f_n\right| < \varepsilon \quad \text{and} \quad |f_n| \leqslant g \quad \text{on } A.$$

(b): Now assume that for all $\varepsilon > 0$ there exists a measurable set A, an integrable function $g \geqslant 0$, and an integer n_0, such that for all $n \geqslant n_0$ one has:

$$\int_{X-A} |f_n| < \varepsilon \quad \text{and} \quad |f_n| \leqslant g \quad \text{on } A.$$

Show that under these conditions f is integrable and that

$$\int_X f_n \to \int_X f.$$

(c): Show by examples that the conditions of part (a) are not sufficient, and those of part (b) are not necessary in order that f be integrable and

$$\int_X f_n \to \int_X f.$$

Δ∇Δ = ∇Δ∇ = Δ∇Δ = ∇Δ∇ = Δ∇Δ

SOLUTION: (a): Let $\varepsilon > 0$. By part (a) of the preceding exercise there exists a measurable set B with finite measure such that $|f|$ is bounded on B by a constant M, and

$$\int_{X-B} |f| < \varepsilon.$$

By Egoroff's Theorem there exists $A \subset B$ such that $\text{meas}(B - A) < \varepsilon/M$ and $f_n \to f$ uniformly on A. Hence there exists n_0 such that for $n \geqslant n_0$

$$|f_n - f| < \frac{\varepsilon}{\text{meas}(A)} \quad \text{on } A, \quad \text{and} \quad \left|\int_X (f_n - f)\right| < \varepsilon.$$

By writing

$$\int_{X-A} f_n = \int_{B-A} f + \int_{X-B} f + \int_X (f_n - f) - \int_A (f_n - f),$$

if $n \geqslant n_0$ one obtains:

$$\left|\int_{X-A} f_n\right| < M\text{meas}(B - A) + \varepsilon + \varepsilon + \frac{\varepsilon}{\text{meas}(A)}\,\text{meas}(A) < 4\varepsilon.$$

Furthermore, if

$$g = |f| + \frac{\varepsilon}{\text{meas}(A)} \quad \text{on } A \quad \text{and} \quad g = 0 \quad \text{on } X - A,$$

it is clear that g is integrable and that $|f_n| \leqslant g$ on A for $n \geqslant n_0$.

SOLUTION: (b): By Fatou's Lemma,

$$\int_{X-A} |f| \leqslant \varepsilon,$$

and by the Lebesgue Dominated Convergence Theorem,

$$\int_A |f - f_n| \to 0.$$

Then by the inequality

$$\int_X |f - f_n| < \int_{X-A} |f| + \int_{X-A} |f_n| + \int_A |f - f_n|$$

one concludes

$$\limsup_{n\to\infty} \int_X |f - f_n| \leqslant 2\varepsilon,$$

which proves that

$$\int_X f_n \to \int_X f.$$

SOLUTION: (c): Here are three counter-examples where all the functions considered are zero outside $[0,1]$, which comes down to taking this interval as X.

(i): Let

$$f_n(x) = -\frac{(n-2)}{2} \quad \text{if } 1 \leqslant x \leqslant \frac{1}{n} \quad \text{and} \quad f_n(x) = 1 \quad \text{if } \frac{1}{n} < x \leqslant 1$$

for $n \geqslant 2$. Then $f_n \to 1$ almost everywhere. If $A = [\frac{1}{2},1]$ one has $|f_n| \leqslant 1$ on A and

$$\int_{X-A} f_n = 0.$$

Nevertheless,

$$\int_X f_n = \frac{1}{2}.$$

(ii): Let

$$f_n(x) = \frac{1}{x}\sin\frac{1}{x} \quad \text{if } \frac{1}{n} < x \leqslant 1 \quad \text{and} \quad f_n(x) = 0 \quad \text{if } 0 \leqslant x \leqslant \frac{1}{n}.$$

Since

$$\lim_{\substack{\alpha\to 0\\ \alpha>0}}\int_\alpha^1 \frac{1}{x}\sin\frac{1}{x}\,dx$$

exists, for every $\varepsilon > 0$ there is an $\alpha > 0$ such that if $A = [\alpha,1]$ one has

$$\left|\int_{X-A} f_n\right| < \varepsilon.$$

On A the $|f_n|$ are uniformly bounded. Now, $f_n(x) \to (1/x)\sin(1/x)$ almost everywhere and this function is not Lebesgue integrable.

(iii): Finally, let

$$f_n(x) = \frac{1}{x}\sin\frac{1}{x} \quad \text{if } \frac{1}{2n} \leqslant x \leqslant \frac{1}{n} \quad \text{and} \quad f_n(x) = 0 \text{ otherwise.}$$

Then $f_n \to 0$ and

$$\int_X f_n \to 0.$$

If the conditions of part (b) were satisfied by the f_n's they would also be satisfied by the $|f_n|$, and one would have:

$$\lim\int_n^{2n} \frac{|\sin x|}{x}\,dx = \lim_n \int_X |f_n| = 0.$$

Now, by Exercise 3·48,

$$\lim_{n\to\infty}\int_n^{2n} \frac{|\sin nx|}{x}\,dx = \lim_{n\to\infty}\int_1^2 \frac{|\sin nx|}{x}\,dx$$

$$= \frac{1}{\pi}\int_0^\pi \sin x\,dx\int_1^2 \frac{dx}{x} = \frac{2\log 2}{\pi}.$$

EXERCISE 3·54: Let (f_n) be a sequence of measurable functions on a measurable set X of $\mathbb{R}^p$. One says that this *SEQUENCE CONVERGES IN MEASURE* towards a measurable function f if for all $\alpha > 0$

$$\lim_{n\to\infty} \operatorname{meas}(|f - f_n| > \alpha) = 0.$$

(a): Show that if $f_n \to f$ in measure there exists a subsequence (f_{n_s}) which converges to f almost everywhere. Give an example showing that the sequence (f_n), itself, need not converge to f almost everywhere.

(b): Show that if the f_n's are positive, converge in measure to f, and if

$$\int_X f_n \leqslant A$$

for all n, then

$$\int_X f \leqslant A$$

(Fatou's Lemma for convergence in measure).

(c): Show that if the f_n converge in measure to f, and if there exists a positive integrable function g such that $|f_n| \leqslant g$ for all n, then f is integrable and

$$\lim_{n\to\infty}\int_X |f - f_n| = 0$$

(Dominated Convergence Theorem for convergence in measure).

(d): Assume that $\operatorname{meas}(X) < \infty$. Show that if the f_n's converge towards f almost everywhere, they also converge towards f in measure. Give an example showing that the condition $\operatorname{meas}(X) < \infty$ is indispensible.

$$\Delta\nabla\Delta = \nabla\Delta\nabla = \Delta\nabla\Delta = \nabla\Delta\nabla = \Delta\nabla\Delta$$

SOLUTION: (a): Choose a sequence of integers $n_1 < n_2 < \cdots$ such that

$$\text{meas}\left(|f - f_{n_s}| > \frac{1}{s}\right) < 2^{-s}.$$

Then the set of points that belong to an infinite number of sets $(|f - f_{n_s}| > 1/s)$ has measure zero (cf., Exercise 1·3). Now, this set contains the one formed of the points where (f_{n_s}) does not converge to f. For every integer $n \geqslant 1$ let f_n be the characteristic function of the interval $[r2^{-s},(r + 1)2^{-s}]$, where r and s are the integers such that $n = 2^s + r$, $0 \leqslant r < 2^s$. The sequence (f_n) converges in measure on $[0,1]$ but does not converge at any point of this interval.

SOLUTION: (b): By part (a) there exists a subsequence which converges to f almost everywhere, and the classic Fatou's Lemma immediately furnishes the result.

SOLUTION: (c): Let $\varepsilon > 0$. There exists an integrable set $A \subset X$ such that

$$\int_{X-A} g < \varepsilon.$$

Set $A_n = (|f - f_n| > \varepsilon/\text{meas}(A))$ and note that by part (a) one has $|f| \leqslant g$ almost everywhere; then, by decomposing the integral on X into a sum of integrals on $X - A$, $A - A_n$, and $A \cap A_n$, one obtains:

$$\limsup_{n\to\infty}\int_X |f - f_n| \leqslant 2\varepsilon + \varepsilon + \limsup_{n\to\infty}\int_{A_n} 2g.$$

Now, $\text{meas}(A_n) \to 0$ by hypothesis, so (cf., Exercise 3·53),

$$\lim_{n\to\infty}\int_{A_n} 2g = 0.$$

SOLUTION: (d): By Egoroff's Theorem, for all $\varepsilon > 0$ there exists a measurable set E such that $\text{meas}(X - E) < \varepsilon$ and $f_n \to f$ uniformly on E. Then, for all $\alpha > 0$ one has $(|f - f_n| > \alpha) \subset X - E$ whenever n is large enough, which proves that $f_n \to f$ in measure. The functions $f_n = \mathbb{1}_{[n,n+1]}$ converge to zero almost everywhere, but not in measure.

EXERCISE 3·55: Let $\alpha > 0$. Prove that if f and the f_i's are positive measurable functions and if $f_i \to f$ in measure, then $f_i^\alpha \to f^\alpha$ in measure.

Δ∇Δ = ∇Δ∇ = Δ∇Δ = ∇Δ∇ = Δ∇Δ

SOLUTION: If $0 < \alpha \leqslant 1$ this results immediately from the inequality:

$$|f^\alpha - f_i^\alpha| \leqslant |f - f_i|^\alpha.$$

When $\alpha > 1$ one has (by the mean value theorem):

$$|f^\alpha - f_i^\alpha| \leqslant \alpha|f - f_i|(f \vee f_i)^{\alpha-1},$$

where $f \vee f_i$ denotes the maximum of f and f_i. Let $\varepsilon > 0$ and let $M > 0$. Then

$$\text{meas}(|f^\alpha - f_i^\alpha| > \varepsilon) \leqslant \text{meas}\left(|f - f_i| > \frac{\varepsilon}{\alpha M^{\alpha-1}}\right)$$
$$+ \text{meas}(f \vee f_i > M).$$

Note that if $f \vee f_i > M$, then

$$f > \tfrac{1}{2}M \quad \text{or} \quad |f - f_i| > \tfrac{1}{2}M;$$

for otherwise one would have

$$f_i = f + (f_i - f) \leqslant \tfrac{1}{2}M + \tfrac{1}{2}M = M.$$

Thus:

$$\text{meas}(|f^\alpha - f_i^\alpha| > \varepsilon) < \text{meas}\left(|f - f_i| > \frac{\varepsilon}{\alpha M^{\alpha-1}}\right)$$

$$+ \text{meas}(|f - f_i| > \tfrac{1}{2}M) + \text{meas}(f > \tfrac{1}{2}M),$$

so that

$$\limsup_{i\to\infty} \text{meas}(|f^\alpha - f_i^\alpha| > \varepsilon) \leqslant \text{meas}(f > \tfrac{1}{2}M).$$

Since $\lim_{M\to\infty} \text{meas}(f > M/2) = 0$,

$$\lim_{i\to\infty} \text{meas}(|f^\alpha - f_i^\alpha| > \varepsilon) = 0.$$

EXERCISE 3·56: Let X be a measurable set of $\mathbb{R}^p$ such that $0 < \text{meas}(X) < \infty$. Denote by M the vector space of measurable complex functions on X. If $f \in \mathsf{M}$, set

$$\rho(f) = \int_X |f|(1 + |f|)^{-1}.$$

(a): Show that $f_n \to 0$ in measure on X if and only if $\rho(f_n) \to 0$.

(b): Show that $(f,g) \mapsto \rho(f - g)$ is a metric on M if one agrees to identify two functions that are equal almost everywhere.

(c): Show that ρ is not a norm on M, but that the sum and the product of two functions of M are continuous operations in the metric defined by ρ.

(d): Show that M is complete in the metric defined by ρ.

Δ∇Δ = ∇Δ∇ = Δ∇Δ = ∇Δ∇ = ΔVΔ

SOLUTION: (a): This follows from the fact that if $A_\varepsilon = (|f| \geqslant \varepsilon)$,

$$\frac{\varepsilon}{1+\varepsilon}\operatorname{meas}(A_\varepsilon) \leqslant \rho(f) \leqslant \operatorname{meas}(A_\varepsilon) + \frac{\varepsilon}{1+\varepsilon}\operatorname{meas}(X).$$

SOLUTION: (b): If $f,g \in M$, from the inequality

$$\frac{f+g}{1+|f+g|} \leqslant \frac{f}{1+|f|} + \frac{g}{1+|g|}$$

one deduces that $\rho(f+g) \leqslant \rho(f) + \rho(g)$. From this it results that $\rho(f-g)$ satisfies the Triangle Inequality. Furthermore, if $\rho(f-g) = 0$ then $f - g = 0$ almost everywhere, that is to say that $f = g$ in M.

SOLUTION: (c): ρ is not a norm, for in general $\rho(\lambda f) \neq |\lambda|\rho(f)$. Nevertheless, if $f_n \to f$ and $g_n \to g$ in measure, then

$$\rho(f+g-f_n-g_n) \leqslant \rho(f-f_n) + \rho(g-g_n) \to 0,$$

so $f + g \to f_n + g_n$ in measure.

Assume now that $f_n \to 0$ in measure and that for all $\varepsilon > 0$ there exists $\alpha > 0$ such that

$$\limsup_{n\to\infty} \operatorname{meas}(|g_n| \geqslant \alpha) < \varepsilon$$

(which is the case if $g_n \to 0$ in measure, or if $g_n = g$ for all n). Since, for all $\beta > 0$,

$$(|f_n g_n| \geqslant \beta) \subset (|g_n| \geqslant \alpha) \cup \left(|f_n| \geqslant \frac{\beta}{\alpha}\right),$$

one will have

$$\limsup_{n\to\infty} \text{meas}(|f_n g_n| \geqslant \beta) \leqslant \varepsilon,$$

which shows that $f_n g_n \to 0$ in measure. From this it follows immediately that if $f_n \to f$ and $g_n \to g$ in measure, then $f_n g_n \to fg$ in measure.

SOLUTION: (d): Let (f_n) be a Cauchy sequence in M. There exists a subsequence (f_{n_s}) such that

$$\rho(f_{n_{s+1}} - f_{n_s}) < 4^{-s}.$$

By the inequality from the answer to part (a), for all s one has:

$$\text{meas}(|f_{n_{s+1}} - f_{n_s}| \geqslant 2^{-s}) \leqslant (1 + 2^s)4^{-s}.$$

By exercise 1·3, for almost all x one has

$$|f_{n_{s+1}}(x) - f_{n_s}(x)| < 2^{-s},$$

for s sufficiently large. In other words, f_{n_s} converges almost everywhere to a function f. Applying Lebesgue's Bounded Convergence Theorem to the integral which gives $\rho(f - f_{n_s})$ shows that $\rho(f - f_{n_s}) \to 0$. Thus the Cauchy sequence (f_n) possesses a convergent subsequence; it is therefore itself convergent.

EXERCISE 3·57: Let (f_n) be a sequence of positive integrable functions that converge in measure to an integrable function f.

Show that if

$$\lim_{n\to\infty}\int f_n = \int f$$

then

$$\lim_{n\to\infty}\int |f - f_n| = 0.$$

(See Exercise 6·114 for a generalisation of this result to L^p spaces).

Δ∇Δ = ∇Δ∇ = Δ∇Δ = ∇Δ∇ = Δ∇Δ

SOLUTION: It is clear that

$$(f - \min(f,f_n) > \alpha) \subset (|f - f_n| > \alpha),$$

so

$$\min(f,f_n) \to f \quad \text{in measure.}$$

On the other hand,

$$0 \leqslant \min(f,f_n) \leqslant f.$$

Hence, by the Dominated Convergence Theorem for convergence in measure (cf., Exercise 3·54)

$$\int \min(f,f_n) \to \int f.$$

Then

$$\int \max(f,f_n) = \int f + \int f_n - \int \min(f,f_n) \to \int f,$$

so that:

$$\int |f - f_n| = \int \max(f,f_n) - \int \min(f,f_n) \to 0.$$

EXERCISE 3·58: Let H be a family of positive integrable functions on a measurable set X of $\mathbb{R}^m$.

(a): Show that the following conditions are equivalent:

(i): $\lim_{c\to\infty} \int_{(f>c)} f(x)dx = 0$ uniformly for $f \in H$;

(ii): $\lim_{\text{meas}(E)\to 0} \int_E f(x)dx = 0$ uniformly for $f \in H$.

(A family satisfying these conditions is called *UNIFORMLY INTEGRABLE*).

(b): Show that if H is uniformly integrable and additionally satisfies the condition:

(iii): For any $\varepsilon > 0$ there exists a measurable set

$A \subset X$ such that $\text{meas}(A) < \infty$,

and

$$\int_{X-A} f(x)dx < \varepsilon \quad \text{for all } f \in H;$$

then

$$\sup_{f\in H} \int_X f(x)dx < \infty.$$

(It is said that H *CONSERVES MASS* if the Condition (iii) is satisfied).

(c): Show that if there exists a positive measurable func-

G defined on $\mathbb{R}_+$ such that

$$\lim_{t\to\infty} t^{-1}G(t) = \infty \quad \text{and} \quad \sup_{f\in H}\int_X G(f(x))dx < \infty,$$

then the family H is uniformly integrable. Conversely, if H is uniformly integrable there exists a function G that satisfies the above conditions and that can further be chosen to be convex and increasing. (We refer the reader to Exercise 6•113 for an application of the notion of uniform integrability to the problem of convergence in L^p spaces).

ΔVΔ = VΔV = ΔVΔ = VΔV = ΔVΔ

SOLUTION: (a): (i) => (ii): For every set E of finite measure,

$$\int_E f \leqslant \int_{(f>c)} f + c\,\text{meas}(E).$$

For every $\varepsilon > 0$ one can choose c such that the integral of f on $(f > c)$ is less than $\varepsilon/2$ for all $f \in H$. Then, if meas$(E) < \varepsilon/2c$,

$$\int_E f < \varepsilon \quad \text{for any } f \in H.$$

(ii) => (i): First let us prove that

$$\lim_{c\to\infty} \text{meas}(f > c) = 0 \quad \text{uniformly for } f \in H. \tag{*}$$

Otherwise, there would exist $\alpha > 0$, a sequence $c_k \to \infty$, and a sequence $f_k \in H$, such that

$$\text{meas}(f_k > c_k) > \alpha \quad \text{for all } k.$$

Let $\beta > 0$ be such that meas$(E) \leqslant \beta$ implies for all $f \in H$:

$$\int_E f < 1.$$

For every k one can find $E_k \subset (f_k > c_k)$ such that:

$$\text{meas}(E_k) = \min(\alpha,\beta).$$

One would then have:

$$1 > \int_{E_k} f_k \geqslant c_k \min(\alpha,\beta),$$

which contradicts $c_k \to \infty$. Now let $\varepsilon > 0$ and $\delta > 0$ be such that $\text{meas}(E) < \delta$ implies

$$\int_E f < \varepsilon$$

for all $f \in H$. There exists c_0 such that $c \geqslant c_0$ implies that $\text{meas}(f > c) < \delta$ for all $f \in H$, and consequently:

$$\int_{(f>c)} f < \varepsilon.$$

SOLUTION: (b): In fact the conclusion still follows if Condition (iii) is weakened to requiring only the existence of a set A of finite measure such that

$$\sup_{f \in H} \int_{X-A} f < \infty,$$

which in essence reduces us to the case where $\text{meas}(X) = a < \infty$. Assume the conclusion is false, i.e.,

$$\sup_{f \in H} \int_X f = \infty.$$

Let (X_1, X_2) be a partition of X into two measurable sets such that $\text{meas}(X_1) = \text{meas}(X_2) = a/2$. Then

$$\sup_{f \in H} \int_{X_i} f = \infty$$

for at least one of the integers $i = 1,2$. Repeating this partition argument, one constructs a sequence of measurable sets $(E_n)_{n>0}$ such that

$$\text{meas}(E_n) = a2^{-n},$$

$$\sup_{f \in H} \int_{E_n} f = \infty.$$

But this manifestly contradicts Condition (ii).

SOLUTION: (c): Assume first that there exists such a function G. Let $\varepsilon > 0$ and set

$$M = \sup_{f \in H} \int_X G(f).$$

There exists c_0 such that $t^{-1}G(t) > M/\varepsilon$ if $t > c_0$. Then for all $c > c_0$ and all $f \in H$

$$\int_{(f>c)} f \leqslant \frac{\varepsilon}{M} \int_X G(f) \leqslant \varepsilon.$$

Now assume that H is uniformly integrable and look for G in the form

$$G(t) = \int_0^t g(u)\,du,$$

where $g(0) = g_n$ = constant on $[n,n + 1[$, with $0 = g_0 \leqslant g_1 \leqslant \cdots$, and $g_n \to \infty$. Then G will be increasing (and hence measurable), convex (since g is increasing), and will satisfy

$$\lim_t t^{-1}G(t) = \infty,$$

for, $g(u) \to \infty$ as $u \to \infty$, and this implies that the mean of g on $[0,t]$ tends to infinity when $t \to \infty$. It remains to choose the g_n suitably in order that

$$\sup_{f\in H}\int_X \{G(f) < \infty. \tag{1}$$

Define

$$a_n(f) = \text{meas}(f > n);$$

by noting that $G(f) = 0$ if $0 \leqslant f \leqslant 1$ (since $g_0 = 0$) one obtains:

$$\int_X G(f) = \sum_{n=1}^{\infty}\int_{(n<f\leqslant n+1)} G(f) \leqslant \sum_{n=1}^{\infty} G(n+1)(a_n(f) - a_{n+1}(f))$$

(as G is increasing). Since $G(1) = 0$ and $G(n+1) - G(n) = g_n$, one has:

$$\sum_{n=1}^{N} G(n+1)(a_n(f) - a_{n+1}(f)) = \sum_{n=1}^{N} g_n a_n(f) - G(N+1)a_{N+1}(f),$$

and consequently:

$$\int_X G(f) \leqslant \sum_{n=1}^{\infty} g_n a_n(f). \tag{2}$$

Let (c_s) be an increasing sequence of integers tending to infinity, and such that for all $f \in H$:

$$\int_{(f>c_s)} f \leqslant 2^{-s}.$$

Then

$$1 \geq \sum_{s=1}^{\infty} \int_{(f>c_s)} f = \sum_{s=1}^{\infty} \sum_{n=c_s}^{\infty} \int_{(n<f\leq n+1)} f$$

$$\geq \sum_{s=1}^{\infty} \sum_{n=c_s}^{\infty} n(a_n(f) - a_{n+1}(f)).$$

Noting that $na_n(f) \to 0$ (cf. Exercise 3·39), an Abel transformation on the summation over n leads to the inequality, valid for all $f \in H$:

$$\sum_{s=1}^{\infty} \sum_{n=c_s}^{\infty} a_n(f) \leq 1,$$

which can also be written, letting g_n denote the number of integers s such that $c_s \leq n$,

$$\sum_{n=1}^{\infty} g_n a_n(f) \leq 1. \tag{3}$$

This choice of the g_n's in the definition of the function g clearly makes Equation (1) hold, on account of Equations (2) and (3).

REMARK: Condition (*) does not imply the uniform integrability of H. In fact, if:

$$f_n = n\mathbb{1}_{[0,1/n]},$$

the sequence (f_n) is not uniformly integrable, since

$$\int_{[0,1/n]} f_n = 1$$

contradicts condition (ii). Nevertherless,

$$\lim_{c\to\infty} \text{meas}(f_n > c) = 0$$

uniformly in n, because if $\varepsilon > 0$ and $c > 1/\varepsilon$ one has $\text{meas}(f_n > c) = 0$ if $n \leqslant c$, and when $n > c$

$$\text{meas}(f_n > c) = \frac{1}{n} < \frac{1}{c} < \varepsilon.$$

EXERCISE 3·59: Let A be a measurable set of $\mathbb{R}^k$ such that $0 < \text{meas}(A) < \infty$, and let $f_1,\ldots,f_n$ be real functions integrable on A. Assume that

$$\int_A f_i = 0, \qquad 1 \leqslant i \leqslant n.$$

Show that for every number α, $0 \leqslant \alpha \leqslant 1$, there exists a measurable set B such that $B \subset A$, $\text{meas}(B) = \alpha\text{meas}(A)$ and

$$\int_B f_i = 0, \qquad 1 \leqslant i \leqslant n.$$

(First of all consider the case $\alpha = \frac{1}{2}$).

ΔVΔ = VΔV = ΔVΔ = VΔV = ΔVΔ

SOLUTION: *First Step: Assume* $\alpha = \frac{1}{2}$: Let us denote the property to be proved by P_n ($n \geqslant 1$). Note that for every measurable set A there exists a measurable set $B \subset A$ such that $\text{meas}(B) = \frac{1}{2}\text{meas}(A)$; call this property P_0. Now show that $P_n \Rightarrow P_{n+1}$ for $n \geqslant 0$. Noticing that if B has the property P_n (relative to A) then the same holds for $A - B$ (since $\alpha = \frac{1}{2}$), one can construct step by step some measurable sets $A_{p,q}$ ($p \geqslant 0$, $0 \leqslant q < 2^p$) such that:

(i): $A_{0,0} = A$;

(ii): $A_{p,2q} \cap A_{p,2q+1} = \emptyset$, $\quad A_{p,2q} \cup A_{p,2q+1} = A_{p-1,q}$,

if $p \geqslant 1$, $0 \leqslant q < 2^{p-1}$;

(iii): $\operatorname{meas}(A_{p,q}) = 2^{-p}\operatorname{meas}(A)$;

(iv): $\displaystyle\int_{A_{p,q}} f_i = 0, \quad 1 \leqslant i \leqslant n,\ p \geqslant 0,\ 0 \leqslant q < 2^p$;

(if $n = 0$ Condition (iv) is absent).

Set:

$$I_{p,q} = [q2^{-p}, (q+1)2^{-p}[, \quad p \geqslant 0,\ 0 \leqslant q < 2^p;$$

then $\operatorname{meas}(A_{p,q}) = \ell(I_{p,q})\operatorname{meas}(A)$, where $\ell(I_{p,q})$ denotes the length of the interval $I_{p,q}$. Finally, for $0 \leqslant t \leqslant 1$ and $p \geqslant 1$ set:

$$I_t = \left[\frac{t}{2}, \frac{t+1}{2}\right[,$$

$$A_p(t) = \bigcup_{I_{p,q} \subset I_t} A_{p,q},$$

$$A(t) = \bigcup_{p \geqslant 1} A_p(t).$$

We are going to show that the $A(t)$ realise some sort of "homotopy" between $A_{1,0}$ and $A_{1,1}$ in the sense that they satisfy the following properties:

(1): $A(0) = A_{1,0}$ and $A(1) = A_{1,1}$;

(2): $\operatorname{meas}(A(t)) = \frac{1}{2}\operatorname{meas}(A)$;

(3): $\displaystyle\int_{A(t)} f_i = 0, \quad 1 \leqslant i \leqslant n$;

(4): For every function f integrable on A the function

$$t \mapsto \int_{A(t)} f$$

is continuous.

This will imply the property P_{n+1}, for by virtue of the Intermediate Value Theorem there exists a t $(0 \leqslant t \leqslant 1)$ such that

$$\int_{A(t)} f_{n+1} = \tfrac{1}{2}\Big(\int_{A(0)} f_{n+1} + \int_{A(1)} f_{n+1}\Big)$$

$$= \tfrac{1}{2}\Big(\int_{A_{1,0}} f_{n+1} + \int_{A_{1,1}} f_{n+1}\Big)$$

$$= \tfrac{1}{2}\int_{A} f_{n+1} = 0.$$

It remains to prove properties (1)-(4). First of all,

$$A_1(0) = A_{1,0},$$

and if $p \geqslant 1$,

$$A_{p+1}(0) = \bigcup_{0 \leqslant q < 2^p} A_{p+1,q} = \bigcup_{0 \leqslant q < 2^{p-1}} (A_{p+1,2q} \cup A_{p+1,2q+1})$$

$$= \bigcup_{0 \leqslant q < 2^{p-1}} A_{p,q} = A_p(0),$$

which shows that $A(0) = A_{1,0}$. Similarly one sees that $A(1) = A_{1,1}$. From the equation

$$\text{meas}(A_p(t)) = \sum_{I_{p,q} \subset I_t} \text{meas}(A_{p,q}) = \text{meas}(A) \sum_{I_{p,q} \subset I_t} \ell(I_{p,q}),$$

one deduces that

$$\lim_{p\to\infty} \text{meas}(A_p(t)) = \ell(I_t)\text{meas}(A) = \tfrac{1}{2}\text{meas}(A).$$

Since $A_p(t) \subset A_{p+1}(t)$ one clearly has property (2). If $1 \leqslant i \leqslant n$,

$$\int_{A_p(t)} f_i = \sum_{I_{p,q} \subset I_t} \int_{A_{p,q}} f_i = 0,$$

whence

$$\int_{A(t)} f_i = \lim_{p\to\infty} \int_{A_p(t)} f_i = 0.$$

It remains to prove (4). Let $\varepsilon > 0$; there exists $\delta > 0$ such that $E \subset A$ and $\text{meas}(E) < \delta$ imply

$$\int_E |f| < \varepsilon$$

(cf., Exercise 3·52). If $0 \leqslant t < \tau \leqslant 1$,

$$\left|\int_{A_p(\tau)} f - \int_{A_p(t)} f\right| \leqslant \int_{E_p(t,\tau)} |f|,$$

where $E_p(t,\tau)$ is the union of the $A_{p,q}$'s for the q's such that $I_{p,q} \subset I_t$ or $I_{p,q} \subset I_\tau$, but $I_{p,q} \not\subset I_t \cap I_\tau$. The sum of the lengths of these intervals is less than $\tau - t + 2^{1-p}$, so that:

$$\text{meas}(E_p(t,\tau)) \leqslant (\tau - t + 2^{1-p})\text{meas}(A).$$

If p_0 is the smallest integer such that $2^{1-p}\text{meas}(A) < \delta/2$, then for $\tau - t < \delta/2\text{meas}(A)$ and $p \geqslant p_0$

$$\left|\int_{A_p(\tau)} f - \int_{A_p(t)} f\right| \leqslant \varepsilon,$$

and on taking $p \to \infty$,

$$\left|\int_{A(\tau)} f - \int_{A(t)} f\right| \leqslant \varepsilon.$$

Second Step: α Is Arbitrary: Let us again use the notations above, and also set:

$$B_p(\alpha) = \bigcup_{I_{p,q} \subset [0,\alpha[} A_{p,q}, \qquad B(\alpha) = \bigcup_{p \geq 1} B_p(\alpha).$$

As before, one sees easily that:

$$\int_{B_p(\alpha)} f_i = 0 \quad (1 \leq i \leq n), \qquad \lim_{p\to\infty} \text{meas}(B_p(\alpha)) = \alpha\text{meas}(A),$$

whence:

$$\int_{B(\alpha)} f_i = 0 \quad (1 \leq i \leq n), \qquad \text{meas}(B(\alpha)) = \alpha\text{meas}(A).$$

REMARK: It will be noted that $B(0) = \emptyset$, $B(1) = A$, $B(\alpha_1) \subset B(\alpha_2)$ if $\alpha_1 \geq \alpha_2$, and that if f is integrable on A, the function

$$\alpha \mapsto \int_{B(\alpha)} f$$

is continuous.

EXERCISE 3·60: Let X be a measurable set of $\mathbb{R}^p$ such that meas(X) = 1 and let f be an integrable mapping of X into $\mathbb{R}^q$.

Show that:

$$\{\int_E f(x)dx : E \subset X \text{ and } E \text{ measurable}\}$$

is a convex set of $\mathbb{R}^q$ (Lyapunov's Theorem).

Δ∇Δ = ∇Δ∇ = ΔVΔ = ∇Δ∇ = ΔVΔ

SOLUTION: Let E_1 and E_2 be two measurable sets of X, let

$$y_i = \int_{E_i} f_i \qquad (i = 1,2),$$

and $0 < \alpha < 1$. Since

$$\int_X (\mathbb{1}_{E_i} f - y_i) = 0 \qquad (i = 1,2),$$

by Exercise 3·59, there exists a measurable set E of X such that $\text{meas}(E) = \alpha$,

$$\int_X \mathbb{1}_E(\mathbb{1}_{E_i} f - y_i) = 0,$$

which implies

$$\int_{E \cap E_i} f = \alpha y_i.$$

Set $F = (E \cap E_1) \cup ((X - E) \cap E_2)$. Then

$$\int_F f = \int_{E \cap E_1} f + \int_{E_2} f - \int_{E \cap E_2} f$$

$$= \alpha y_1 + (1 - \alpha) y_2.$$

EXERCISE 3·61: Let X be a measurable set of $\mathbb{R}^p$ such that $\text{meas}(X) = 1$, and for all $x \in X$ let $A(x)$ be a set of $\mathbb{R}^q$. Denote by

$$\int_X A(x)\,dx$$

the subset of $\mathbb{R}^q$ formed by the integrals

$$\int_X f(x)\,dx,$$

where f is an integrable mapping of X into $\mathbb{R}^q$ such that $f(x) \in A(x)$ for all $x \in X$.

Show that this subset is convex (Richter's Theorem).

Δ∇Δ = ∇Δ∇ = Δ∇Δ = ∇Δ∇ = Δ∇Δ

SOLUTION: Let f_i $(i = 1,2)$ be integrable mappings at X into $\mathbb{R}^q$ with $f_i(x) \in A(x)$ for all $x \in X$. Let

$$y_i = \int_X f_i$$

and let $0 < \alpha < 1$. Since

$$\int_X (f_i - y_i) = 0,$$

there is a measurable subset E of X, with $\text{meas}(E) = \alpha$, such that

$$\int_E (f_i - y_i) = 0$$

(cf., exercise 3·59). Hence

$$\int_E f_i = \alpha y_i$$

Set

$$f(x) = \begin{cases} f_1(x) & x \in E, \\ f_2(x) & x \in X - E. \end{cases}$$

Then $f(x) \in A(x)$ for all $x \in X$ and

$$\int_X f = \int_E f_1 + \int_X f_2 - \int_E f_2 = \alpha y_1 + (1 - \alpha) y_2.$$

EXERCISE 3·62: Let X be a measurable subset of $\mathbb{R}^p$ such that $\text{meas}(X) = 1$ and let A be a subset of $\mathbb{R}^q$.

Show that the set

$$\int_X A\,dx$$

is equal to the convex hull of A. (See the previous exercise for the notation).

$$\Delta\nabla\Delta = \nabla\Delta\nabla = \Delta\nabla\Delta = \nabla\Delta\nabla = \Delta\nabla\Delta$$

SOLUTION: We shall argue by induction on the dimension of the affine subspace generated by A. If this dimension is zero, that is to say, if A is a point, the proposition is trivial. Let $\Gamma(A)$ denote the convex hull of A. If $a_1,\ldots,a_n$ belong to A and $\alpha_1,\ldots,\alpha_n$ are positive numbers adding up to one, there exists a partition of X into measurable subsets $X_1,\ldots,X_n$ such that $\text{meas}(X_i) = \alpha_i$. The function which takes the value a_i on X_i has $\alpha_1 a_1 + \cdots + \alpha_n a_n$ as its integral, which shows that

$$\Gamma(A) \subset \int_X A\,dx.$$

If ℓ is an affine function on $\mathbb{R}^q$ such that $\ell(a) \geqslant 0$ for all $a \in A$, then for every integrable mapping f of X into A

$$\ell\left(\int_X f(x)\,dx\right) = \int_X \ell(f(x))\,dx \geqslant 0,$$

which proves that:

$$\int_X A\,dx \subset \overline{\Gamma(A)}.$$

Now assume that, with f being as above,

$$y = \int_X f(x)\,dx \in \overline{\Gamma(A)} - \Gamma(A).$$

Since the interior of $\overline{\Gamma(A)}$ is equal to that of $\Gamma(A)$ (a classical result about convex sets), y must be a boundary point of $\overline{\Gamma(A)}$. There would then exist a non-constant affine function such that

$$\ell(y) = \int_X \ell(f(x))\mathrm{d}x = 0,$$

and $\ell(a) \geqslant 0$ for all $a \in \overline{\Gamma(A)}$, and in particular $\ell(f(x)) \geqslant 0$ for all $x \in X$. By modifying f on a negligeable set, one would then have $\ell(f(x)) = 0$ for all $x \in X$, or $f(x) \in \ell^{-1}(0) \cap A$, so then $y \notin \Gamma(\ell^{-1}(0) \cap A)$. Since it can always be assumed that A affinely generates $\mathbb{R}^q$, and consequently that the affine dimension of $\ell^{-1}(0) \cap A$ is strictly less than that of A, one is led to a contradiction.

REMARK: More generally, one can show analogously that if g is measurable and positive on $\mathbb{R}^n$, and

$$\int g(x)\mathrm{d}x = 1,$$

the set of points

$$\int f(x)g(x)\mathrm{d}x,$$

where $f: \mathbb{R}^n \mapsto A$ is integrable, is equal to $\Gamma(A)$.

EXERCISE 3·63: Let f be a mapping of $X = \mathbb{R}^m$ into $Y = \mathbb{R}^n$.

(a): Show that there exists a smallest closed set A_f such that $f(x) \in A_f$ for almost all $x \in X$, and that A_f is non-empty. Next show that if f is measurable, then $y \in A_f$ if and only if $\operatorname{meas}(f^{-1}(V)) > 0$ for every open neighbourhood V of y.

(b): Now assume that f is integrable (or measurable and bounded) and consider the set I_f formed of points of Y of the type:

$$\frac{1}{\text{meas}(E)}\int_E f(x)\mathrm{d}x,$$

where E is an arbitrary measurable part of X such that

$$0 < \text{meas}(E) < \infty.$$

Show that I_f is convex. What are the relations between A_f and I_f?

ΔVΔ = VΔV = ΔVΔ = VΔV = ΔVΔ

SOLUTION: (a): In order to prove the existence of A_f it suffices to prove that of a largest open set V_f of Y such that $f^{-1}(V_f)$ has measure zero; A_f will then be the complement of V_f. To do that it suffices to prove that if $(V_\alpha)_{\alpha\in I}$ is a family of open sets of Y such that $f^{-1}(V_\alpha)$ has measure zero for all $\alpha\in I$, then the union V of the V_α's possesses the same property. By virtue of Lindelöf's Theorem (cf., the end of the Exercise) there exists a sequence $(\alpha_n)_{n\,1}$ of elements of I such that

$$V = \bigcup_{n=1}^{\infty} V_{\alpha_n},$$

and consequently,

$$f^{-1}(V) = \bigcup_{n=1}^{\infty} f^{-1}(V_{\alpha_n})$$

certainly has measure zero. Furthermore, as $X = f^{-1}(Y)$, one has $Y_f \neq Y$, and consequently $A_f \neq \emptyset$.

Now assume that f is measurable. If $y\in A_f$ and if V is an open neighbourhood of y, $V_f \cup V$ is an open set which strictly contains V_f, so that:

$$f^{-1}(V_f\cup V) = f^{-1}(V_f)\cup f^{-1}(V)$$

has positive measure; since $f^{-1}(V_f)$ has measure zero, $\text{meas}(f^{-1}(V)) > 0$. Conversely, if $y \notin A_f$ and $V = V_f$ one has $\text{meas}(f^{-1}(V)) = 0$.

SOLUTION: (b): Let E_i ($i = 0,1$) be two measurable sets of X such that $0 < \text{meas}(E_i) < \infty$, $E = E_0 \cup E_1$, and

$$y_i = \frac{1}{\text{meas}(E_i)} \int_{E_i} f,$$

which may also be written as:

$$\int_E \mathbb{1}_{E_i}(f - y_i) = 0.$$

Assume first that $\text{meas}(E_0 \cap E_1) > 0$. By the remark at the end of Exercise 3·59 there exists a family $(F_t)_{0 \leq t \leq 1}$ of measurable sets such that

$$F_t \subset E, \qquad F_0 = \emptyset, \qquad F_1 = E,$$

$$t \mapsto \text{meas}(E_i \cap F_t) \quad \text{is continuous},$$

and

$$\int_{F_t} \mathbb{1}_{E_i}(f - y_i) = 0,$$

which we may write as:

$$\int_{E_i \cap F_t} f = \text{meas}(E_i \cap F_t) y_i.$$

Set:

$$G_t = (E_0 - F_t) \cup (E_1 \cap F_t);$$

then

$$\int_{G_t} f = \text{meas}(E_0 - F_t)y_0 + \text{meas}(E_1 \cap F_t)y_1.$$

Now note that $G_t \supset E_0 \cap E_1$, so that $\text{meas}(G_t) > 0$ for all t; the function

$$t \mapsto \frac{\text{meas}(E_0 - F_t)}{\text{meas}(G_t)}$$

is therefore continuous and varies from 1 to 0. For all α, $0 \leqslant \alpha \leqslant 1$, there therefore exists a t for which it takes the value α, and then

$$\frac{1}{\text{meas}(G_t)}\int_{G_t} f = \alpha y_0 + (1 - \alpha)y_1.$$

When $\text{meas}(E_0 \cap E_1) = 0$ one can assume that $E_0 \cap E_1 = \emptyset$. It is possible to determine two families $(F_{i,t})_{0 \leqslant t \leqslant 1}$ such that:

$$\left.\begin{array}{ll} F_{i,t} \subset E_i, & \text{meas}(F_{i,t}) = t\,\text{meas}(E_i) \\ \displaystyle\int_{F_{i,t}} (f - y_i) = 0, & \end{array}\right\} \quad (i = 0,1),$$

and then if $G_t = F_{0,t} \cup (E_1 - F_{1,t})$,

$$\frac{1}{\text{meas}(G_t)}\int_{G_t} f = \frac{t\,\text{meas}(E_0)y_0 + (1 - t)\text{meas}(E_1)y_1}{t\,\text{meas}(E_0) + (1 - t)\text{meas}(E_1)},$$

so that the first member runs over the segment $[y_1, y_0]$ when t varies from 0 to 1. We have thus proved that I_f is convex.

Let us now show that $A_f \subset \bar{I}_f$. If $y \in A_f$ and $\varepsilon > 0$, the set $\{x: \|f(x) - y\| \leqslant \varepsilon\}$ does not have measure zero; it therefore contains a measurable set E such that $0 < \text{meas}(E) < \infty$. But then

$$\left\| y - \frac{1}{\text{meas}(E)}\int_E f \right\| = \left\| \frac{1}{\text{meas}(E)}\int_E (f - y) \right\| \leqslant \varepsilon.$$

Finally, let us show that $I_f \subset \Gamma(A_f)$, where $\Gamma(A_f)$ denotes the convex hull of A_f. Proceed as in exercise 3·62, showing first of all that if ℓ is an affine function on Y such that $\ell(y) \geqslant 0$ for all $y \in A_f$, then:

$$\ell\left(\frac{1}{\text{meas}(E)}\int_E f\right) = \frac{1}{\text{meas}(E)}\int_E \ell(f) \geqslant 0,$$

which proves that $I_f \subset \overline{\Gamma(A_f)}$. Next, if there existed a non-constand affine function ℓ such that $\ell(y) \geqslant 0$ for $y \in A_f$, and

$$\int_E \ell(f) = 0,$$

one would be able to assume that $A_f \subset \ell^{-1}(0)$, and one would end by arguing by induction on the affine dimension of A_f. Let us note that if f is bounded, then A_f is compact, and consequently $\bar{I}_f = \Gamma(A_f)$.

== == = $\overset{=}{\underset{=}{}}$ = == ==

LINDELOF'S THEOREM: *In a topological space possessing a countable basis of open sets* (U_n) *— which is the case for* $\mathbb{R}^m$ *— the union of an arbitrary family of open sets* (V_α) *is the union of a* countable *subfamily of them.*

This is proved by considering initially the subset J of integers n for which U_n is contained in at least one of the V_α. Then, with every $n \in J$ one associates an index α_n such that $U_n \subset V_{\alpha_n}$. Then,

$$\bigcup_\alpha V_\alpha = \bigcup_{n \in J} V_{\alpha_n}.$$

Indeed, if $x \in V_\alpha$ there exists n such that $x \in U_n \subset V_\alpha$ (definition of a basis of open sets). But then $n \in J$ and $x \in V_{\alpha_n}$, which proves the Theorem.

EXERCISE 3·64: Let f be an integrable function on $\mathbb{R}$.

Show that if

$$\int_a^b f(x)\mathrm{d}x = 0$$

for all real numbers a and b, then $f = 0$ almost everywhere.

ΔVΔ = VΔV = ΔVΔ = VΔV = ΔVΔ

SOLUTION: Every open set V of $\mathbb{R}$ being the countable union of intervals, for all such V one has:

$$\int_V f = 0.$$

If E is measurable there exists a decreasing sequence of open sets (V_n) such that

$$E \subset \bigcap_n V_n \quad \text{and} \quad \operatorname{meas}(E) = \lim_n \operatorname{meas}(V_n),$$

so that:

$$\int_E f = \lim_n \int_{V_n} f = 0.$$

Thus the set I_f of all the numbers

$$\frac{1}{\operatorname{meas}(E)}\int_E f, \qquad 0 < \operatorname{meas}(E) < \infty,$$

reduces to $\{0\}$. Now $\bar{I}_f$ contains the smallest closed set A_f such that $f(x) \in A_f$ for almost all x (cf. the preceding exercise). Hence one has $A_f = \{0\}$, or in other terms $f = 0$ almost everywhere.

EXERCISE 3·65: Let p be a norm on $\mathbb{R}^n$.

(a): Show that if E is a subset of measure zero of $\mathbb{R}_+$, then $\{x:p(x) \in E\}$ is a subset of measure zero of $\mathbb{R}^n$.

(b): Let $V = \text{meas}(p \leqslant 1)$ and f a measurable function on $\mathbb{R}_+$. Show that:

$$\int_{\mathbb{R}^n} f(p(x))\mathrm{d}x = nV\int_0^\infty t^{n-1}f(t)\mathrm{d}t$$

if $f \geqslant 0$, or if

$$\int_0^\infty t^{n-1}|f(t)|\mathrm{d}t < \infty.$$

ΔVΔ = VΔV = ΔVΔ = VΔV = ΔVΔ

SOLUTION: Note first that:

$$\text{meas}(p = r) = \lim_{\varepsilon \downarrow 0} \text{meas}(r \leqslant p < r + \varepsilon)$$

$$= \text{meas}(p < 1)\lim_{\varepsilon \downarrow 0}((r + \varepsilon)^n - r^n) = 0.$$

This allows us to write

$$\text{meas}(a \leqslant p \leqslant b) = V(b^n - a^n) = nV\int_a^b t^{n-1}\mathrm{d}t,$$

and consequently the formula

$$\int_{\mathbb{R}^n} f(p(x))\mathrm{d}x = nV\int_0^\infty t^{n-1}f(t)\mathrm{d}t \tag{1}$$

is valid for every step function.

Let E be a subset of measure zero of $\mathbb{R}_+$ contained in $[0,R]$. There exists a seqeunce (f_i) of step functions, zero on $[2R,\infty[$, such that

$$0 \leqslant f_i \leqslant f_{i+1}, \qquad \sup_i \int_0^\infty f_i < \infty, \qquad \text{and} \quad f_i \to \infty \text{ on } E.$$

Then $f_i op \leqslant f_{i+1} op$,

$$\sup_i \int f_i op = \sup_i (nV \int_0^\infty t^{n-1} f_i(t) dt)$$

$$< nV(2R)^{n-1} \sup_i \int_0^\infty f_i(t) dt < \infty,$$

and $f_i op \to \infty$ on $\{x : p(x) \in E\}$, which proves that this set has measure zero. If E has measure zero without being bounded one has

$$(p \in E) = \bigcup_i (p \in E \cap [0,R_i]) \quad \text{if } R_i \to \infty,$$

and therefore has measure zero. Hence if $f_i \to f$ almost everywhere on $\mathbb{R}_+$, then $f_i op \to fop$ almost everywhere on $\mathbb{R}^n$.

Now assume that

$$\int_0^\infty t^{n-1} |f(t)| dt < \infty$$

and let (f_i) be a sequence of step functions such that

$$f_i \to f \quad \text{(a.e.)}, \qquad \int_0^\infty t^{n-1} |f(t) - f_i(t)| dt \to 0. \tag{2}$$

Then

$$\lim_{i,j\to\infty} \int_{\mathbb{R}^n} |f_i op - f_j op| = nV \lim_{i,j\to\infty} \int_0^\infty t^{n-1} |f_i(t) - f_j(t)| dt \to 0,$$

and consequently $f_i \circ p$ converges in $L^1(\mathbb{R}^n)$. Since $f_i \circ p \to f \circ p$ almost everywhere, $f_i \circ p \to f \circ p$ in $L^1(\mathbb{R}^n)$. Replacing f by f_i in Equation (1), and passing to the limit, one obtains the desired result. Finally, if f is measurable and greater than or equal to zero, then $f_K = \inf(f,k)\mathbb{1}_{[0,k]} \to f$ on $\mathbb{R}_+$, $0 \leqslant f_K \leqslant f_{K+1}$. Passing to the limit in Equation (1) where f is replaced by f_K, it is seen that this Formula is still true even if $t^{n-1}f(t)$ is not integrable.

REMARK: To obtain Equation (2) we have used the following proposition: *If $g \geqslant 0$ is locally integrable, if f is measurable, and if*

$$\int |f| g < \infty,$$

there exists a sequence of step functions (f_i) such that

$$(f - f_i)g \to 0 \quad (a.e.), \qquad \int |f - f_i| g \to 0.$$

This is a particular case of a general proposition in measure theory. It can be proved directly by observing that if $h \in L^\infty$ and

$$\int_P gh = 0$$

for every rectangle P, then $gh = 0$ almost everywhere. In other words the annihilator in L^∞ of the set of functions $\mathbb{1}_P g$ (which belongs to L^1 because g is locally integrable) is the vector subspace $\{h : h \in L^\infty, hg = 0 \text{ a.e.}\}$. Now $fg \in L^1$ and is annihilated by this subspace. Hence fg is the limit in L^1 of functions $f_i g$, where the f_i's are step functions (by the Hahn-Banach Theorem), and by taking a subsequence one can assume that $f_i g \to fg$ almost everywhere.

If one wants to avoid using the Hahn-Banach Theorem (as well as the property $(L^1)' = L^\infty$) one can argue in the Hilbert space

L^2, and note that if $f \geqslant 0$ then $\sqrt{fg} \in L^2$. If $\sqrt{fg}$ were not to belong to the closed vector subspace generated by the functions $\mathbb{1}_P\sqrt{g}$, the projection theorem would assure us of the existence of a function $h \in L^2$ such that

$$\int h\sqrt{fg} > 0, \qquad \int_P h\sqrt{g} = 0 \quad P \text{ a rectangle of } \mathbb{R}^n.$$

This is absurd, for the second condition implies that $h\sqrt{g} = 0$ almost everywhere. Hence there exists a sequence (f_i) of step functions such that $f_i\sqrt{g} \to \sqrt{fg}$ in L^2; but then the functions f_i^2 are also step functions, and

$$\int |f - f_i^2|g < \|(\sqrt{f} - f_i)\sqrt{g}\|_2\|(\sqrt{f} + f_i)\sqrt{g}\|_2 \to 0.$$

If f is not positive, write $f = u + iv$, $u = u_+ - u_-$, $v = v_+ - v_-$. This result generalises to the case of a function f such that:

$$\int |f|^p g < \infty, \qquad p \geqslant 1,$$

for, on reducing to the case where $f \geqslant 0$ there exists a sequence (f_i) of positive step functions (cf., the second proof of the case $p = 1$) such that $f_i g \to f^p g$ in L^1, and consequently $f_i^{1/p}g^{1/p} \to fg^{1/p}$ in L^p (cf., Exercise 6·105).

Instead of considering sequences of step functions one can, for example, consider infinitely differentiable functions with compact support.

EXERCISE 3·66: Let V be a convex bounded set of $\mathbb{R}^n$.

Show that:

$$\operatorname{meas}(V - V) \leqslant \binom{2n}{n}\operatorname{meas}(V).$$

(If $p(x) = \inf(\lambda > 0 : x \in \lambda(V - V))$ is the gauge of the symmetric convex set $V - V$, p is a norm on $\mathbb{R}^n$. Show that for all $x \in V - V$ there exists $y \in \mathbb{R}^n$ such that:

$$(1 - p(x))V + y \subset V \cap (V + x)$$

and use the preceding exercise).

$$\Delta\nabla\Delta = \nabla\Delta\nabla = \Delta\nabla\Delta = \nabla\Delta\nabla = \Delta\nabla\Delta$$

SOLUTION: If $x \in V - V$ then $p(x) < 1$ and $x = p(x)z$, where z belongs to the boundary of $V - V$. Let (z_i^1) and (z_i^2) be two sequences of points of V such that $z_i^1 - z_i^2 \to z$. As the open set V is bounded one can assume that $z_i^1 \to z^1$, $z_i^2 \to z^2$, so that:

$$x = p(x)(z^1 - z^2), \qquad z^1 \in \bar{V},\ z^2 \in \bar{V}.$$

Let $y = p(x)z^1$ and observe that as V is open and convex $(1 - t)a + tb \in V$ if $0 \leqslant t < 1$, $a \in V$, $b \in \bar{V}$. Consequently, for $a \in V$

$$(1 - p(x))a + y = (1 - p(x))a + p(x)z^1 \in V,$$

$$(1 - p(x))a + y = (1 - p(x))a + p(x)z^2 + x \in V + x,$$

which proves that

$$(1 - p(x))V + y \subset V \cap (V + x).$$

Denoting the characteristic function of V by φ one obtains:

$$\begin{aligned}(1 - p(x))^n \text{meas}(V) &= \text{meas}\{(1 - p(x))V + y\} \\ &\leqslant \text{meas}(V \cap (V + x)) = (\varphi * \check{\varphi})(x),\end{aligned}$$

whence

$$\text{meas}(V)\int_{V-V} (1 - p(x))^n dx \leqslant \int_{\mathbb{R}^n} (\varphi * \check{\varphi})(x) = \text{meas}(V)^2,$$

and consequently, because meas(V) > 0,

$$\int_{V-V} (1 - p(x))^n dx \leqslant \text{meas}(V).$$

By the preceding exercise

$$\int_{V-V} (1 - p(x))^n dx = n\,\text{meas}(V - V)\int_0^1 t^{n-1}(1 - t)^n dt$$

$$= \frac{n\Gamma(n)\Gamma(n + 1)}{\Gamma(2n + 1)} \text{meas}(V - V)$$

$$= \frac{(n!)^2}{(2n)!} \text{meas}(V - V),$$

whence, at last,

$$\text{meas}(V - V) \leqslant \binom{2n}{n} \text{meas}(V).$$

EXERCISE 3·67: Let p be an even, positive function on $\mathbb{R}$, decreasing on $[0,\infty[$, and such that:

$$\int_{-\infty}^{+\infty} p(t)dt = 1, \qquad \int_{-\infty}^{+\infty} t^2 p(t)dt = \sigma^2 < \infty.$$

(a): Set:

$$G(x) = 2\int_{\sqrt{x}}^{\infty} p(t)dt, \qquad x \geqslant 0.$$

Show that G is convex, and calculate:

$$\int_0^{\infty} G(x)dx.$$

(b): Let OAB be a right triangle, M a point on its hypotenuse, and P and Q the orthogonal projections of this point onto the other two sides of the triangle. Find the maximum area of the rectangle OPMQ as M varies along AB.

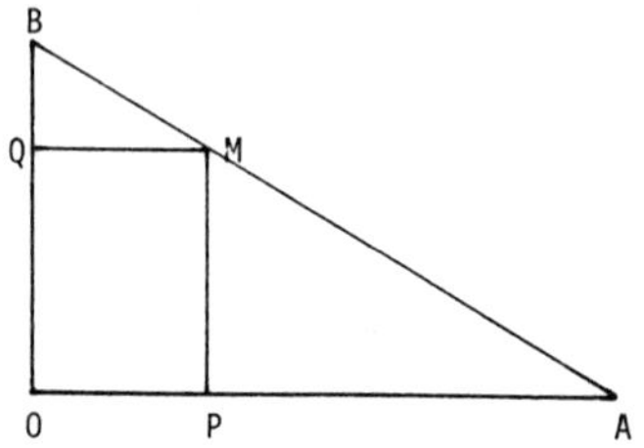

(c): From the preceding considerations deduce that for all $\lambda > 0$

$$\int_{|t|\geqslant\lambda\sigma} p(t)dt \leqslant \frac{1}{2\lambda^2} .$$

(d): Prove the following result (F. Gauss (1821): an improvement, under the given condition, of the Bienyamé-Chebycheff Inequality):

$$\int_{|t|\geqslant\lambda\sigma} p(t)dt \leqslant \begin{cases} \dfrac{1}{2\lambda^2} & \text{if } \lambda \geqslant \sqrt{\dfrac{3}{2}} , \\ 1 - \dfrac{2\lambda}{3}\sqrt{\dfrac{2}{3}} & \text{if } 0 \leqslant \lambda \leqslant \sqrt{\dfrac{3}{2}} \end{cases}$$

Δ∇Δ = ∇Δ∇Δ= ΔVΔ = ∇Δ∇= Δ∇Δ

SOLUTION: (a): Because p is decreasing and positive on $(0,\infty)$, the function

$$x \mapsto \int_x^\infty p(t)dt$$

is convex and decreasing on this interval. Since $\sqrt{x}$ is concave, it follows that the functions $x \mapsto G(x)$ and $x \mapsto G(x^2)$ are convex and decreasing. Furthermore,

$$\int_0^\infty G(x)\mathrm{d}x = 2\int_0^\infty \mathrm{d}x\int_{\sqrt{x}}^\infty p(t)\mathrm{d}t = 2\int_0^\infty p(t)\int_0^{t^2}\mathrm{d}x$$

$$= 2\int_0^\infty t^2 p(t)\mathrm{d}t = \sigma^2.$$

SOLUTION: (b): If OA = a, OB = b, OP = x, OQ = y, then

$$\frac{x}{a} + \frac{y}{b} = 1,$$

and consequently:

$$xy = \frac{ab}{4}\left[1 - \left(\frac{x}{a} - \frac{y}{b}\right)^2\right] .$$

The maximum area of the rectangle OPMQ is achieved when $\frac{x}{a} = \frac{y}{b}$, that is to say when $x = a/2$, $y = b/2$, and this maximum is equal to half the area of the triangle OAB.

SOLUTION: (c): Consider the area A of the triangle bounded by the coordinate axes and a line of support for the curve $y = G(x)$ at the point $(\lambda^2\sigma^2, G(\lambda^2\sigma^2))$. By the preceding,

$$\lambda^2\sigma^2 G(\lambda^2\sigma^2) \leqslant \tfrac{1}{2}A \leqslant \tfrac{1}{2}\sigma^2,$$

whence:

$$G(\lambda^2\sigma^2) \leqslant \frac{1}{2\lambda^2},$$

which is the desired inequality.

(d): Since the function $\lambda \mapsto G(\lambda^2\sigma^2)$ is convex and $G(0) = 1$, for $0 \leqslant \lambda \leqslant \lambda_0$

$$G(\lambda^2\sigma^2) \leqslant \left(1 - \frac{\lambda}{\lambda_0}\right) + \frac{\lambda}{\lambda_0}\, G(\lambda_0^2\sigma^2) \leqslant 1 - \left(1 - \frac{1}{2\lambda_0^2}\right)\frac{\lambda}{\lambda_0} .$$

The best approximation will be obtained by choosing λ_0 so that the line

$$\mu = 1 - \left(1 - \frac{1}{2\lambda_0^2}\right)\frac{\lambda}{\lambda_0}$$

is tangent at the point $(\lambda_0, 1/2\lambda_0^2)$ to the curve

$$\mu = \frac{1}{2\lambda^2}.$$

For that to happen it is necessary that:

$$-\frac{1}{\lambda_0}\left(1 - \frac{1}{2\lambda_0^2}\right) = -\frac{1}{\lambda_0^3},$$

i.e.,

$$\lambda_0 = \sqrt{\frac{3}{2}}.$$

EXERCISE 3·68: Let X be a measurable set of $\mathbb{R}^m$ such that meas(X) = 1, and f a real measurable function on X such that:

$$V = \int_X f^2 < \infty, \qquad \int_X f = 0.$$

Show that if $0 < \alpha < 1$,

$$\liminf_{\varepsilon \to 0} \varepsilon^{-2}\left(1 - \int_X |1 + \varepsilon f|^\alpha\right) = \tfrac{1}{2}\alpha(1 - \alpha)V.$$

Δ∇Δ = ∇Δ∇ = ΔVΔ = ∇Δ∇ = Δ∇Δ

SOLUTION: Now,

$$\varepsilon^{-2}\left(1 - \int_X |1 + \varepsilon f|^\alpha\right) = \int_X \frac{1 + \alpha\varepsilon f - |1 + \varepsilon f|^\alpha}{\varepsilon^2}.$$

Note that

$$\lim_{u\to 0}\left(\frac{1 + \alpha u - |1 + u|^{\alpha}}{u^2}\right) = \tfrac{1}{2}\alpha(1 - \alpha),$$

and also that

$$\lim_{|u|\to\infty}\left(\frac{1 + \alpha u - |1 + u|^{\alpha}}{u^2}\right) = 0,$$

so that there exists a constant B such that

$$\left|1 + \alpha u - |1 + u|^{\alpha}\right| \leqslant Bu^2 \text{ for all } u.$$

Then

$$\lim_{\varepsilon\to 0}\left(\frac{1 + \alpha\varepsilon f - |1 + \varepsilon f|^{\alpha}}{\varepsilon^2}\right) = \tfrac{1}{2}\alpha(1 - \alpha)f^2,$$

$$\left|\frac{1 + \alpha\varepsilon f - |1 + \varepsilon f|^{\alpha}}{\varepsilon^2}\right| \leqslant Bf^2.$$

One can therefore apply Lebesque's Theorem, which gives the result.

EXERCISE 3·69: Let X be a measurable set of $\mathbb{R}^n$, g a positive measurable function on X such that

$$\int_X g(x)\mathrm{d}x = 1,$$

and let φ be a measurable convex real function on a convex set I of $Y = \mathbb{R}^m$.

(a): Let f be a measurable mapping of X into I and such that fg is integrable.

Show that:

$$\varphi\left(\int_X f(x)g(x)\mathrm{d}x\right) \leqslant \int_X \varphi(f(x))g(x)\mathrm{d}x. \qquad \text{(\textit{JENSEN'S FORMULA})}$$

(b): Further assume that $g(x) > 0$ for all $x \in X$ and that φ is strictly convex.

Show that the inequality in Jensen's Formula is strict unless f is equal to a constant almost everywhere.

(c): Show that Jensen's Formula is still true if, with I an interval of $\mathbb{R}$,

$$\int_X f(x)g(x)\mathrm{d}x > -\infty.$$

(If I is not bounded from above, one sets

$$\varphi(\infty) = \lim_{y\to\infty}\varphi(y)).$$

Δ∇Δ = ∇Δ∇ = Δ∇Δ = ∇Δ∇ = Δ∇Δ

SOLUTION: Recall that for a real measurable function h it is said that

$$\int h > -\infty$$

if

$$\int h_- < +\infty$$

and one then sets

$$\int h = \int h_+ - \int h_- \in \,]-\infty,+\infty].$$

It is easy to see that if

$$h_1 \leqslant h_2, \qquad \int h_1 > -\infty,$$

then

$$\int h_1 \leqslant \int h_2.$$

$\Delta\nabla\Delta = \nabla\Delta\nabla = \Delta\nabla\Delta = \nabla\Delta\nabla = \Delta\nabla\Delta$

SOLUTION: (a): By exercise 3·62

$$y_0 = \int_X f(x)g(x)\mathrm{d}x$$

is a point of I. Assume first that y_0 is an interior point of I. There then exists a linear form u on Y such that:

$$\varphi(y) \geqslant \varphi(y_0) + u(y - y_0), \qquad y \in I. \tag{*}$$

In particular,

$$\varphi(f(x)) \geqslant \varphi(y_0) + u(f(x) - y_0), \qquad x \in X.$$

Multiplying both sides by $g(x)$ and then integrating yield Jensen's Formula, since

$$\int_X u(f(x) - y_0)g(x)\mathrm{d}x = u\left(\int_X (f(x) - y_0)g(x)\mathrm{d}x\right) = 0.$$

If y_0 belongs to the boundary of I there exists a (non-constant) affine function ℓ on Y such that $\ell(y_0) = 0$ and $\ell(y) \geqslant 0$ for all $y \in I$. Then

$$\int_X \ell(f(x))g(x)\mathrm{d}x = \ell(y_0) = 0,$$

and $\ell(f(x)) \geqslant 0$ for all $x \in X$. Since it can be assumed that $g(x) > 0$

for all $x \in X$ (otherwise replace X by the set of $x \in X$ such that $g(x) > 0$), one therefore has $f(x) \in \ell^{-1}(0) \cap I$ for all $x \in X$ (after modifying f if necessary on a set of measure zero). One then argues by induction on the affine dimension of I.

SOLUTION: (b): Let us assume that the equality holds in Jensen's Formula. Using again the proof above it is seen first of all that one can assume that y_0 is interior to I, and next that

$$\varphi(f(x)) = \varphi(y_0) + u(f(x) - y_0)$$

for almost all x. Now, if φ is strictly convex this equality can only hold if $f(x) = y_0$.

SOLUTION: (c): It suffices to examine the case where I is not bounded from above, $\varphi(\infty) > -\infty$ and

$$\int_X f(x)g(x)\mathrm{d}x = +\infty.$$

If $\varphi(\infty) = \infty$ there exist $\alpha > 0$ and $\beta \in \mathbb{R}$ such that $\varphi(y) \geqslant \alpha y + \beta$ for all $y \in I$, whence

$$\int_X \varphi(f(x))g(x)\mathrm{d}x \geqslant \int_X (\alpha f(x) + \beta)g(x)\mathrm{d}x = +\infty.$$

If $\varphi(\infty) = a < \infty$, then $\varphi(y) \geqslant a$ for all $y \in I$, and consequently

$$\int_X \varphi(f(x))g(x)\mathrm{d}x \geqslant \int_X ag(x)\mathrm{d}x = a.$$

== == = = = == ==

Here, briefly expounded, are the proofs of the properties of convex functions used above.

Let us, in $Y \times \mathbb{R}$, consider the set,

$$J = \{(y,t):y \in I,\ t \geqslant \varphi(y)\}.$$

It is easy to see that J is convex, and that if $y_0 \in I$, the point $(y_0,\varphi(y_0))$ belongs to the boundary of J. Hence there exists an affine function on $Y \times \mathbb{R}$ which vanishes at this point, is positive on J, and is not constant. This is expressed by the existence of a linear form u on Y and of two numbers α,β such that u and α are not simultaneously zero, and

$$u(y_0) + \alpha\varphi(y_0) + \beta = 0,$$

$$u(y) + \alpha t + \beta \geqslant 0, \qquad y \in I \text{ and } t \geqslant \varphi(y).$$

The second condition is equivalent to $\alpha \geqslant 0$ and

$$u(y) + \alpha\varphi(y) + \beta \geqslant 0, \qquad y \in I,$$

so

$$\alpha(\varphi(y) - \varphi(y_0)) + u(y - y_0) \geqslant 0. \tag{1}$$

Now note that if y_0 is an interior point of I then necessarily $\alpha > 0$; otherwise one would have $u \neq 0$, $u(y_0) + \beta = 0$ and $u(y) + \beta \geqslant 0$ for all $y \in I$, which would contradict the fact that y_0 is an interior point. Dividing both members of Formula (1) by α and replacing $-u/\alpha$ by u yields Formula (*).

If for $y \in I$, $y \neq y_0$, one has

$$\varphi(y) = \varphi(y_0) + u(y - y_0);$$

for all $0 \leqslant \alpha \leqslant 1$,

$$\alpha\varphi(y_0) + (1 - \alpha)\varphi(y) \geqslant \varphi(\alpha y_0 + (1 - \alpha)y)$$

$$\geqslant \varphi(y_0) + (1 - \alpha)u(y - y_0) = \alpha\varphi(y_0) + (1 - \alpha)\varphi(y),$$

so

$$\varphi(\alpha y_0 + (1 - \alpha)y) = \alpha\varphi(y_0) + (1 - \alpha)\varphi(y),$$

and φ is not strictly convex.

If I is an interval of $\mathbb{R}$ not bounded from above, there exists an increasing function g on I such that:

$$\varphi(z) - \varphi(y) = \int_y^z g(t)\mathrm{d}t$$

when $y \in \mathring{I}$ and $z \geqslant y$. If $\varphi(\infty)$ is finite, the integral:

$$\int_y^\infty g(t)\mathrm{d}t = \varphi(\infty) - \varphi(y)$$

is convergent, and consequently $g \leqslant 0$; but then $\varphi(\infty) - \varphi(y) \leqslant 0$. This inequality is still valid if $y = \inf I \in I$, for then

$$\varphi(y) \geqslant \lim_{\substack{z\to y\\ z>y}} \varphi(z).$$

EXERCISE 3·70: Let X be a measurable set of $\mathbb{R}^n$ such that meas(X) = 1 and f be a positive measurable function on X.

Show that if one sets

$$A = \int_X f(x)\mathrm{d}x,$$

then:

$$\sqrt{1 + A^2} \leqslant \int_X \sqrt{1 + f^2(x)}\mathrm{d}x \leqslant 1 + A.$$

If $X = [0,1]$ and $f = F'$, where F is continuously differentiable, the preceding inequalities have a simple geometric interpretation. Use this interpretation to discover under what conditions equality holds, and then prove your answer.

ΔVΔ = VΔV = ΔVΔ = VΔV = ΔVΔ

SOLUTION: If $\varphi(x) = (1 + x^2)^{\frac{1}{2}}$, φ is continuous on $\mathbb{R}$ and $\varphi''(x) = (1 + x^2)^{-3/2}$; φ is therefore convex. If f is integrable, then by Jensen's Inequality

$$\sqrt{1 + A^2} \leqslant \int_X \sqrt{1 + f^2}. \qquad (*)$$

If $A = \infty$ equality holds, for $\sqrt{1 + f^2} \geqslant f$. Finally, the inequality

$$\int_X \sqrt{1 + f^2} \leqslant 1 + A \qquad (**)$$

follows from $\sqrt{1 + f^2} \leqslant 1 + f$. If $X = [0,1]$ and $f = F' \geqslant 0$, one has $A = F(1) - F(0)$, and the integral of $\sqrt{1 + f^2}$ is the length of the arc $\widehat{AC}$ (cf. the Figure).

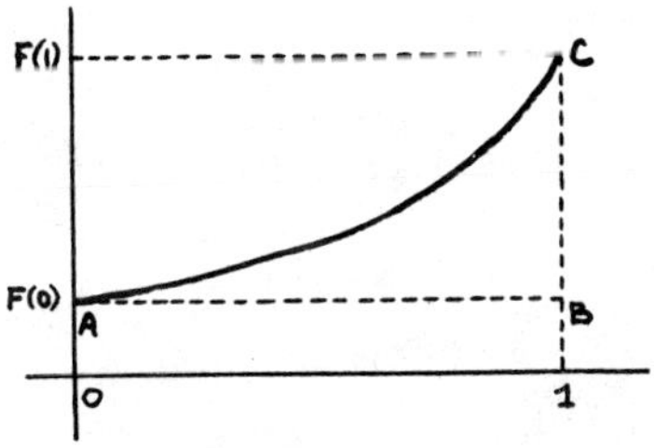

Inequalities (*) and (**) express that:

$$AC \leqslant \widehat{AC} \leqslant AB + BC.$$

One will therefore have equality in (*) if f is constant, and in (**) if f is zero. In fact, if:

$$\int_0^1 \sqrt{1 + f^2} = \sqrt{1 + A^2},$$

then on the one hand

$$\sqrt{1 + f^2} \geqslant \sqrt{1 + A^2} + \frac{A(f - A)}{\sqrt{1 + A^2}},$$

and on the other hand the integrals of both sides are equal; since f is continuous, it folllows that $f = A$. Similarly, if

$$\int_0^1 \sqrt{1 + f^2} = 1 + A = \int_0^1 (1 + f),$$

then since $\sqrt{1 + f^2} \leqslant 1 + f$, one has

$$\sqrt{1 + f^2} = 1 + f,$$

and so

$$f = 0.$$

EXERCISE 3·71: Let X be a measurable set of $\mathbb{R}^n$ such that meas(X) = 1, and let f,g be two positive measurable functions on X.

Show that if $fg \geqslant 1$, then:

$$\int_X f.\int_X g \geqslant 1.$$

Δ∇Δ = ∇Δ∇ = ΔVΔ = ∇Δ∇ = Δ∇Δ

SOLUTION: The inequality $fg \geqslant 1$ in fact implies that $f(x) > 0$ and $g(x) > 0$ for all $x \in X$, so that

$$\int_X f > 0, \qquad \int_X g > 0.$$

We therefore need only consider the case when

$$\int_X f < \infty.$$

Then Jensen's Inequality applied to f and to $\varphi(x) = 1/x$, $x > 0$, gives

$$\left(\int_X f\right)^{-1} \leqslant \int_X f^{-1} \leqslant \int_X g.$$

EXERCISE 3·72: Let f be a bounded measurable function on $E = \mathbb{R}^p$ such that $f(x) > 1$ for all $x \in E$.

(a): Show that if g is integrable on E, and

$$\left|\int_E f^n g\right| \leqslant M, \qquad n = 0,1,2,\ldots ,$$

then:

$$\int_E f^n g = 0, \qquad n = 0,1,2,\ldots .$$

(b): From this deduce that if g is integrable on $[0,a]$ and

$$\left|\int_0^a e^{nt} g(t)dt\right| \leqslant M, \qquad n = 0,1,2,\ldots ,$$

then $g = 0$ almost everywhere.

Δ∇Δ = ∇Δ∇ = ΔΔ = ∇Δ∇ = ΔΔ

SOLUTION: (a): Dividing g by M reduced us to the case where $M = 1$. There exists a constant A such that $1 < f \leqslant A$; for all $x \geqslant 0$, one therefore has, uniformly on E:

$$e^{xf} = \sum_{n=0}^{\infty} \frac{x^n}{n!} f^n,$$

and consequently

$$\left|\int e^{xf}g\right| = \left|\sum_{n=0}^{\infty} \frac{x^n}{n!}\int f^n g\right| \leqslant e^x.$$

Setting $F = f - 1$, one therefore has $0 < F \leqslant B = A - 1$, and

$$\left|\int e^{xF}g\right| \leqslant 1, \qquad x \geqslant 0. \tag{1}$$

For all $z = x + iy \in \mathbb{C}$ set

$$\phi(z) = \int e^{zF}g. \tag{2}$$

Then $|e^{zF}g| \leqslant e^{Bx^+}|g|$, which proves that ϕ is an entire function and that

$$|\phi(z)| \leqslant \|g\|_1 e^{Bx^+}. \tag{3}$$

By Inequality (1) it is also true that

$$|\phi(x)| \leqslant 1, \qquad x \geqslant 0. \tag{4}$$

(3) and (4) imply, on setting $C = \max(\|g\|_1, 1)$, that

$$|\phi(z)| \leqslant C \quad \text{if} \quad \mathrm{Re}(z) \leqslant 0 \quad \text{or} \quad z \in \mathbb{R}_+. \tag{5}$$

We will show that in fact:

$$|\phi(z)| \leqslant C, \qquad z \in \mathbb{C}. \tag{6}$$

It will then follow from Liouville's Theorem that ϕ is constant. Since it is clear that $\phi(-x) \to 0$ as $x \to +\infty$, this constant is zero. Differentiating relation (2) n times and setting $z = 0$ gives

$$\int F^n g = \int (f - 1)^n g = 0, \qquad n \geqslant 0.$$

But

$$f^n = \sum_s C^n_s (f - 1)^s$$

which implies the desired result.

We now next prove (6). This follows from the Phrägmen-Lindelöf Theorem. We shall give the proof of it in this particular case. Without loss of generality g is real, so that $\phi(\bar{z}) = \overline{\phi(z)}$. It is therefore sufficient to prove that (6) holds when $\mathrm{Re}(z) > 0$, $\mathrm{Im}(z) > 0$.

When $\mathrm{Re}(z) \geqslant 0$, $\mathrm{Im}(z) \geqslant 0$, define

$$G(z) = \exp(-\varepsilon e^{-i\alpha\pi/4} z^\alpha)\phi(z)$$

where $\varepsilon > 0$ and $1 < \alpha < 2$. If $z = \rho e^{i\vartheta}$, $\rho \geqslant 0$, $0 \leqslant \vartheta \leqslant \pi/2$, then

$$|G(z)| = \exp(-\varepsilon\rho^\alpha \cos\alpha(\vartheta - \tfrac{1}{4}\pi))|\phi(z)|,$$

so that by (5):

$$|G(x)| \leqslant C, \qquad |G(iy)| \leqslant C, \qquad x \geqslant 0,\ y \geqslant 0. \tag{7}$$

By (3), since $x^+ \leqslant \rho$ one has

$$|G(z)| \leqslant C\exp(-\varepsilon\rho^\alpha \cos(\vartheta - \tfrac{1}{4}\pi) + B\rho). \tag{8}$$

Since $-\alpha\pi/4 \leqslant \alpha(\vartheta - \pi/4) \leqslant \alpha\pi/4 < \pi/2$, one deduces from (8) that:

$$|G(z)| \leqslant C\exp\left(-\varepsilon\rho^\alpha \cos\left(\frac{\alpha\pi}{4}\right) + B\rho\right). \tag{9}$$

Because $\alpha > 1$ the right side of (9) tends to zero as $\rho \to \infty$; therefore there exists ρ_0 such that

$$|G(z)| \leqslant C; \quad \mathrm{Re}(z) \geqslant 0, \quad \mathrm{Im}(z) \geqslant 0, \quad |z| \geqslant \rho_0. \tag{10}$$

Now by (7) and (10) $|G| \leqslant C$ on the boundary of the domain

$\mathrm{Re}(z) > 0$, $\mathrm{Im}(z) > 0$, $|z| < \rho_0$, by the Maximum Principle one also has $|G| \leqslant C$ in this domain. Finally, if $z = \rho e^{i\vartheta}$, $\rho \geqslant 0$, $0 \leqslant \vartheta \leqslant \pi/2$,

$$|G(z)| = \exp(-\varepsilon\pi^{\alpha}\cos\alpha(\vartheta - \tfrac{1}{4}\pi))|\phi(z)| \leqslant C.$$

On taking $\varepsilon \to 0$ this yields:

$$|\phi(z)| \leqslant C, \qquad \mathrm{Re}(z) \geqslant 0, \qquad \mathrm{Im}(z) \geqslant 0.$$

SOLUTION: (b): By the preceding,

$$\int_0^a e^{nt} g(t)\,\mathrm{d}t = 0, \qquad n \geqslant 0.$$

It follows that for every polynomial P,

$$\int_1^{e^a} P(u) g(\log u) \frac{\mathrm{d}u}{u} = 0.$$

If φ is a continuous function on $[1, e^a]$ and if (P_i) is a sequence of polynomials which tends uniformly to φ on this interval, then since $u^{-1}g(\log u)$ is integrable, one obtains

$$\int_1^{e^a} \varphi(u) g(\log u) \frac{\mathrm{d}u}{u} = 0.$$

But then $u^{-1}g(\log u) = 0$ almost everywhere on $[1, e^a]$, i.e., $g(t) = 0$ almost everywhere on $[0, a]$.

CHAPTER 4

Asymptotic Evaluation of Integrals

EXERCISE 4·74: Let φ be a continuous real function on $[0,\alpha]$. Assume that $\varphi(x) > 0$ if $0 < x \leqslant \alpha$ and that $\varphi(x) \sim Ax^r$ as $x \to 0$ ($A > 0, r \geqslant 0$). For all $t > 0$ set

$$F(t) = \int_0^\alpha \frac{dx}{t + \varphi(x)} .$$

(a): Show that if $0 \leqslant r < 1$,

$$\lim_{t \to 0} F(t) = \int_0^\alpha \frac{dx}{\varphi(x)} .$$

(b): Show that if $r > 1$, then as $t \to 0$

$$F(t) \sim \frac{\pi}{rA^{1/r}\sin(\pi/r)} \cdot \frac{1}{t^{1 - 1/r}} .$$

(c): Show that if $r = 1$, then as $t \to 0$

$$F(t) \sim \frac{1}{A} \log \frac{1}{t} .$$

(For part (c), show that this can be reduced to the study of the integral

$$\int_{\alpha t}^{\alpha} (t + \varphi(x))^{-1} dx,$$

and make the change of variable $x = \alpha t^{y}$).

ΔVΔ = VΔV = ΔVΔ = VΔV = ΔVΔ

SOLUTION: (a): As t decreases to zero, $(t + \varphi(x))^{-1}$ increases to $1/\varphi(x)$ for all $x \in]0,\alpha]$, whence the result.

SOLUTION: (b): Make the change of variables $x = (ty)^{1/r}$. Then

$$F(t) = \frac{1}{rt^{1 - 1/r}} \int_0^{\alpha^r/t} \frac{y^{1/r - 1}}{1 + t^{-1}\varphi(t^{1/r}y^{1/r})} dy.$$

Set:

$$f_t(y) = \begin{cases} \dfrac{y^{1/r - 1}}{1 + t^{-1}\varphi(t^{1/r}y^{1/r})} & \text{if } 0 \leqslant y \leqslant \alpha^r/t, \\ 0 & \text{if } y > \alpha^r/t. \end{cases}$$

Since, for $y > 0$ fixed and $t \to 0$, $t^{-1}\varphi(t^{1/r}y^{1/r}) \to Ay$,

$$\lim_{t \to 0} f_t(y) = \frac{y^{1/r - 1}}{1 + Ay}.$$

Furthermore, $x^{-r}\varphi(x)$ extends to a strictly positive continuous function on $[0,\alpha]$; hence there exists a constant $B > 0$ such that $\varphi(x) \geqslant Bx^r$. Therefore

$$f_t(y) \leqslant \frac{y^{1/r - 1}}{1 + By}.$$

The Dominated Convergence Theorem can therefore be applied, yielding

$$\lim_{t\to 0}\int_0^\infty f_t(y)\,dy = \int_0^\infty \frac{y^{1/r-1}}{1+Ay}\,dy = \frac{1}{A^{1/r}}\cdot\frac{\pi}{\sin(\pi/r)}$$

SOLUTION: (c): Note that the relation proved in part (a) is true for all $r \geqslant 0$, so that $\lim_{t\to 0} F(t) = \infty$ if $r \geqslant 1$. Now,

$$\int_0^{\alpha t} \frac{dx}{t+\varphi(x)} \leqslant \int_0^{\alpha t} \frac{dx}{t+Bx} = \frac{1}{B}\log(1+B\alpha),$$

so that

$$F(t) \sim \int_{\alpha t}^{\alpha} \frac{dx}{t+\varphi(x)} .$$

The change of variable $x = \alpha t^y$ gives

$$\int_{\alpha t}^{\alpha} \frac{x}{t+\varphi(x)} = \alpha\log\frac{1}{t}\int_0^1 \frac{t^y}{t+\varphi(\alpha t^y)}\,dy.$$

When $0 < y < 1$ and $t \to 0$ one has $\varphi(\alpha t^y) \sim \alpha A t^y$ and $t = o(t^y)$. Consequently

$$\lim_{t\to 0} \frac{t^y}{t+\varphi(\alpha t^y)} = \frac{1}{\alpha A} .$$

Furthermore, for $0 \leqslant y \leqslant 1$ and $t > 0$,

$$\frac{t^y}{t+\varphi(\alpha t^y)} \leqslant \frac{t^y}{t+\alpha B t^y} < \frac{1}{\alpha B} .$$

from this it follows that

$$\lim_{t\to 0}\int_0^1 \frac{t^y}{t+\varphi(\alpha t^y)}\,dy = \frac{1}{\alpha A}.$$

EXERCISE 4·74: Let p,q,r be three positive real numbers. Find the necessary and sufficient condition that

$$\int_0^\infty \frac{x^p}{1+x^q|\sin\pi x|^r}\,dx < \infty.$$

What type of counter-example does this furnish?

ΔVΔ = VΔV = ΔVΔ = VΔV = ΔVΔ

SOLUTION: The integral is equal to

$$\sum_{n=0}^{\infty} I_n,$$

where

$$I_n = \int_n^{n+1} \frac{x^p}{1+x^q|\sin\pi x|^r}\,dx = \int_0^1 \frac{(x+n)^p}{1+(x+n)^q(\sin\pi x)^r}\,dx.$$

For $t > 0$, set

$$\varphi(t) = \int_0^{\frac{1}{2}} \frac{dx}{t+(\sin\pi x)^r};$$

then

$$\frac{2n^p}{(n+1)^q}\varphi((n+1)^{-q}) < I_n < \frac{2(n+1)^p}{n^q}\varphi(n^{-q}).$$

By the results of the preceding exercise, as $t \to 0$ one has

$$\varphi(t) \to \int_0^{\frac{1}{2}} (\sin \pi x)^{-r} dx = A_r \qquad \text{if } 0 \leqslant r < 1,$$

$$\varphi(t) \sim \frac{1}{\pi} \log \frac{1}{t} \qquad \text{if } r = 1,$$

$$\varphi(t) \sim \frac{1}{r\sin(\pi/r)} \left(\frac{1}{t}\right)^{1 - 1/r} \qquad \text{if } r > 1.$$

Consequently, if $q > 0$,

$$I_n \sim 2A_r n^{p-q} \qquad \text{if } 0 \leqslant r < 1,$$

$$I_n \sim \frac{2q}{\pi} n^{p-q} \log n \qquad \text{if } r = 1,$$

$$I_n \sim \frac{2}{r\sin(\pi/r)} n^{p - q/r} \qquad \text{if } r > 1.$$

It follows that if $q > 0$ the integral converges if and only if

$$\frac{q}{p + 1} > \max(r, 1).$$

For $q = 0$ the integral is never convergent, so that the preceding condition is valid in all cases.

If $p > 0, r > 0$ and $q > (p + 1)\max(1,r)$ then for $x = n$ the integrand has the value n^p; it is therefore not bounded as $x \to \infty$ although its integral is finite.

EXERCISE 4·75: Let f,g be two real functions on $]0,a[$. Assume that

(i): $f(x) \sim Ax^\alpha$ and $g(x) \sim x^\beta$ when $x \to 0$ ($\alpha > 0, A > 0, \beta > -1$);

(ii): f is strictly positive and increasing on $]0,a[$;

(iii): $\displaystyle\int_0^a |g(x)| e^{-f(x)} dx < \infty$.

Show that as $t \to \infty$

$$\int_0^a g(x)e^{-tf(x)}dx \sim \frac{1}{\alpha}\Gamma\left(\frac{\beta+1}{\alpha}\right)(At)^{-(\beta+1)/\alpha} .$$

APPLICATIONS: (1): Prove that $\Gamma(\alpha+1) \sim (\alpha/e)^{\alpha}\sqrt{2\pi\alpha}$ as $\alpha \to \infty$.

(2): Prove that

$$B(p,q) = \frac{\Gamma(p)\Gamma(q)}{\Gamma(p+q)} .$$

(Begin by showing that

$$B(p,q) = \frac{p+q}{q} B(p+1,q).)$$

SOLUTION: Let $0 < B < A$, $\gamma > 1$; there exists b such that $0 < b < a$ and

$$f(x)) \geqslant Bx^{\alpha}, \quad 0 \leqslant g(x) \leqslant \gamma x^{\beta} \quad \text{if } 0 < x \leqslant b. \tag{*}$$

Since $f(x) \geqslant f(b) > 0$ for $b \leqslant x < a$, when $t > 1$ one has

$$\left| t^{(\beta+1)/\alpha}\int_b^a g(x)e^{-tf(x)}dx \right|$$

$$\leqslant t^{(\beta+1)/\alpha}e^{-(t-1)f(b)}\int_0^a |g(x)|e^{-f(x)}dx,$$

which proves that the left hand side tends to zero as $t \to \infty$. Moreover, the change of variable $x \mapsto xt^{-1/\alpha}$ gives

$$t^{(\beta+1)/\alpha}\int_0^b g(x)e^{-tf(x)}dx = \int_0^{\infty} F_t(x)dx$$

where

$$F_t(x) = \begin{cases} t^{\beta/\alpha} g(xt^{-1/\alpha}) e^{-tf(xt^{-1/\alpha})} & \text{if } 0 < x \leqslant bt^{1/\alpha}, \\ 0 & \text{if } x > bt^{1/\alpha}. \end{cases}$$

By (i),

$$\lim_{t\to\infty} F_t(x) = x^\beta e^{-Ax^\alpha},$$

and by (*),

$$0 \leqslant F_t(x) \leqslant \gamma x^\beta e^{-Bx^\alpha}.$$

Applying Lebesgue's Theorem, and recalling that the integral from b to a tends to zero

$$\lim_{t\to\infty} t^{(\beta+1)/\alpha} \int_0^a g(x) e^{-tf(x)} dx = \int_0^\infty x^\beta e^{-Ax^\alpha} dx$$

$$= \frac{1}{\alpha} A^{-(\beta+1)/\alpha} \Gamma\left(\frac{\beta + 1}{\alpha}\right).$$

FIRST APPLICATION: Setting $x = \alpha(u + 1)$ one obtains

$$\Gamma(\alpha + 1) = \int_0^\infty x^\alpha e^{-x} dx = \alpha^{\alpha+1} e^{-\alpha} \int_{-1}^\infty e^{-\alpha(u-\log(u+1))} du.$$

The above result can be applied to each of the integrals $\int_{-1}^0$ and $\int_0^\infty$ with $\alpha = 2$, $A = \frac{1}{2}$, $\beta = 0$, giving

$$\int_{-1}^\infty e^{-\alpha(u-\log(u+1))} du \sim 2 \times \tfrac{1}{2}\Gamma(\tfrac{1}{2}) \left(\frac{\alpha}{2}\right)^{-\frac{1}{2}} = \sqrt{\frac{2\pi}{\alpha}}$$

whence:

$$\Gamma(\alpha + 1) \sim \alpha^{\alpha+\frac{1}{2}} e^{-\alpha} \sqrt{2\pi}.$$

SECOND APPLICATION: For $p > 0, q > 0$,

$$B(p + 1,q) = \int_0^1 \left(\frac{x}{1 - x}\right)^p (1 - x)^{p+q-1} dx$$

$$= \frac{p}{p + q} \int_0^1 (1 - x)^{p+q} \left(\frac{x}{1 - x}\right)^{p-1} \frac{dx}{(1 - x)^2}$$

$$= \frac{p}{p + q} B(p,q).$$

From this it follows that for every integer $n \geqslant 1$,

$$B(p,q) = \frac{(p + q)(p + q + 1)\cdots(p + q + n)}{p(p + 1)\cdots(p + n)} B(p + n + 1,q).$$

Now it is known that as $n \to \infty$,

$$\frac{\alpha(\alpha + 1)\cdots(\alpha + n)}{n!} \sim \frac{n^\alpha}{\Gamma(\alpha)},$$

So

$$\frac{(p + q)\cdots(p + q + n)}{p\cdots(p + n)} \sim \frac{\Gamma(p)}{\Gamma(p + q)} n^q.$$

On the other hand,

$$B(p + n + 1,q) = \int_0^1 x^{q-1}(1 - x)^p e^{n\log(1-x)} dx.$$

Applying the first part with $\alpha = A = 1$ and $\beta = q - 1$,

$$B(p + n + 1,q) \sim \Gamma(q) n^{-q},$$

So

$$B(p,q) = \frac{\Gamma(p)\Gamma(q)}{\Gamma(p+q)} .$$

EXERCISE 4·76: Prove that for $n \geqslant 0$,

$$\int_0^\infty x^n e^{-x^{\frac{1}{4}}} \sin x^{\frac{1}{4}} dx = 0,$$

and then that as $t \to \infty$,

$$\int_0^\infty e^{itx - x^{\frac{1}{4}}} \sin x^{\frac{1}{4}} dx \sim \frac{i}{4}\,\Gamma(\tfrac{1}{4}) e^{i\pi/8} t^{-5/4} .$$

Δ∇Δ = ∇Δ∇ = Δ∇Δ = ∇Δ∇ = ΔΔ∇Δ

SOLUTION: The function

$$F(z) = \int_0^\infty x^{4n+3} e^{-zx} dx$$

is holomorphic for $\mathrm{Re}(z) > 0$. When z is real one has:

$$F(z) = \frac{(4n+3)!}{z^{4n+4}} .$$

By analytic continuation this formula is valid for every z with $\mathrm{Re}(z) > 0$. In particular, if $z = 1 + i$,

$$\int_0^\infty x^{4n+3} e^{-(1+i)x} dx = (-1)^{n+1} \frac{(4n+3)!}{2^{2n+2}} .$$

Taking the imaginary parts of both sides yields:

$$\int_0^\infty x^{4n+3} e^{-x} \sin x dx = 0,$$

and carrying the change of variable $x \to x^{\frac{1}{4}}$ gives

$$\int_0^\infty x^n e^{-x^{\frac{1}{4}}} \sin x^{\frac{1}{4}} dx = 0.$$

Now set:

$$f(z,t) = \exp(itz - z^{\frac{1}{4}}) \sin z^{\frac{1}{4}},$$

which for fixed t is holomorphic for $\mathrm{Re}(z) > 0$ and continuous for $\mathrm{Re}(z) \geqslant 0$. (One chooses the principal branch of $z^{\frac{1}{4}}$ in this half-plane). Taking into account the majorisation

$$|\sin(x + iy)| \leqslant e^y,$$

which is valid for $y \geqslant 0$, then for $R \geqslant 0$ and $t \geqslant 0$ one has:

$$|f(Re^{i\vartheta},t)| \leqslant \exp[-tR\sin\vartheta - R^{\frac{1}{4}}(\cos\frac{\vartheta}{4} - \sin\frac{\vartheta}{4})],$$

and consequently when $0 \leqslant \vartheta \leqslant \pi/2$:

$$|f(Re^{i\vartheta},t)| \leqslant \exp\left(-\frac{2tR\vartheta}{\pi} - \sqrt{2}R^{\frac{1}{4}}\sin\frac{\pi}{8}\right). \tag{1}$$

Moreover, for $z = x + iy$, $y \geqslant 0$:

$$|z^{-\frac{1}{4}} e^{ty} f(z,t)| \leqslant \min(|z|^{-\frac{1}{4}}, |z^{-\frac{1}{4}} \sin z^{\frac{1}{4}}|),$$

and there therefore exists a constant M such that

$$|f(z,t)| < M|z|^{\frac{1}{4}} e^{-ty}, \quad y \geqslant 0. \tag{2}$$

Therefore, using (1), the integral of $f(z,t)$ along the circular quadrant $\vartheta \to z = Re^{i\vartheta}$ $(0 \leqslant \vartheta \leqslant \pi/2)$ is majorised in modulus by

$$R\exp(-\sqrt{2}R^{\frac{1}{4}}\sin\frac{\pi}{8}) \int_0^\infty \exp\left(-\frac{2tR\vartheta}{\pi}\right) d\vartheta \leqslant \frac{\pi}{2t} \exp(-\sqrt{2}R^{\frac{1}{4}}\sin\frac{\pi}{8}),$$

and consequently tends to zero as $R \to \infty$. By Cauchy's Theorem one therefore has

$$\int_0^\infty f(x,t)\mathrm{d}x = i\int_0^\infty f(iy,t)\mathrm{d}y$$

$$= it^{-5/4}e^{i\pi/8}\int_0^\infty t^{\frac{1}{4}}e^{-i\pi/8}f(it^{-1}y,t)\mathrm{d}y.$$

It is easily verified that

$$\lim_{t\to\infty} t^{\frac{1}{4}}e^{-i\pi/8}f(it^{-1}y,t) = y^{\frac{1}{4}}e^{-y},$$

and by (2)

$$|t^{\frac{1}{4}}e^{-i\pi/8}f(it^{-1}y,t)| \leqslant My^{\frac{1}{4}}e^{-y}.$$

Therefore by Lebesgue's Theorem

$$\lim_{t\to\infty} t^{5/4}\int_0^\infty f(x,t)\mathrm{d}x = ie^{i\pi/8}\int_0^\infty y^{\frac{1}{4}}e^{-y}\mathrm{d}y$$

$$= ie^{i\pi/8}\Gamma\left(\frac{5}{4}\right) = \frac{i}{4}\,\Gamma(\tfrac{1}{4})e^{i\pi/8}.$$

EXERCISE 4·77: For every integer $n \geqslant 1$ denote by d_n the number of partitions of a set with n elements. Set $d_0 = 1$.

(a): Show that for all complex numbers z:

$$\sum_{n=0}^{\infty} d_n \frac{z^n}{n!} = \exp(e^z - 1).$$

(b): Deduce from this that for every real number $u > 0$:

$$d_n = \frac{n!}{2\pi i}\int_{u-i\infty}^{u+i\infty} z^{-(n+1)}\exp(e^z)dz.$$

(c): Let u_n be the unique real root of the equation $ze^z = n + 1$.

Show that

$$d_n \sim \frac{n!}{e\sqrt{2\pi}} \exp(e^{u_n} - u_n e^{u_n}\log u_n - \tfrac{1}{2}u_n),$$

and deduce from this that

$$\log d_n \sim n\log n.$$

Δ∇Δ = ∇Δ∇ = ΔVΔ = ∇Δ∇ = ΔVΔ

SOLUTION: (a): To determine a partition of $E_{n+1} = \{1,2,\ldots,n + 1\}$ one may first fix the part of E_{n+1} which contains $n + 1$; if this part contains $p + 1$ elements $(0 \leqslant p \leqslant n)$ there are $\binom{n}{p}$ possible choices. Next it remains to choose a partition of the $n - p$ remaining elements, which gives, taking account of the convention $d_0 = 1$,

$$d_{n+1} = \sum_{p=0}^{n} \binom{n}{p} d_{n-p}. \tag{1}$$

This relation can be written as

$$(n + 1)\frac{d_{n+1}}{(n + 1)!} = \sum_{p=0}^{n} \frac{1}{p!}\frac{d_{n-p}}{(n - p)!}.$$

If the series

$$f(z) = \sum_{n=0}^{\infty} d_n \frac{z^n}{n!} \tag{2}$$

has a radius of convergence $R > 0$, then taking the derivative and using the above relation shows that for $|z| < R$

$$f'(z) = e^z f(z),$$

and consequently, since $f(0) = 1$,

$$f(z) = \exp(e^z - 1).$$

This function is entire, and if its Taylor expansion is written in the form of Equation (2) the coefficients d_n have to satisfy Equation (1), so, since $d_0 = 1$, the proposition follows.

SOLUTION: (b): By Cauchy's Theorem:

$$d_n = \frac{n!}{2\pi i e}\int_\Gamma z^{-(n+1)}\exp(e^z)dz$$

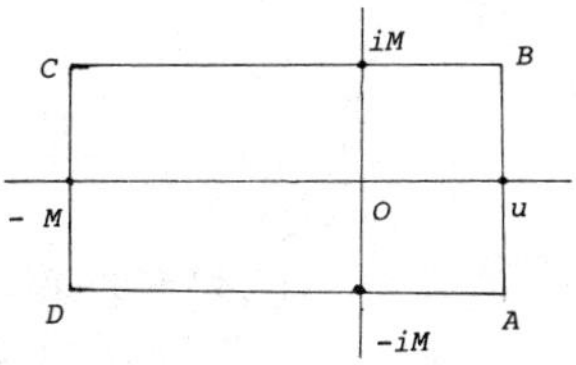

where Γ denotes the rectangle indicated by the figure. Note that, on this rectangle, if $z = x + iy$,

$$|\exp(e^z)| = \exp(e^x \cos y) \leqslant \exp(e^u),$$

and consequently

$$\left|\int_{BCDA}\right| \leqslant \exp(e^u)\ \frac{2u + 4M}{M^{n+1}} .$$

On the other hand:

$$|(u + iy)^{-(n+1)}\exp(e^{u+iy})| = (u^2 + y^2)^{-(n+1)/2}\exp(e^u).$$

From this one deduces that for $n \geqslant 1$:

$$d_n = \frac{n!}{2\pi e}\int_{-\infty}^{+\infty}(u + iy)^{-(n+1)}\exp(e^{u+iy})dy, \tag{3}$$

the integral being absolutely convergent.

SOLUTION: (c): Introducing the principal branch of log z in the half-plane $\mathrm{Re}(z) > 0$, and if u is chosen so that $ue^{u} = n + 1$,

$$d_n = \frac{n!}{2\pi e}\exp(e^{u} - ue^{u}\log u)\int_{-\infty}^{+\infty} g(y,u)\,dy, \tag{4}$$

where

$$g(y,u) = \exp\{e^{u}[e^{iy} - 1 - u\log\left(1 + \frac{iy}{u}\right)]\}.$$

Note that

$$|g(y,u)| = \exp\{-e^{u}[2\sin^2\frac{y}{2} + \frac{u}{2}\log\left(1 + \frac{y^2}{u^2}\right)]\}.$$

Therefore

$$\left|\int_{|y|\geqslant\pi} g(y,u)\,dy\right| \leqslant 2\int_{\pi}^{\infty}(1 + u^{-2}y^2)^{-\frac{1}{2}ue^{u}}\,dy$$

$$\leqslant 2\int_{\pi}^{\infty}\frac{dy}{1 + \frac{1}{2}u^{-1}e^{u}y^2}.$$

The last integral has the value

$$e^{-u/2}\sqrt{2u}\tan^{-1}\left(\frac{1}{\pi}e^{-u/2}\sqrt{2u}\right),$$

whence

$$\int_{|y|\geqslant\pi} g(y,u)\,dy = O(ue^{-u}). \tag{5}$$

If $|y| \leqslant \pi$ then $|\sin(y/2)| \geqslant |y|/\pi$, so that:

$$|g(y,u)| < \exp\left(-\frac{2}{\pi^2}e^{u}y^2\right).$$

Setting $y = te^{-u/2}$ yields:

$$\int_{-\pi}^{+\pi} g(y,u)dy = e^{-u/2}\int_{-\pi e^{u/2}}^{\pi e^{u/2}} g(te^{-u/2},u)dt,$$

and for $t > 0$,

$$\lim_{u\to\infty} g(te^{-u/2},u) = e^{-t^2/2}.$$

Since for $0 \leqslant |t| \leqslant \pi e^{u/2}$

$$|g(te^{-u/2},u)| \leqslant e^{-2t^2/\pi^2},$$

Lebesgue's Theorem implies that:

$$\int_{-\pi}^{\pi} g(y,u)dy \sim e^{-4/2}\int_{-\infty}^{\infty} e^{-t^2/2}dt = e^{-4/2}\sqrt{2\pi}.$$

Taking account of Equations (4) and (5),

$$d_n \sim \frac{n!}{e\sqrt{2\pi}} \exp(e^{u_n} - u_n e^{un}\log u_n - \tfrac{1}{2}u_n).$$

Hence

$$\log d_n = \log n! - (u_n e^{u_n}\log u_n - e^{u_n} + \tfrac{1}{2}u_n) + O(1).$$

Now $\log n! \sim n\log n$ and

$$u_n e^{u_n}\log u_n - e^{u_n} + \tfrac{1}{2}u_n \sim n\log u_n = o(n(u_n + \log u_n))$$

$$= o(n\log n),$$

which proves that $\log d_n \sim n\log n$.

EXERCISE 4·78: Show that for every real number t the integral

$$\phi(t) = \frac{1}{\sqrt{\pi}}\int_0^\infty \cos\left(tx + \frac{x^3}{3}\right)dx$$

is convergent, and that the function ϕ so defined (called the *AIRY FUNCTION*) is a solution of the differential equation,

$$\phi'' - t\phi = 0.$$

Next prove that as $t \to \infty$

$$\phi(t) = \frac{1}{2t^{\frac{1}{4}}}\exp\left(-\frac{2}{3}t^{3/2}\right)(1 + O(t^{-3/4})),$$

$$\phi(-t) = \frac{1}{t^{\frac{1}{4}}}\sin\left(\frac{2}{3}t^{3/2} + \frac{\pi}{4}\right) + O\left(\frac{1}{t}\right).$$

ΔVΔ = VΔV = ΔVΔ = VΔV = ΔVΔ

SOLUTION: Note

$$2\sqrt{\pi}\phi(t) = \int_{-\infty}^{+\infty}\exp i\left(tx + \frac{x^3}{3}\right)dx.$$

If $z = x + iy$,

$$\left|\exp i\left(tz + \frac{z^3}{3}\right)\right| = \exp(-ty - x^2y + \frac{1}{3}y^3),$$

so that if $0 \leqslant y \leqslant \alpha$ and $|t| \leqslant A$,

$$\left|\exp i\left(tz + \frac{z^3}{3}\right)\right| \leqslant Be^{-x^2y},$$

where $B = \exp\left(\alpha A + \frac{1}{3}\alpha^3\right)$. Therefore

$$\left|\int_{x+i0}^{x+i\alpha} \exp i\left(tz + \frac{z^3}{3}\right)dz\right| \leqslant B\int_0^{\alpha} e^{-x^2y}dy \leqslant \frac{B}{x^2}.$$

This shows that the integral which defines ϕ is convergent, and that if L_α is the straight line $(-\infty + i\alpha, \infty + i\alpha)(\alpha \geqslant 0)$ then

$$2\sqrt{\pi}\phi(t) = \int_{L_\alpha} \exp i\left(tz + \frac{z^3}{3}\right)dz. \tag{1}$$

From this it follows that

$$2\sqrt{\pi}\phi''(t) = -\int_{L_\alpha} z^2 \exp i\left(tz + \frac{z^3}{3}\right)dz$$

the differentiation under the integral sign being legitimate because on L_α one has for $|t| \leqslant A$

$$\left|z^2 \exp i\left(tz + \frac{z^3}{3}\right)\right| \leqslant B(x^2 + \alpha^2)e^{-\alpha x^2}.$$

Hence

$$2\sqrt{\pi}(\phi''(t) - t\phi(t)) = -\int_{L_\alpha} (t + z^2)\exp i\left(tz + \frac{z^3}{3}\right)dz$$

$$= \left[i\exp i\left(tz + \frac{z^3}{3}\right)\right]\Big|_{z=-\infty+i\alpha}^{\infty+i\alpha} = 0.$$

== == = = = == ==

When $t > 0$, carrying out the change of variable $z \mapsto t^{\frac{1}{2}}z$ in Equation (1) and setting $\beta = t^{-\frac{1}{2}}\alpha, \lambda = t^{3/2}$ yields

$$2\sqrt{\frac{\pi}{t}}\,\phi(t) = \int_{L_\beta} \exp i\lambda\left(z + \frac{z^3}{3}\right)dz.$$

Since the derivative of $z + z^3/3$ vanishes for $z = i$, let us take $\beta = 1$ in the formula above. This yields

$$2\sqrt{\frac{\pi}{t}}\,\phi(t) = e^{-2\lambda/3}\int_{-\infty}^{+\infty}\exp\lambda\left(-u^2 + \frac{iu^3}{3}\right)du$$

$$= \frac{1}{\sqrt{\lambda}}\,e^{-2\lambda/3}\int_{-\infty}^{+\infty}\exp\left(-u^2 + \frac{iu^3}{3\sqrt{\lambda}}\right)du.$$

The latter integral can be written

$$\sqrt{\pi} + \int_{-\infty}^{+\infty} e^{-u^2}(e^{iu^3/3\sqrt{\lambda}} - 1)du.$$

The absolute value of integral above is majorised by

$$\frac{1}{3\sqrt{\lambda}}\int_{-\infty}^{+\infty}|u|^3 e^{-u^2}du.$$

From this, one deduces that as $t \to \infty$

$$\phi(t) = \tfrac{1}{2}t^{-\frac{1}{4}}\exp\left(-\frac{2}{3}\,t^{3/2}\right)(1 + O(t^{-3/4})).$$

== == = $\overline{\overline{=}}$ = == ==

Setting $\lambda = t^{3/2}$ as before,

$$\sqrt{\frac{\pi}{t}}\,\phi(-t) = \int_0^{\infty}\cos\lambda\left(\frac{x^3}{3} - x\right)dx.$$

As the derivative of $x^3/3 - x$ vanishes for $x = 1$, one is led to the change of variable $x = 1 + u/\sqrt{\lambda}$, which gives

$$\sqrt{\frac{\pi}{t}}\,\phi(-t) = \frac{1}{\sqrt{\lambda}}\,\mathrm{Re}\{e^{-2i\lambda/3}\int_{-\sqrt{\lambda}}^{\infty}\exp i\left(u^2 + \frac{u^3}{3\sqrt{\lambda}}\right)du\}.$$

Set

$$f(z) = \exp i\left(z^2 + \frac{z^3}{3\sqrt{\lambda}}\right) = Re^{i\psi},$$

where, as $z = \rho e^{i\vartheta}$,

$$R = \exp(-\rho^2\sin 2\vartheta - \frac{\rho^3}{3\sqrt{\lambda}}\sin 3\vartheta),$$

$$\psi = \rho^2\cos 2\vartheta + \frac{\rho^3}{3\sqrt{\lambda}}\cos 3\vartheta.$$

Assume that $0 \leqslant \vartheta \leqslant \pi/4$ and $\rho \geqslant -\sqrt{\lambda}$. Then $\sin 2\vartheta \geqslant 0$ and $\sin 3\vartheta \geqslant 0$. From this it follows that for every number ξ such that $0 \leqslant \xi \leqslant 1$:

$$\sin 2\vartheta + \frac{\xi\rho}{3\sqrt{\lambda}}\sin 3\vartheta \geqslant \frac{4\vartheta}{\pi} - \vartheta = \alpha\vartheta,$$

where $\alpha = 4/\pi - 1 > 0$. Therefore

$$|R| \leqslant e^{-\alpha\rho^2\vartheta}. \tag{2}$$

By the Mean Value Theorem

$$R - e^{-\rho^2\sin 2\vartheta} = -\frac{\rho^3}{3\sqrt{\lambda}}\sin 3\vartheta\exp\left(-\rho^2\sin 2\vartheta - \frac{\xi\rho^3}{3\sqrt{\lambda}}\sin 3\vartheta\right),$$

where $0 < \xi < 1$. Consequently:

$$|R - e^{-\rho^2\sin 2\vartheta}| \leqslant \frac{|\rho|^3}{3\sqrt{\lambda}} e^{-\alpha\rho^2\vartheta}. \tag{3}$$

Let L be the half-line $\rho e^{i\pi/4}$, $\rho \geqslant -\sqrt{\lambda}$, and Γ the circular arc $-\sqrt{\lambda}e^{i\vartheta}$, $0 \leqslant \vartheta \leqslant \pi/4$. Then (2) shows

$$\int_{-\sqrt{\lambda}}^{\infty} f(u)du = \int_{\Gamma+L} f(z)dz,$$

and then that

$$\left|\int_\Gamma f(z)dz\right| \leqslant \sqrt{\lambda}\int_0^{\pi/4} e^{-\alpha\lambda\vartheta}d\vartheta = O\left(\frac{1}{\sqrt{\lambda}}\right) .$$

Note next that for $\vartheta = \pi/4$, $\psi = -\rho^3/3\sqrt{2\lambda}$. Then (2) and (3) show that

$$|f(\rho e^{i\pi/4}) - e^{-\rho^2}| \leqslant |Re^{i\psi} - R| + |R - e^{-\rho^2}|$$

$$\leqslant \left(\frac{|\rho|^3}{3\sqrt{2\lambda}} + \frac{|\rho|^3}{3\sqrt{\lambda}}\right)e^{-\alpha\pi\rho^2/4}.$$

Consequently:

$$\int_L f(z)dz = e^{i\pi/4}\int_{-\sqrt{\lambda}}^{\infty} e^{-\rho^2}d\rho + H,$$

with

$$H \leqslant \text{constant} \times \frac{1}{\sqrt{\lambda}}\int_{-\sqrt{\lambda}}^{\infty} |\rho|^3 e^{-\alpha\pi\rho^2/4}d\rho = O\left(\frac{1}{\sqrt{\lambda}}\right) .$$

Since, furthermore

$$\int_{-\infty}^{-\sqrt{\lambda}} e^{-\rho^2}d\rho \sim \frac{e^{-\lambda}}{2\sqrt{\lambda}} ,$$

one has

$$\int_{-\sqrt{\lambda}}^{\infty} f(u)du = e^{i\pi/4}\sqrt{\pi} + O\left(\frac{1}{\sqrt{\lambda}}\right) .$$

Finally,

$$\sqrt{\frac{\pi}{t}}\,\phi(-t) = \sqrt{\frac{\pi}{\lambda}}\,\mathrm{Re}\{e^{-2i\lambda/3}(e^{i\pi/4} + O\left(\frac{1}{\sqrt{\lambda}}\right))\},$$

i.e.,

$$\phi(-t) = t^{-\frac{1}{4}}\cos\left(-\frac{2}{3}\,t^{3/2} + \frac{\pi}{4}\right) + O\left(\frac{1}{t}\right)$$

$$= t^{-\frac{1}{4}}\sin\left(\frac{2}{3}\,t^{3/2} + \frac{\pi}{4}\right) + O\left(\frac{1}{t}\right) .$$

CHAPTER 5

Fubini's Theorem

EXERCISE 5·79: For each of the functions f below, calculate

$$\int_0^1 dx \int_0^1 f(x,y)dy, \qquad \int_0^1 dy \int_0^1 f(x,y)dx, \qquad \iint_{0 \leqslant x,y \leqslant 1} |f(x,y)|dxdy.$$

(a):

$$f(x,y) = \frac{x^2 - y^2}{(x^2 + y^2)^2};$$

(b):

$$f(x,y) = \begin{cases} (x - \frac{1}{2})^{-3} & \text{if } 0 < y < |x - \frac{1}{2}|, \\ 0 & \text{otherwise.} \end{cases}$$

(c):

$$f(x,y) = \frac{x - y}{(x^2 + y^2)^{3/2}};$$

(d):

$$f(x,y) = (1 - xy)^{-p}, \qquad p > 0.$$

$$\Delta\nabla\Delta = \nabla\Delta\nabla = \Delta\nabla\Delta = \nabla\Delta\nabla = \Delta\nabla\Delta$$

SOLUTION: (a):

$$\int_0^1 dx \int_0^1 \frac{x^2 - y^2}{(x^2+y^2)^2}\, dy = \int_0^1 dx \left[\frac{y}{x^2+y^2}\right]_{y=0}^{y=1} = \int_0^1 \frac{dx}{x^2+1} = \frac{\pi}{4}.$$

$$\int_0^1 dy \int_0^1 \frac{x^2 - y^2}{(x^2+y^2)^2}\, dx = \int_0^1 dy \left[-\frac{x}{x^2+y^2}\right]_{x=0}^{x=1} = \int_0^1 \frac{(-dy)}{y^2+1} = -\frac{\pi}{4}.$$

$$\iint_{0\leqslant x,y\leqslant 1} \left|\frac{x^2 - y^2}{(x^2+y^2)^2}\right| dx dy = 2\int_0^1 dx \int_0^x \frac{x^2 - y^2}{(x^2+y^2)^2}\, dy$$

$$= 2\int_0^1 dx \left[\frac{y}{x^2+y^2}\right]_{y=0}^{y=x} = \int_0^1 \frac{dx}{x} = \infty.$$

SOLUTION: (b):

$$\int_0^1 f(x,y)dy = \frac{|x - \frac{1}{2}|}{(x - \frac{1}{2})^3},$$

and this function of x is not integrable on $(0,1)$. Moreover, if $0 < y \leqslant \frac{1}{2}$,

$$\int_0^1 f(x,y)dx = \int_0^{\frac{1}{2}-y} (x - \tfrac{1}{2})^{-3} dx + \int_{\frac{1}{2}+y}^1 (x - \tfrac{1}{2})^{-3} dx = 0,$$

and if $\frac{1}{2} \leqslant y \leqslant 1$ this integral is again trivially zero. Thus:

$$\int_0^1 dy \int_0^1 f(x,y)dx = 0.$$

Finally,

$$\int_0^1 |f(x,y)|dy = |x - \tfrac{1}{2}|^{-2},$$

so

$$\int_0^1\int_0^1 |f(x,y)|\,dx\,dy = \infty.$$

SOLUTION: (c):

$$\int_0^1 dx\int_0^1 \frac{x - y}{(x^2 + y^2)^{3/2}}\,dy = \int_0^1 dx\left[\frac{1}{x}\,\frac{y + x}{(x^2 + y^2)^{\frac{1}{2}}}\right]_{y=0}^{y=1}$$

$$= \int_0^1\left(\frac{x + 1}{\sqrt{x^2 + 1}} - 1\right)\frac{dx}{x} = \left[\log\frac{x + \sqrt{x^2 + 1}}{1 + \sqrt{x^2 + 1}}\right]_{x=0}^{x=1} = \log 2.$$

Clearly

$$\int_0^1 dy\int_0^1 \frac{x - y}{(x^2 + y^2)^{3/2}}\,dx = -\log 2,$$

and

$$\frac{1}{2}\int_0^1 dx\int_0^1\left|\frac{x - y}{(x^2 + y^2)^{3/2}}\right|dy = \int_0^1 dx\int_0^x \frac{x - y}{(x^2 + y^2)^{3/2}}\,dy$$

$$= \int_0^1 dx\left[\frac{1}{x}\,\frac{y + x}{(x^2 + y^2)^{\frac{1}{2}}}\right]_{y=0}^{y=x} = (\sqrt{2} - 1)\int \frac{du}{x} = \infty.$$

SOLUTION: (d): As the function is positive the three integrals are equal. For example, if $p \neq 1$:

$$\int_0^1 dx\int_0^1 (1 - xy)^{-p}\,dy = \frac{1}{p - 1}\int_0^1 \frac{(1 - x)^{1-p} - 1}{x}\,dx,$$

an integral which converges only if $p < 2$. When $p = 1$:

$$\int_0^1 dx\int_0^1 (1 - xy)^{-1}\,dy = -\int_0^1 \frac{\log(1 - x)}{x}\,dx < +\infty.$$

EXERCISE 5·80: Let $0 < a < b$. By applying Fubini's Theorem to the double integral:

$$\iint_{\substack{0\leqslant x\leqslant 1\\ a\leqslant y\leqslant b}} x^y\,dx\,dy,$$

prove that:

$$\int_0^1 \frac{x^b - x^a}{\log x}\,dx = \log\frac{1+b}{1+a}.$$

ΔVΔ = VΔV = ΔVΔ = VΔV = ΔVΔ

SOLUTION: On the one hand

$$\int_a^b dy\int_0^1 x^y\,dx = \int_a^b \frac{dy}{y+1} = \log\frac{1+b}{1+a},$$

and on the other hand,

$$\int_0^1 dx\int_a^b x^y\,dy = \int_0^1 dx\left[\frac{x^y}{\log x}\right]_{y=a}^{y=b} = \int_0^1 \frac{x^b - x^a}{\log x}\,dx.$$

EXERCISE 5·81: Let R be the region in the quadrant $x \geqslant 0, y \geqslant 0$ bounded by the curves $y - x = 0$, $y^2 - x^2 = 1$, $xy = a$, $xy = b$ $(0 < a < b)$. Calculate the integral

$$\iint_R (y^2 - x^2)^{xy}(x^2 + y^2)\,dx\,dy.$$

by using a change of variables that transforms R into a rectangle.

ΔVΔ = VΔV = ΔVΔ = VΔV = ΔVΔ

SOLUTION: Set $u = y^2 - x^2$ and $v = xy$; R is then transformed into a rectangle $0 \leqslant u \leqslant 1$, $a \leqslant v \leqslant b$. Now

$$\frac{D(u,v)}{D(x,y)} = -2(x^2 + y^2),$$

so

$$\iint_R (x^2 - y^2)^{xy}(x^2 + y^2)\mathrm{d}x\mathrm{d}y = \tfrac{1}{2}\int_a^b \mathrm{d}v\int_0^1 u^v \mathrm{d}u = \tfrac{1}{2}\int_a^b \frac{\mathrm{d}v}{v+1}$$

$$= \tfrac{1}{2}\log\frac{1+b}{1+a} = \tfrac{1}{2}\log\frac{1+b}{1+a}.$$

EXERCISE 5·82: Let T be the interior of the tetrahedron defined by $x \geqslant 0, y \geqslant 0, z \geqslant 0$, and $x + y + z \leqslant 1$. Calculate the integral:

$$\iiint_T xyz(1 - x - y - z)\mathrm{d}x\mathrm{d}y\mathrm{d}z$$

by carrying the change of variable $x + y + z = X$, $y + z = XY$, and $z = XYZ$.

Δ∇Δ = ∇Δ∇ = ΔVΔ = ∇Δ∇ = Δ∇Δ

SOLUTION: We have

$$x = X(1 - Y), \qquad y = XY(1 - Z), \qquad z = XYZ.$$

The open tetrahedron T is thus transformed into the open rectangle $0 < X,Y,Z < 1$. Also,

$$\mathrm{d}x\wedge\mathrm{d}y\wedge\mathrm{d}z = (\mathrm{d}x + \mathrm{d}y + \mathrm{d}z)\wedge(\mathrm{d}y + \mathrm{d}z)\wedge\mathrm{d}z$$

$$= \mathrm{d}X\wedge(Y\mathrm{d}X + X\mathrm{d}Y)\wedge(YZ\mathrm{d}X + ZX\mathrm{d}Y + XY\mathrm{d}Z)$$

$$= X^2Y\mathrm{d}X\wedge\mathrm{d}Y\wedge\mathrm{d}Z,$$

so

$$\iiint_T xyz(1 - x - y - z)\mathrm{d}x\mathrm{d}y\mathrm{d}z =$$

$$= \iiint_{0<X,Y,Z<1} X^5Y^3Z(1 - X)(1 - Y)(1 - Z)\mathrm{d}X\mathrm{d}Y\mathrm{d}Z =$$

$$= \frac{\Gamma(6)\Gamma(2)\Gamma(4)\Gamma(2)\Gamma(2)\Gamma(2)}{\Gamma(8)\Gamma(6)\Gamma(4)} = \frac{1}{7!}.$$

EXERCISE 5·83: Calculate:

$$\iint_{x>0,y>0} \frac{\mathrm{d}x\mathrm{d}y}{(1 + y)(1 + x^2y)}.$$

Deduce from this the value of

$$\int_0^\infty \frac{\log x}{x^2 - 1}\,\mathrm{d}x.$$

ΔVΔ = VΔV = ΔVΔ = VΔV = ΔVΔ

SOLUTION: As the integrand is positive:

$$\iint_{x>0,y>0} = \int_0^\infty \frac{\mathrm{d}y}{1 + y} \cdot \int_0^\infty \frac{\mathrm{d}x}{1 + x^2y} = \frac{\pi}{2}\int_0^\infty \frac{y^{-\frac{1}{2}}}{1 + y}\,\mathrm{d}y$$

$$= \frac{\pi}{2}\,\frac{\pi}{\sin(\pi/2)} = \frac{\pi^2}{2}.$$

From this one deduces that

$$\frac{\pi^2}{2} = \int_0^\infty \mathrm{d}x\int_0^\infty \frac{\mathrm{d}y}{(1 + y)(1 + x^2y)} =$$

(Contd)

$$= \int_0^\infty \frac{dx}{x^2 - 1} \int_0^\infty \left(\frac{x^2}{1 + x^2 y} - \frac{1}{1 + y}\right) dy$$

$$= \int_0^\infty \frac{dx}{x^2 - 1} \left[\log \frac{1 + x^2 y}{1 + y}\right]_{y=0}^{y=\infty}$$

$$= 2\int_0^\infty \frac{\log x}{x^2 - 1}\, dx,$$

so

$$\int_0^\infty \frac{\log x}{x^2 - 1}\, dx = \frac{\pi^2}{4} .$$

EXERCISE 5·84: Calculate

$$\iint_{\substack{x>0, y>0\\ x+y<a}} \frac{3y\,dx\,dy}{\sqrt{1 + (x + y)^3}} .$$

$\Delta\nabla\Delta = \nabla\Delta\nabla = \Delta\nabla\Delta = \nabla\Delta\nabla = \Delta\nabla\Delta$

SOLUTION: Make the change of variables $u = x + y$, $v = x - y$, for which $D(u,v)/D(x,y) = -2$ and $y = \frac{1}{2}(u - v)$. The domain of integration becomes $0 < u < a$, $-u < v < u$. The integral is therefore equal to:

$$\frac{3}{4}\int_0^a \frac{du}{\sqrt{1 + u^3}} \int_{-u}^{u} (u - v)\,dv = \frac{3}{2}\int_0^a \frac{u^2\,du}{\sqrt{1 + u^3}}$$

$$= \sqrt{1 + a^3} - 1.$$

EXERCISE 5·85: Calculate

$$\iiint_{\substack{0<x,y<1\\ z>0}} \frac{dx\,dy\,dz}{(1+x^2z^2)(1+y^2z^2)},$$

and deduce from this the value of

$$\int_0^\infty \left(\frac{\tan^{-1}z}{z}\right)^2 dz.$$

Δ∇Δ = ∇Δ∇ = Δ∇Δ = ∇Δ∇ = Δ∇Δ

SOLUTION:

$$\iiint_{\substack{0<x,y<1\\ z>0}} \frac{dx\,dy\,dz}{(1+x^2z^2)(1+y^2z^2)} =$$

$$= \iint_{0<x,y<1} dx\,dy \int_0^\infty \frac{dz}{(1+x^2z^2)(1+y^2z^2)}$$

$$= \iint_{0<x,y<1} \frac{dx\,dy}{x^2-y^2}\int_0^\infty \left(\frac{x^2}{1+x^2z^2} - \frac{y^2}{1+y^2z^2}\right) dz$$

$$= \frac{\pi}{2}\iint_{0<x,y<1} \frac{dx\,dy}{x+y} = \frac{\pi}{2}\int_0^1 dx \int_0^1 \frac{dy}{x+y}$$

$$= \frac{\pi}{2}\int_0^1 (\log(x+1) - \log x)\,dx = \pi\log 2.$$

It follows that

$$\pi\log 2 = \int_0^\infty dz \iint_{0<x,y<1} \frac{dx\,dy}{(1+x^2z^2)(1+y^2z^2)}$$

$$= \int_0^\infty \left(\int_0^1 \frac{dx}{1 + x^2 z^2}\right)^2 dz$$

$$= \int_0^\infty \left(\frac{\tan^{-1} z}{z}\right)^2 dz.$$

EXERCISE 5·86: Let q be a positive definite quadratic form on $\mathbb{R}^n$. Prove the formula

$$\int_{\mathbb{R}^n} e^{-q(x)} dx = \frac{\pi^{n/2}}{\sqrt{\Delta}},$$

where Δ is the determinant of q.

$$\Delta\nabla\Delta = \nabla\Delta\nabla = \Delta\nabla\Delta = \nabla\Delta\nabla = \Delta\nabla\Delta$$

SOLUTION: There exists an orthonormal basis with respect to which

$$q(x) = \lambda_1 x_1^2 + \cdots + \lambda_n x_n^2, \qquad \lambda_i > 0.$$

Then

$$\Delta = \lambda_1 \cdots \lambda_n,$$

so

$$\int_{\mathbb{R}^n} e^{-q(x)} dx = \prod_{s=1}^n \int_{-\infty}^{+\infty} e^{-\lambda_s u^2} du = \prod_{s=1}^n \sqrt{\frac{\pi}{\lambda_s}} = \frac{\pi^{n/2}}{\sqrt{\Delta}}.$$

EXERCISE 5·87: Determine the values of α for which:

$$\iint_{0<x,y<\pi/2} \left(\frac{\cos x}{1 - \sin x \sin y}\right)^\alpha dx dy < \infty.$$

$$\Delta\nabla\Delta = \nabla\Delta\nabla = \Delta\nabla\Delta = \nabla\Delta\nabla = \Delta\nabla\Delta$$

SOLUTION: By carrying the change of variables $x \to \frac{1}{2}\pi - x$, $y \to \frac{1}{2}\pi - y$, one is led to study

$$I = \iint_{0<x,y<\pi/2} \left(\frac{\sin x}{1 - \cos x \cos y}\right)^{\alpha} dx dy.$$

The integrand is continuous on the square $0 \leqslant x,y \leqslant \pi/2$, except at the point (0,0). If $r = (x^2 + y^2)^{\frac{1}{2}}$, then in a neighbourhood of (0,0)

$$\frac{\sin x}{1 - \cos x \cos y} = \frac{2x + O(x^3)}{r^2 + O(r^4)} = \frac{2x}{r^2}(1 + O(r^2)),$$

whence:

$$\left(\frac{\sin x}{1 - \cos x \cos y}\right)^{\alpha} = \frac{2^{\alpha} x^{\alpha}}{r^{2\alpha}} + O\left(\frac{x^{\alpha}}{r^{2\alpha-2}}\right).$$

Note that changing to polar coordinates gives

$$\iint_{\substack{0<r\leqslant 1\\ x>0,y>0}} x^{\alpha} r^{-\beta} dx dy = \int_0^1 r^{\alpha-\beta+1} dr \, . \int_0^{\pi/2} \cos^{\alpha} x dx,$$

a quantity which is finite if and only if $\alpha > -1$ and $\beta < \alpha + 2$. Hence one has $I < \infty$ only in the case where $-1 < \alpha < 2$.

EXERCISE 5·88: Calculate:

$$I_p(a) = \int \cdots \iint_{x_i>0, x_1+\cdots+x_p<a} x_1 x_2 \cdots x_p dx_1 dx_2 \ldots dx_p$$

Δ∇Δ = ∇Δ∇ = ∇Δ∇ = Δ∇Δ = ∇Δ∇

SOLUTION: The change of variables $x_i \to a x_i$ shows:

$$I_p(a) = a^{2p} I_p(1).$$

If $p \geq 2$:

$$I_p(1) = \int_0^1 x_1 \, dx_1 \int_{x_i>0,\, x_2+\cdots+x_p<1-x_1} x_2 \cdots x_p \, dx_2 \ldots dx_p$$

$$= I_{p-1}(1) \int_0^1 x_1 (1 - x_1)^{2p-2} \, dx_1$$

$$= I_{p-1}(1) \frac{\Gamma(2)\Gamma(2p - 1)}{\Gamma(2p + 1)} .$$

Since

$$I_1(1) = \int_0^1 x \, dx = \tfrac{1}{2},$$

it follows that

$$I_p(1) = \frac{1}{2} \frac{\Gamma(3)}{\Gamma(2p + 1)} = \frac{1}{(2p)!} ,$$

so that

$$I_p(a) = \frac{a^{2p}}{(2p)!} .$$

EXERCISE 5·89: Prove the formula (*DIRICHLET'S INTEGRAL*):

$$\iint \cdots \int_{x_i>0,\ \left(\frac{x_1}{a_1}\right)^{p_1} + \cdots + \left(\frac{x_n}{a_n}\right)^{p_n} \leq 1} x_1^{\alpha_1 - 1} \cdots x_n^{\alpha_n - 1} \, dx_1 \ldots dx_n = \frac{a_1^{\alpha_1} \cdots a_n^{\alpha_n}}{p_1 \cdots p_n} \cdot \frac{\Gamma\left(\frac{\alpha_1}{p_1}\right) \cdots \Gamma\left(\frac{\alpha_n}{p_n}\right)}{\Gamma\left(\frac{\alpha_1}{p_1} + \cdots + \frac{\alpha_n}{p_n} + 1\right)} .$$

$$\Delta\nabla\Delta = \nabla\Delta\nabla = \Delta\nabla\Delta = \nabla\Delta\nabla = \Delta\nabla\Delta$$

SOLUTION: The change of variable $X_i = (x_i/a_i)^{p_i}$ transforms the integral into

$$\frac{a_1^{\alpha_1}\cdots a_n^{\alpha_n}}{p_1\cdots p_n}\iint\cdots\int_{\substack{X_i>0\\ X_1+\cdots+X_n\leqslant 1}} X_1^{\frac{\alpha_1}{p_1}-1}\cdots X_n^{\frac{\alpha_n}{p_n}-1}\,dX_1\ldots dX_n.$$

Thus it suffices to prove the formula when $a_i = p_i = 1$. For $1 \leqslant s \leqslant n$ and $\lambda \geqslant 0$ set

$$I_s(\lambda) = \iint\cdots\int_{\substack{x_i>0\\ x_i+\cdots+x_s\leqslant\lambda}} x_1^{\alpha_1-1}\cdots x_s^{\alpha_s-1}\,dx_1\ldots dx_s.$$

It is clear that $I_s(\lambda) = \lambda^{\alpha+\cdots+\alpha_s} I_s(1)$. Consequently,

$$I_n(1) = I_{n-1}(1)\int_0^1 x_n^{\alpha_n-1}(1-x_n)^{\alpha_1+\cdots+\alpha_{n-1}}dx_n$$

$$= I_{n-1}(1)\,\frac{\Gamma(\alpha_n)\Gamma(\alpha_1+\cdots+\alpha_{n-1}+1)}{\Gamma(\alpha_1+\cdots+\alpha_n+1)}.$$

Since

$$I_1(1) = \int_0^1 x_1^{\alpha_1-1}\,dx_1 = \frac{1}{\alpha_1},$$

it follows that

$$I_n(1) = \frac{1}{\alpha_1}\,\frac{\Gamma(\alpha_n)\cdots\Gamma(\alpha_2)\Gamma(\alpha_1+1)}{\Gamma(\alpha_1+\cdots+\alpha_n+1)}.$$

$$= \frac{\Gamma(\alpha_1)\Gamma(\alpha_2)\cdots\Gamma(\alpha_n)}{\Gamma(\alpha_1 + \cdots + \alpha_n + 1)}.$$

EXERCISE 5·90: Determine the values of the real numbers α, β, γ for which:

$$\iiint_{\substack{x>0\\y>0\\z>0}} \frac{dx\,dy\,dz}{1 + x^\alpha + y^\beta + z^\gamma} < \infty,$$

and then calculate this integral.

Δ∇Δ = ∇Δ∇ = Δ∇Δ = ∇Δ∇ = ΔCΔ

SOLUTION: First of all it is clear that one must have $\alpha > 1$, $\beta > 1$, and $\gamma > 1$, for if, for example, $\alpha \leqslant 1$:

$$\int_0^\infty \frac{dx}{1 + x^\alpha + y^\beta + z^\gamma} = \infty$$

for all $y > 0$ and all $z > 0$. Set

$$x = u^{2/\alpha}, \qquad y = v^{2/\beta}, \qquad z = w^{2/\gamma}.$$

The integral may be written

$$I = \frac{8}{\alpha\beta\gamma} \iiint_{\substack{u>0\\v>0\\w>0}} \frac{u^{2/\alpha - 1} v^{2/\beta - 1} w^{2/\gamma - 1}}{1 + u^2 + v^2 + w^2}\, du\,dv\,dw.$$

Now change to spherical coordinates

$$u = r\sin\vartheta\cos\varphi, \qquad v = r\sin\vartheta\sin\varphi, \qquad w = r\cos\vartheta,$$

which yields

$$I = \frac{8}{\alpha\beta\gamma}\int_0^\infty \frac{r^{2(1/\alpha+1/\beta+1/\gamma)-1}}{r^2+1}\,dr\int_0^{\pi/2}\cos^{2/\alpha-1}\varphi\sin^{2/\beta-1}\varphi d\varphi \times$$

$$\int_0^{\pi/2}\sin^{2(1/\alpha+1/\beta)-1}\vartheta\cos^{2/\gamma-1}\vartheta d\vartheta$$

$$= \frac{1}{\alpha\beta\gamma}\frac{\Gamma\left(\frac{1}{\alpha}\right)\Gamma\left(\frac{1}{\beta}\right)}{\Gamma\left(\frac{1}{\alpha}+\frac{1}{\beta}\right)}\frac{\Gamma\left(\frac{1}{\alpha}+\frac{1}{\beta}\right)\Gamma\left(\frac{1}{\gamma}\right)}{\Gamma\left(\frac{1}{\alpha}+\frac{1}{\beta}+\frac{1}{\gamma}\right)}\int_0^\infty \frac{u^{1/\alpha+1/\beta+1/\gamma-1}}{1+u}\,du$$

Thus if

$$\frac{1}{\alpha}+\frac{1}{\beta}+\frac{1}{\gamma} < 1$$

then

$$I = \frac{\pi}{\alpha\beta\gamma}\frac{\Gamma\left(\frac{1}{\alpha}\right)\Gamma\left(\frac{1}{\beta}\right)\Gamma\left(\frac{1}{\gamma}\right)}{\Gamma\left(\frac{1}{\alpha}+\frac{1}{\beta}+\frac{1}{\gamma}\right)\sin\pi\left(\frac{1}{\alpha}+\frac{1}{\beta}+\frac{1}{\gamma}\right)},$$

which can also be written

$$= \frac{1}{\alpha\beta\gamma}\Gamma\left(\frac{1}{\alpha}\right)\Gamma\left(\frac{1}{\beta}\right)\Gamma\left(\frac{1}{\gamma}\right)\Gamma\left(1-\frac{1}{\alpha}-\frac{1}{\beta}-\frac{1}{\gamma}\right).$$

EXERCISE 5·91: Calculate:

$$\iiint_{0<x,y,z<\pi}\frac{dx\,dy\,dz}{1-\cos x\cos y\cos z}.$$

Δ∇Δ = ∇Δ∇ = Δ∇Δ = ∇Δ∇ = Δ∇Δ

SOLUTION: Set

$$u = \tan\tfrac{1}{2}x, \qquad v = \tan\tfrac{1}{2}y, \qquad z = \tan\tfrac{1}{2}z,$$

so that the integral becomes:

$$I = 4\iiint_{\substack{u>0\\ v>0\\ w>0}} \frac{du\,dv\,dw}{u^2 + v^2 + w^2 + u^2v^2w^2}\,.$$

Changing to spherical coordinates yields

$$I = 4\int_0^{\pi/2} d\vartheta \int_0^{\pi/2} d\varphi \int_0^{\infty} \frac{\sin\vartheta\, dr}{1 + r^4\sin^4\vartheta\cos^2\vartheta\cos^2\varphi\sin^2\varphi}\,.$$

Finally, setting

$$r = t^{\frac{1}{4}}(\sin\vartheta)^{-1}\cos^{-\frac{1}{2}}\vartheta\cos^{-\frac{1}{2}}\varphi\sin^{-\frac{1}{2}}\varphi,$$

we have

$$I = \int_0^{\pi/2} \cos^{-\frac{1}{2}}\vartheta\, d\vartheta \int_0^{\pi/2} \cos^{-\frac{1}{2}}\varphi\sin^{-\frac{1}{2}}\varphi\, d\varphi \int_0^{\infty} \frac{u^{\frac{1}{4}-1}}{1 + u}\, du$$

$$= \frac{1}{4}\,\frac{\Gamma(\frac{1}{4})\Gamma(\frac{1}{2})}{\Gamma(\frac{3}{4})}\,\frac{\Gamma(\frac{1}{4})\Gamma(\frac{1}{4})}{\Gamma(\frac{1}{2})}\,\frac{\pi}{\sin\frac{1}{4}\pi} = \frac{\pi\sqrt{2}}{4}\,\frac{\Gamma(\frac{1}{4})^3}{\Gamma(\frac{3}{4})}\,.$$

Noticing that

$$\Gamma(\tfrac{1}{4})\Gamma(\tfrac{3}{4}) = \frac{\pi}{\sin\frac{1}{4}\pi} = \pi\sqrt{2},$$

the answer can be written

$$I = \tfrac{1}{4}\Gamma(\tfrac{1}{4})^4.$$

EXERCISE 5·92: Use a double integral to represent the difference:

$$\int_{-\infty}^{+\infty} f(x)^2 dx \int_{-\infty}^{+\infty} g(x)^2 dx - \left(\int_{-\infty}^{+\infty} f(x)g(x)dx\right)^2.$$

Δ∇Δ = ∇Δ∇ = Δ∇Δ = ∇Δ∇ = Δ∇Δ

SOLUTION: It will be found that

$$\frac{1}{2}\iint_{\mathbb{R}^2} (f(x)g(y) - f(y)g(x))^2 dx dy$$

satisfies the requirement.

EXERCISE 5·93: Show that the centre of gravity G of a homogeneous cone satisfies:

$$\overrightarrow{OG} = \tfrac{3}{4}\overrightarrow{OG}_0,$$

where O denotes the vertex of the cone, and G_0 the centre of gravity of its base.

Δ∇Δ = ∇Δ∇ = Δ∇Δ = ∇Δ∇ = Δ∇Δ

FIRST SOLUTION: On placing the coordinate origin at O and making the plane xOy parallel to the base of the cone C, the coordinates ξ,η,ζ of G are given by:

$$\xi = \frac{\iiint_C x dx dy dz}{\iiint_C dx dy dz},$$

and two analogous formulae. If ξ_0,η_0,ζ_0 are the coordinates of G_0, S_0 the area of the base, and C_z the cross-section of the cone cut by the plane parallel to xOy with height z, then:

$$\iint_{C_z} dx\,dy = \left(\frac{z}{\zeta_0}\right)^2 S_0,$$

$$\iint_{C_z} x\,dx\,dy = \left(\frac{z}{\zeta_0}\right)^3 \xi_0 S_0,$$

$$\iint_{C_z} y\,dx\,dy = \left(\frac{z}{\zeta_0}\right)^3 \eta_0 S_0,$$

$$\iint_{C_z} z\,dx\,dy = z\left(\frac{z}{\zeta_0}\right)^2 S_0 = \left(\frac{z}{\zeta_0}\right)^3 \zeta_0 S_0,$$

so that

$$\xi = \frac{\int_0^{\zeta_0}\left(\frac{z}{\zeta_0}\right)^3 \xi_0 S_0\,dz}{\int_0^{\zeta_0}\left(\frac{z}{\zeta_0}\right)^2 S_0\,dz} = \frac{3}{4}\,\xi_0, \qquad \text{etc. \dots .}$$

SECOND SOLUTION: G is the centre of mass of the segment OG_0 weighted with density ku^2, where $u = \overline{OM}$ and k is a constant. It follows immediately that $\overrightarrow{OG} = \frac{3}{4}\overrightarrow{OG_0}$.

EXERCISE 5·94: Show that the volume bounded by a ruled surface and two parallel planes is equal to

$$V = \frac{h}{6}(S_1 + S_2 + 4S_3),$$

where h denotes the distance between the two planes, S_1 and S_2 are the areas of the cross-sections cut out by these two planes, and S_3 is the area of the cross-section cut out by the plane parallel to the other two and located at a distance $h/2$ from each of them. (The *Pile of Sand Formula*).

$$\Delta\nabla\Delta = \nabla\Delta\nabla = \Delta\nabla\Delta = \nabla\Delta\nabla = \Delta\nabla\Delta$$

SOLUTION: As the z-axis is perpendicular to the planes under consideration, the cross-section cut out by the horizontal plane with height z is bounded by a curve given parametrically by equations of the type

$$\Gamma_z : \begin{cases} x = a(t) + zb(t), \\ y = c(t) + zd(t), \end{cases}$$

where a,b,c,d are periodic functions that we shall assume to be piecewise continuously differentiable. The area $S(z)$ of the corresponding section, given by

$$S(z) = \frac{1}{2}\int_{\Gamma_z} x\,dy - y\,dx$$

is therefore a second-degree polynomial in z. From this it follows that (Simpson's Formula)

$$V = \int_0^h S(z)dz = \frac{h}{6}(S_1 + S_2 + 4S_3).$$

EXERCISE 5·95: Show that if

$$\int_0^1 |f(y)|dy < \infty, \qquad \int_1^\infty |f(y)|y^{-2}dy < \infty,$$

then the order of the integration can be inverted in

$$\int_0^\infty \sin ax\,dx \int_0^\infty f(y)e^{-xy}dy.$$

From this deduce that

$$\int_0^\infty \frac{\sin ax}{\sqrt{1+x^2}}\,dx = a\int_0^\infty \frac{J_0(y)}{a^2+y^2}\,dy,$$

where J_0 is the Bessel function:

$$J_0(y) = \frac{2}{\pi}\int_0^{\pi/2} \cos(y\cos\vartheta)\,d\vartheta.$$

$$\Delta\nabla\Delta = \nabla\Delta\nabla = \Delta\nabla\Delta = \nabla\Delta\nabla = \Delta\nabla\Delta$$

SOLUTION: If $0 < \varepsilon < X < \infty$ then

$$\int_\varepsilon^X \sin ax\,dx \int_0^\infty f(y)e^{-xy}dy = \int_0^\infty f(y)\,dy \int_\varepsilon^X e^{-xy}\sin ax\,dx,$$

since

$$\int_\varepsilon^X dx \int_0^\infty |\sin ax f(y)e^{-xy}|\,dy < (X-\varepsilon)\int_0^\infty e^{-\varepsilon y}|f(y)|\,dy < \infty.$$

The Second Mean Value Theorem gives

$$\left|\int_\varepsilon^X e^{-xy}\sin ax\,dx\right| \leqslant \frac{2}{a},$$

and the inequality $|\sin ax| \leqslant ax$ implies

$$\left|\int_\varepsilon^X e^{-xy}\sin ax\,dx\right| \leqslant \frac{a}{y^2}.$$

Consequently,

$$\left|f(y)\int_\varepsilon^X e^{-xy}\sin ax\,dx\right| \leqslant \frac{2}{a}\mathbb{1}_{(0,1)}(y)|f(y)| + a\mathbb{1}_{(1,\infty)}(y)|f(y)|y^{-2}.$$

The Dominated Convergence Theorem leads to the formula

$$\int_0^\infty \sin ax\,dx \int_0^\infty f(y)e^{-xy}dy = \int_0^\infty f(y)\,dy \int_0^\infty e^{-xy}\sin ax\,dx.$$

Now,

$$\int_0^\infty e^{-xy}\sin ax\,dx = \frac{a}{y^2 + a^2},$$

so

$$a\int_0^\infty \frac{f(y)}{a^2 + y^2}\,dy = \int_0^\infty \sin ax\,dx \int_0^\infty f(y)e^{-xy}dy.$$

Since J_0 is continuous and $|J_0| \leqslant 1$, it is clear that the conditions of the problem are satisfied if $f = J_0$. Moreover:

$$\int_0^\infty J_0(y)e^{-xy}dy = \int_0^\infty e^{-xy}dy\, \frac{2}{\pi}\int_0^{\pi/2} \cos(y\cos\vartheta)d\vartheta$$

$$= \frac{2}{\pi}\int_0^{\pi/2} d\vartheta \int_0^\infty e^{-xy}\cos(y\cos\vartheta)dy$$

$$= \frac{2}{\pi}\int_0^{\pi/2} \frac{x\,d\vartheta}{x^2 + \cos^2\vartheta} = \frac{2x}{\pi}\int_0^\infty \frac{dt}{1 + x^2 + x^2t^2}$$

$$= \left[\frac{2}{\pi\sqrt{1 + x^2}} \tan^{-1} \frac{xt}{\sqrt{1 + x^2}}\right]_{t=0}^{t=\infty} = \frac{1}{\sqrt{1 + x^2}}.$$

(Switching order of integration is legitimate, since

$$\frac{2}{\pi}\int_0^\infty dy \int_0^{\pi/2} e^{-xy}|\cos(y\cos\vartheta)|d\vartheta \leqslant \frac{1}{x} < \infty).$$

EXERCISE 5·96: Let H be a continuously differentiable function on $[0,\infty[$. For all $r > 0$ set

$$m(r) = \sup_{x \geqslant 0}(x \log r - H(x)).$$

show that if

$$\int_0^\infty e^{-H'(x)} dx < \infty$$

then

$$\int_0^\infty \frac{m(r)}{1 + r^2} dr < \infty.$$

Δ∇Δ = ∇Δ∇ = ΔVΔ = ∇Δ∇ = Δ∇Δ

SOLUTION: We shall first prove the following property: *If* φ is a measurable function on $[0,\infty[$ such that $\varphi(t) > 0$ for $t \geqslant 0$ and:

$$\int_0^\infty \varphi(t) dt < \infty,$$

and if for $r > 0$ one sets

$$\mu(r) = \sup_{x \geqslant 0} \int_0^x \log(r\varphi(t)) dt,$$

then

$$\int_0^\infty \frac{\mu(r)}{r^2} dr < \infty.$$

Notice that $\mu(r) \geqslant 0$ for all $r > 0$. On the other hand,

$$\int_0^x \log(r\varphi(t)) dt \leqslant \int_0^x \log^+(r\varphi(t)) dt,$$

so

$$0 \leqslant \mu(r) \leqslant \int_0^\infty \log^+(r\varphi(t))dt.$$

Consequently,

$$\int_0^\infty \frac{\mu(r)}{r^2}\, dr \leqslant \int_0^\infty dt \int_0^\infty \frac{\log^+(r\varphi(t))}{r^2}\, dr$$

$$= \left(\int_0^\infty \varphi(t)dt\right)\left(\int_0^\infty r^{-2}\log^+ r dr\right)$$

$$< \infty.$$

=== == === == ==

To return to the problem, put $\varphi = e^{-H'}$. Then:

$$x\log r - H(x) = - H(o) + \int_0^x \log(r\varphi(t))dt,$$

so

$$m(r) = - H(o) + \mu(r),$$

which implies the stated result.

EXERCISE 5·97: Let X be a measurable set of $\mathbb{R}^n$ with $0 < \text{meas}(X) < \infty$, and let f be an integrable function on X. Show that if for every complex number z

$$\int_X \log|1 + zf| = 0,$$

then $f(x) = 0$ for almost all x.

Show that

$$\int_X \log^+|\rho f| = 0$$

for all $\rho > 0$ by using the formula

$$\frac{1}{2\pi}\int_0^{2\pi} \log|1 + ze^{it}|dt = \log^+|z|.$$

ΔVΔ = ∇Δ∇ = ΔVΔ = ∇Δ∇ = ΔVΔ

SOLUTION: It may be assumed that meas$(X) = 1$. Let $\rho > 0$. Then:

$$\int_0^{2\pi} dt\int_X \log|1 + \rho e^{it}f(x)|dx = 0.$$

If Fubini's Theorem can be applied then

$$0 = \int_X dx\int_0^{2\pi} \log|1 + \rho e^{it}f(x)|dt - 2\pi\int_X \log^+|\rho f|.$$

From this it follows that $|f| < \rho^{-1}$ almost everywhere, hence that $f = 0$ almost everywhere. It remains to be seen that Fubini's Theorem actually can be applied. By hypothesis

$$\int_X \log^+|1 + zf| = \int_X \log^-|1 + zf|$$

for every complex number z. Therefore

$$\int_0^{2\pi} dt\int_X \Big|\log|1 + \rho e^{it}f(x)|\Big|dx = 2\int_0^{2\pi} dt\int_X \log^+|1 + \rho e^{it}f(x)|dx$$

$$\leqslant 4\pi(1 + \rho\int_X |f|) < \infty.$$

(We have used the inequality $\log^+|1 + z| \leqslant 1 + |z|$).

EXERCISE 5·98: With every function f that is positive on $E = \mathbb{R}^n$ associate the set $D_f \subset E \times \mathbb{R}$ formed by the points (x,t) such that $0 \leqslant t \leqslant f(x)$.

(a): Show that f is measurable if and only if D_f is measurable.

(b): Show that if f is measurable and $p > 0$:

$$\int_E f(x)^p dx = p\int_0^\infty t^{p-1}\text{meas}(f > t)dt.$$

(c): Show that if f is measurable its graph is a set of measure zero in $E \times \mathbb{R}$.

ΔVΔ = VΔV = ΔVΔ = VΔV = ΔVΔ

SOLUTION: (a): If f is measurable, the function $\varphi(x,t) = f(x) - t$ is also. Since $D_f = (\varphi \geqslant 0) \cap (t \geqslant 0)$ it follows that D_f is measurable.

Now assume that D_f is measurable. Then $x \mapsto \text{meas}((D_f)_x) = f(x)$ is measurable (Fubini's Theorem; here, for $A \subset E \times \mathbb{R}$ and $x \in E$, A_x denotes the set of t's such that $(x,t) \in A$).

SOLUTION: (b): Fubini's Theorem also shows that

$$\int_E f^p = p\int_E dx\int_0^{f(x)} t^{p-1}dt = p\int_0^\infty t^{p-1}dt\int_{(f>t)} dx$$

$$= p\int_0^\infty t^{p-1}\text{meas}(f > t)dt.$$

SOLUTION: (c): One can show, as in part (a), that the set D'_f

of (x,t)'s such that $0 \leqslant t < f$ is measurable, and the calculation carried out in part (b) proves that

$$\int_E f = \text{meas}(D_f) = \text{meas}(D'_f).$$

Consequently $D_f - D'_f$, which is the graph of f, has measure zero.

EXERCISE 5·99: Let (D_n) be a sequence of closed discs, containing in the unit disc D, of radii $r_n > 0$, and mutually disjoint. show that if $\text{meas}(D - \bigcup_n D_n) = 0$ then $\sum r_n = \infty$.

$\Delta\nabla\Delta = \nabla\Delta\nabla = \Delta\nabla\Delta = \nabla\Delta\nabla = \Delta\nabla\Delta$

SOLUTION: Let I_n be the orthogonal projection of D_n on the x-axis. Then $\sum \text{meas}(I_n) = 2 \sum r_n$, and consequently if $\sum r_n < \infty$ then almost all x's belong to only a finite number of the I_n's; that is to say that the vertical line L_x with abscissae x meets only a finite number of discs D_n, say $D_{n_1}, \ldots, D_{n_k}$. If $|x| < 1$ none of the intervals $L_x \cap D_{n_i}$ can be equal to $L_x \cap D$, for in that case one would have $D_{n_i} = D$, which is absurd, because $r_n > 0$ for all n and because the D_n's are mutually disjoint and contained in D. It is then clear that:

$$\sum_{i=1}^{k} \text{meas}(L_x \cap D_{n_i}) < \text{meas}(L_x \cap D) \tag{1}$$

for almost every $x \in]-1,1[$. If ψ is the characteristic function of $D - \bigcup_n D_n$, then (1) means that for almost all x ($|x| < 1$)

$$\int \psi(x,y)dy > 0,$$

and consequently:

$$\mathrm{meas}(D - \bigcup_n D_n) = \int \mathrm{d}x \int \psi(x,y)\,\mathrm{d}y > 0.$$

CHAPTER 6

The L^p-Spaces

EXERCISE 6·100: Prove Holder's Inequality by using Jensen's Inequality with the function $\varphi(x) = x^p$, $x \geqslant 0$, $p > 1$.

$$\Delta\nabla\Delta = \nabla\Delta\nabla = \Delta\nabla\Delta = \nabla\Delta\nabla = \Delta\nabla\Delta$$

SOLUTION: Since φ is convex, for a positive measurable f and a positive g with integral equal to one,

$$\left(\int fg\right)^p \leqslant \int f^p g \qquad (*)$$

Now assume that $f \geqslant 0$, $g \geqslant 0$, and:

$$\int f^p = \int g^q = 1,$$

where q is such that $\frac{1}{p} + \frac{1}{q} = 1$. Replacing f in (*) by fg^{1-q} and g by g^q yields:

$$\int fg \leqslant 1.$$

EXERCISE 6·101: Prove Minkowski's Inequality using Jensen's Inequality and the function $\varphi(x) = (1 - x^{1/p})^p$, $0 \leqslant x \leqslant 1$, $p \geqslant 1$.

ΔVΔ = VΔV = ΔVΔ = VΔV = ΔVΔ

SOLUTION: The function φ is convex, for it is continuous on $[0,1]$, and on $]0,1[$ its derivative is equal to

$$-\left(\frac{1 - x^{1/p}}{x^{1/p}}\right)^{p-1},$$

which is increasing in x. Consequently, if $0 \leqslant f \leqslant 1$, $g \geqslant 0$, and

$$\int g = 1,$$

then

$$1 \leqslant \left(\int fg\right)^{1/p} + \left(\int (1 - f^{1/p})^p g\right)^{1/p}. \qquad (*)$$

Now assume that $f \geqslant 0$, $g \geqslant 0$, and

$$\int (f + g)^p = 1.$$

Replacing g in (*) by $(f + g)^p$ and f by 0 where $f + g = 0$ and by $f^p(f + g)^{-p}$ otherwise, yields

$$1 \leqslant \left(\int f^p\right)^{1/p} + \left(\int g^p\right)^{1/p}.$$

EXERCISE 6·102: Prove Minkowski's Second Inequality using Jensen's Inequality and the function $\varphi(x) = (1 + x^{1/p})^p$, $x \geqslant 0$, $0 \leqslant p \leqslant 1$.

ΔVΔ = VΔV = ΔVΔ = VΔV = ΔVΔ

SOLUTION: φ is continuous on $[0,\infty[$, and its derivative on $]0,\infty[$ is equal to:

$$\left(\frac{x^{1/p}}{1 + x^{1/p}}\right)^{1-p}$$

which is increasing in x, so φ is convex. If $f \geqslant 0$, $g \geqslant 0$, and

$$\int g = 1,$$

then

$$1 + \left(\int fg\right)^{1/p} < \left(\int (1 + f^{1/p})^p g\right)^{1/p}. \tag{*}$$

If now $f \geqslant 0$, $g \geqslant 0$, and

$$\int g^p = 1,$$

then replacing g by g^p in (*) and f by 0 when $g = 0$ and by $f^p g^{-p}$ otherwise, yields

$$1 + \left(\int f^p\right)^{1/p} \leqslant \left(\int (f + g)^p\right)^{1/p}.$$

EXERCISE 6·103: Let $0 < p < 1$. If $f, g \in L^p$ set:

$$d(f,g) = \int |f - g|^p.$$

Show that this defines a metric on L^p (with the condition that two functions equal almost everywhere are identified), and that, when provided with this metric, L^p is complete. Is the mapping $f \mapsto d(f,0)$ a norm?

$\Delta\nabla\Delta = \nabla\Delta\nabla = \Delta\nabla\Delta = \nabla\Delta\nabla = \nabla\Delta\nabla$

SOLUTION: We have $(x + y)^p \leqslant x^p + y^p$ if $x \geqslant 0$, $y \geqslant 0$, and $0 \leqslant p \leqslant 1$, as can be seen by studying the function

$$x \to x^p + 1 - (x + 1)^p.$$

From this it follows that if $f,g,h \in L$,

$$d(f,g) = \int |f - g|^p \leq \int (|f - h| + |h - g|)^p$$

$$\leq \int |f - h|^p + \int |g - h|^p$$

$$= d(f,h) + d(h,g).$$

Furthermore, $d(f,g) = 0$ implies that $|f - g|^p = 0$, that is to say $f = g$ almost everywhere.

To show that L^p is complete it suffices to prove that if $f_n \in L^p$ and

$$\sum_n d(f_n, f_{n+1}) = A < \infty,$$

then f_n converges to an element of L^p. Set

$$g_N = \sum_{n=1}^{N} |f_n - f_{n+1}|, \qquad g = \sum_{n=1}^{\infty} |f_n - f_{n+1}|.$$

Then $g_N \to g$ and

$$\int g_N^p \leq A$$

By Fatou's Lemma

$$\int g^p \leq A$$

Consequently $g < \infty$ almost everywhere; but then the series

$$\sum_n (f_n - f_{n+1})$$

is almost everywhere absolutely convergent, which implies the existence of a measurable function f such that $f_n \to f$ almost everywhere. If $r < s$ then

$$\int |f_r - f_s|^p \leqslant \sum_{n=r}^{\infty} d(f_n, f_{n+1}).$$

Making $s \to \infty$ and using Fatou's Lemma again yields

$$\int |f_r - f|^p \leqslant \sum_{n=r}^{\infty} d(f_n, f_{n+1}),$$

which proves that $f = f_r + (f - f_r) \in L^p$ and that $f_n \to f$ in L^p. Since $d(\lambda f, 0) = |\lambda|^p d(f, 0)$, the mapping $f \mapsto d(f, 0)$ is not a norm on L^p if $0 < p < 1$.

EXERCISE 6·104: Let K be a compact set of $\mathbb{R}^n$, $\varepsilon > 0$, and $1 \leqslant p < \infty$.

Show that there exists $f \in L^p(\mathbb{R}^n)$ such that $f \geqslant 0$, $\|f\|_p = 1$, and:

$$\|f_a - f\|_p < \varepsilon, \qquad a \in K.$$

ΔVΔ = VΔV = ΔVΔ = VΔV = ΔVΔ

SOLUTION: Let us denote by B_r the ball with centre 0 and radius r, and by χ_r is characteristic function. Assume that $K \subset B_r$, and for $R > r$ let:

$$f = \frac{1}{\text{meas}(B_R)^{1/p}} \chi_R.$$

We have $f \geqslant 0$, $\|f\|_p = 1$ and $|f_a - f|$ is, to within a factor $\text{meas}(B_R)^{-1/p}$, the characteristic function of the union of the set of points x for which $|x - a| \leqslant R$, $|x| > R$, and of that defined by $|x| \leqslant R$, $|x - a| > R$. These two sets are mapped into each other by the symmetry $x \to a - x$, and the first is contained in the 'annulus' $R < |x| < R + a$. Thus when $a \in K$

$$\|f_a - f\|_p \leqslant 2\left(\frac{\text{meas}(B_{R+r})}{\text{meas}(B_R)} - 1\right)^{1/p}.$$

Since $\text{meas}(B_R) = R^n \text{meas}(B_1)$ the right side tends to zero as $R \to \infty$, and consequently can be made less then ε by choosing R large enough.

EXERCISE 6·105: If $p > 0$ denote by L_+^p the set of positive functions whose p-th powers are integrable.

Show that for all $\alpha > 0$ the mapping $f \mapsto f^\alpha$ is a topological isomorphism of L_+^p onto $L_+^{p/\alpha}$.

Δ∇Δ = ∇Δ∇ = Δ∇Δ = ∇Δ∇ = Δ∇Δ

SOLUTION: This is a matter of proving that

$$\int |f - f_i|^p \to 0 \quad \text{implies} \quad \int |f^\alpha - f_i^\alpha|^{p/\alpha} \to 0.$$

When $0 < \alpha \leqslant 1$ we use the inequality

$$|x^\alpha - y^\alpha| \leqslant |x - y|^\alpha, \qquad x \geqslant 0,\ y \geqslant 0,$$

which gives

$$\int |f^\alpha - f_i^\alpha|^{p/\alpha} \leqslant \int |f - f_i|^p.$$

Now let us consider the case $\alpha > 1$. By the Mean Value Theorem,

$$|f^\alpha - f_i^\alpha| \leqslant \alpha|f - f_i|(f \vee f_i)^{\alpha-1}, \qquad f \vee f_i = \max(f, f_i),$$

and by Hölder's Inequality for the pair $\alpha, \alpha(\alpha - 1)^{-1}$

$$\int|f^\alpha - f_i^\alpha|^{p/\alpha} < \alpha^{p/\alpha}\left(\int|f - f_i|^p\right)^{1/\alpha}\left(\int(f\vee f_i)^p\right)^{1 - 1/\alpha}$$

It remains to observe that

$$(f\vee f_i)^p \leqslant (f + |f - f_i|)^p \leqslant C_p(|f|^p + |f - f_i|^p),$$

where $C_p = \max(2^{p-1},1)$.

EXERCISE 6·106: (a): Let $f_1,\ldots,f_n$ be measurable functions on $\mathbb{R}^m$. Show that

$$\|f_1\cdots f_n\|_{p_0} \leqslant \|f_1\|_{p_1}\cdots\|f_n\|_{p_n}$$

if

$$0 < p_i \leqslant \infty, \qquad 0 \leqslant i \leqslant n, \qquad \frac{1}{p_0} = \frac{1}{p_1} + \cdots + \frac{1}{p_n}$$

(as usual, the convention $1/\infty = 0$ is in effect).

(b): Show that if f,g are measurable and positive

$$\int fg < \|f\|_p^{1 - p/r}\|g\|_q^{1 - q/r}\left(\int f^p g^q\right)^{1/r}$$

if

$$1 \leqslant p < \infty, \quad 1 \leqslant q < \infty, \quad 1 \leqslant r \leqslant \infty, \quad \frac{1}{r} = \frac{1}{p} + \frac{1}{q} - 1$$

(the convention is made that $a^0 = 1$, including the cases $a = 0$ or $a = \infty$).

(c): If $f \in L^p(\mathbb{R}^m)$, $g \in L^q(\mathbb{R}^m)$ and

$$1 \leqslant p,q \leqslant \infty, \qquad \frac{1}{p} + \frac{1}{q} - 1 \geqslant 0,$$

show that for almost all x the function

$$t \mapsto f(t)g(x - t)$$

is integrable, and that the function:

$$h(x) = \int f(t)g(x - t)dt$$

is such that

$$\|h\|_r \leq \|f\|_p \|g\|_q \quad \text{if} \quad \frac{1}{r} = \frac{1}{p} + \frac{1}{q} - 1. \quad (\textit{YOUNG'S INEQUALITY})$$

Δ∇Δ = ∇Δ∇ = Δ∇Δ = ∇Δ∇ = Δ∇Δ

SOLUTION: (a): If $I = \{i : p_i < \infty\}$ it is clear that

$$\|f_1 \cdots f_n\|_{p_0} < \|\prod_{i \in I} f_i\|_{p_0} \cdot \prod_{i \notin I} \|f_i\|_\infty$$

and:

$$\frac{1}{p_0} = \sum_{i \in I} \frac{1}{p_i} .$$

It may therefore be assumed that $p_i < \infty$ for all i. First consider the case where $n = 2$. Then

$$1 = \frac{p_0}{p_1} + \frac{p_0}{p_2} .$$

Applying Hölder's Inequality for $p = p_1/p_0$, $q = p_2/p_0$ to the functions $|f|^{p_0}, |g|^{p_0}$ yields

$$\int |fg|^{p_0} \leq \left(\int |f|^{p_1}\right)^{p_0/p_1} \left(\int |g|^{p_2}\right)^{p_0/p_2},$$

which gives the formula in this case. When $n > 2$ one proceeds by induction; in fact, if

$$\frac{1}{\omega} = \frac{1}{p_2} + \cdots + \frac{1}{p_n}$$

then

$$\frac{1}{p_0} = \frac{1}{p_1} + \frac{1}{\omega},$$

whence

$$\|f_1 f_2 \cdots f_n\|_{p_0} \leq \|f_1\|_{p_1} \|f_2 \cdots f_n\|_\omega \leq \|f_1\|_{p_1} \|f_2\|_{p_2} \cdots \|f_n\|_{p_n}.$$

SOLUTION: (b): If $r = \infty$ this is a matter of Hölder's formula. Therefore assume that $r < \infty$ and consider first the case where $p > 1$, $q > 1$. Then $r > p$, $r > q$. Set

$$p_1 = \frac{pr}{r - p}, \quad p_2 = \frac{qr}{r - q}, \quad p_3 = r.$$

Then

$$\frac{1}{p_1} + \frac{1}{p_2} + \frac{1}{p_3} = \frac{1}{p} + \frac{1}{q} - \frac{1}{r} = 1,$$

so that if h_1, h_2, h_3 are three positive measurable functions, by part (a) above one has

$$\int h_1 h_2 h_3 \leq \left(\int h_1^{p_1}\right)^{1/p_1} \left(\int h_2^{p_2}\right)^{1/p_2} \left(\int h_3^{p_3}\right)^{1/p_3}.$$

If

$$h_1^{p_1} = f^p, \quad h_2^{p_2} = g^q, \quad h_3^{p_3} = f^p g^q,$$

that is to say, if

$$h_1 = f^{1 - p/r}, \quad h_2 = g^{1 - q/r}, \quad h_3 = f^{p/r} g^{q/r},$$

the desired formula is obtained.

When $p = q = 1$ one has $r = 1$, and the formula becomes trivial. Lastly, of $p > 1$, $q = 1$, one has $r = p$, and it is a matter of proving that:

$$\int fg \leqslant \left(\int g\right)^{1 - 1/p}\left(\int f^p g\right)^{1/p},$$

which is none other than Hölder's formula, because

$$g^{1 - 1/p}(f^p g)^{1/p} = fg.$$

SOLUTION: (c): Assume for now that $p < \infty$, $q < \infty$, and that

$$\frac{1}{p} + \frac{1}{q} - 1 > 0,$$

so that $r < \infty$. By noticing that

$$\int |g(x - t)|^q dt = \|g\|_q^q,$$

it follows from part (b) that

$$\left(\int |f(t)g(x - t)|dt\right)^r \leqslant \|f\|_p^{r-p}\|g\|_q^{r-q}\int |f(t)|^p|g(x - t)|^q dt,$$

so that

$$\int dx\left(\int |f(t)g(x - t)|dt\right)^r$$

$$\leqslant \|f\|_p^{r-p}\|g\|_q^{r-q}\int dx\int |f(t)|^p|g(x - t)|^q dt$$

$$= \|f\|_p^{r-p}\|g\|_q^{r-q}\int |f(t)|^p dt\int |g(x - t)^q|dx =$$

$$= \|f\|_p^r \|g\|_q^r < \infty.$$

From this it follows that for almost all

$$\int |f(t)g(x - t)|\,dt < \infty,$$

which ensures that the function h is defined almost everywhere. Since

$$|h(x)| \leqslant \int |f(t)g(x - t)|\,dt,$$

one has

$$\int |h(x)|^r dx \leqslant \|f\|_p^r \|g\|_q^r,$$

which accomplishes the proof in this case.

If $q = \infty$ one must have $1/p \geqslant 1$, hence $p = 1$. It therefore remains to examine the cases where $\frac{1}{p} + \frac{1}{q} = 1$, for which $r = \infty$. By Holder's Inequality, for all x one then has

$$|h(x)| \leqslant \|f\|_p \|g\|_q,$$

i.e.

$$\|h\|_\infty \leqslant \|f\|_p \|g\|_q.$$

EXERCISE 6·107: Let $1 \leqslant p \leqslant \infty$ and $\frac{1}{p} + \frac{1}{q} = 1$.

(a): Show that if $f \in L^p$

$$\|f\|_p = \sup\{ |\int fg| : \|g\|_q \leqslant 1\}.$$

(b): Let f be a positive measurable function.

Show that:

$$\|f\|_p = \sup\{\int fg : g \geqslant 0 \text{ and } \|g\|_q \leqslant 1\}.$$

(c): Let f be a measurable function. Assume that there exists a constant M such that:

$$\left|\int fg\right| \leqslant M$$

for every simple function g such that fg is integrable and $\|g\|_q \leqslant 1$.

Show that $\|f\|_p \leqslant M$.

(d): Let f be a measurable function such that fg is integrable for every function $g \in L^q$.

Show that $\|f\|_p < \infty$.

Δ∇Δ = ∇Δ∇ = Δ∇Δ = ∇Δ∇ = Δ∇Δ

SOLUTION: If $f = 0$ almost everywhere all these properties are trivial, therefore it will be assumed that $0 < \|f\|_p \leqslant \infty$.

SOLUTION: (a): By Hölder's Inequality,

$$\sup\{\left|\int fg\right| : \|g\|_q \leqslant 1\} \leqslant \|f\|_p.$$

In order to prove the inequality in the opposite direction let us first of all assume that $p < \infty$, and let:

$$g_0(x) = \begin{cases} \overline{f(x)}\,|f(x)|^{p-2} & \text{if } f(x) \neq 0, \\ 0 & \text{if } f(x) = 0, \end{cases}$$

so that $fg_0 = |f|^p$. If $p > 1$ one has $|g_0|^q = |f|^p$, and then if $g(x) = \|f\|_p^{-p/q} g_0(x)$ we have $\|g\|_q = 1$ and $\int fg = \|f\|_p$. If $p = 1$ we have $\|g_0\|_\infty = 1$ and $\int fg_0 = \|f_1\|$. Finally, when $p = \infty$ let $0 < M < \|f\|_\infty$ and let $E \subset (|f| \geqslant M)$ such that $0 < \text{meas}(E) < \infty$.

Set

$$g(x) = \begin{cases} \dfrac{1}{\text{meas}(E)} \dfrac{\overline{f(x)}}{|f(x)|} & \text{if } x \in E, \\ 0 & \text{if } x \notin E. \end{cases}$$

Then $\|g\|_1 = 1$ and:

$$\int fg = \frac{1}{\text{meas}(E)} \int_E |f(x)|\,dx \geq M.$$

REMARK: In fact we have proved that if $1 \leq p < \infty$

$$\|f\|_p = \max\left(\left|\int fg\right| : \|g\|_q = 1\right).$$

This is nothing astonishing if one knows that in this case L^q is the dual of L^p. Against this, if $0 < f(x) < 1$ for all x, and $\|f\|_\infty = 1$, then for every function g such that $\|g\|_1 \leq 1$ we have $\left|\int fg\right| < 1$, otherwise there would exist $\alpha \in \mathbb{C}$ such that $|\alpha| = 1$ and $\int f(\alpha g) = 1$. If $\alpha g = g_1 + ig_2$, then $\|g_1\|_1 \leq 1$ and $\int fg_1 = 1$; but under these conditions one would have $\|g_1^+\|_1 \leq 1$ and

$$\int fg_1^+ - \int fg_1^- = 1, \qquad 0 \leq \int fg_1^+ \leq 1, \qquad \int fg_1^- \geq 0,$$

whence:

$$\int fg_1^+ = 1 \geq \int g_1^+,$$

that is to say:

$$\int (1 - f)g_1^+ \leq 0,$$

which would imply that $g_1^+ = 0$ almost everywhere, which contradicts $\int fg_1^+ = 1$.

SOLUTION: (b): When $\|f\|_p < \infty$ and $g \in L^q$, one has $\left|\int fg\right| \leqslant \int |f||g|$, whence the result in this case. If $\|f\|_p = \infty$ let us set:

$$f_n(x) = \begin{cases} f(x) & \text{if } |x| \leqslant n \text{ and } f(x) \leqslant n, \\ 0 & \text{otherwise.} \end{cases}$$

By what has preceded, there exists $g_n \geqslant 0$ such that $\|g_n\|_q \leqslant 1$ and $\int f_n g_n \geqslant \|f_n\|_p - 1$. Since $\int fg_n \geqslant \int f_n g_n$ and $\|f_n\|_p \to \|f\|_p = \infty$, the result is again true in this case.

SOLUTION: (c): Let $E_n = \{x : |x| \leqslant n \text{ and } |f(x)| \leqslant n\}$, and let $f_n = f\mathbb{1}_{E_n}$. If g is such that $\|g\|_q \leqslant 1$ there exists a sequence of simple functions (g_i) such that $\|g - g_i\|_q \to 0$. Then, because $f_n \in L^p$,

$$\int f_n g = \lim_i \int f_n g_i = \lim_i \int f(g_i \mathbb{1}_{E_n}).$$

The functions $g_i \mathbb{1}_{E_n}$ are simple, $\|g_i \mathbb{1}_{E_n}\|_q \leqslant 1$, and $f(g_i \mathbb{1}_{E_n})$ is integrable. From this it follows that

$$\left|\int f_n g\right| \leqslant M.$$

By part (a) this implies that $\|f_n\|_p \leqslant M$, and therefore that $\|f\|_p = \lim_n \|f_n\|_p \leqslant M$.

SOLUTION: (d): Let f_n be as above, and set

$$u(g) = \int fg, \quad u_n(g) = \int f_n g, \quad g \in L^q.$$

The u_n's are continuous linear forms on L^q. Moreover, for $g \in L^q$, $f_n g \to fg$ and $|f_n g| \leqslant |fg|$, $\int |fg| < \infty$. By Lebesgue's Theorem $u_n(g) \to u(g)$. By the Banach-Steinhaus Theorem u is a continuous linear form on L^q, that is to say there exists a constant M such that $\left|\int fg\right| \leqslant M$ if $\|g\|_q \leqslant 1$. Using part (c), it follows from this that $\|f\|_p \leqslant M$.

EXERCISE 6 108: Denote by E_n the set of step functions on $\mathbb{R}^n$. Let f be a locally integrable function on $\mathbb{R}^n$.

Show that if $g \in E_n$, $g \geqslant 0$, then

$$\int f^+ g = \sup\left\{\int fh : h \in E_n \text{ and } 0 \leqslant h \leqslant g\right\} \tag{*}$$

if f is real. From this deduce that in this case

$$\int |f| g = \sup\left\{\int fh : h \in E_n \text{ and } -g \leqslant h \leqslant g\right\}. \tag{**}$$

Finally, show that when f is complex

$$\int |f| g = \sup\left\{\left|\int fh\right| : h \in E_n \text{ and } |h| \leqslant g\right\}. \tag{***}$$

Δ∇Δ = ∇Δ∇ = Δ∇Δ = ∇Δ∇ = Δ∇Δ

SOLUTION: By replacing f with zero outside the support of g one can assume f is integrable. Notice that if f is real and $0 \leqslant h \leqslant g$, then since $f^+ \geqslant f$,

$$\int fh \leqslant \int f^+ g.$$

When f is complex and $|h| \leqslant g$,

$$\left|\int fh\right| \leqslant \int |f| g,$$

and, furthermore, if f and h are real:

$$\int fh \leqslant \left|\int fh\right|.$$

All of this shows that the left side of each of the equalities to be proved is greater than or equal to the right side.

Let $\varepsilon > 0$, and let $f_1 \in E_n$ be such that

$$\int |f - f_1| g < \varepsilon.$$

We may further assume that f_1 is real if f is. Set

$$h_1 = \frac{\bar{f}_1}{|f_1|} g$$

(where we have adopted the convention that $\bar{f}_1 |f_1|^{-1} = 0$ when $f_1 = 0$), and if f is real

$$h_2 = \frac{f_1^+}{f_1} g$$

(with an analogous convention). It is clear that $h_1 \in E_n$, $|h_1| \leqslant g$, and that if f is real, $-g \leqslant h_1 \leqslant g$, $h_2 \in E_n$, $0 \leqslant h_2 \leqslant g$. Furthermore:

$$\int f_1 h_1 = \int |f_1| g,$$

and if f is real,

$$\int f_1 h_2 = \int f_1^+ g.$$

When f is real,

$$\int |f| g - \int f h_1 = \int (|f| - |f_1|) g + \int (f_1 - f) h_1$$

$$\leqslant 2 \int |f - f_1| g < 2\varepsilon.$$

Replacing $|f|, |f_1|, h_1$ by f^+, f_1^+, h_2, and noticing that $|f^+ - f_1^+| \leqslant$

$|f - f_1|$, one obtains

$$\int f^+ g - \int f h_2 < 2\varepsilon.$$

Finally, when f is complex,

$$\int |f| g - \left|\int f h_1\right| \leqslant \left|\int |f| g - \int f h_1\right|,$$

and the rest follows as above.

REMARK: E_n can be replaced by K, the space of compactly supported continuous functions.

EXERCISE 6·109: Let $1 \leqslant p \leqslant \infty$ and $\frac{1}{p} + \frac{1}{q} = 1$.

Show that if f is a locally integrable function and if there exists a constant M such that:

$$\left|\int fg\right| \leqslant M$$

for every step function g such that $\|g\|_q \leqslant 1$, then $\|f\|_p \leqslant M$.

Δ∇Δ = ∇Δ∇ = Δ∇Δ = ∇Δ∇ = Δ∇Δ

SOLUTION: By the preceding exercise, if g is a real positive step function such that $\|g\|_q \leqslant 1$, then

$$\int |f| g = \sup\left\{\left|\int fh\right| : h \in E_n \text{ and } |h| \leqslant g\right\} \leqslant M. \tag{1}$$

To begin, assume that $p > 1$, and set

$$\varphi_n(x) = \begin{cases} |f(x)| & \text{if } |x| \leqslant n \text{ and } |f(x)| \leqslant n, \\ 0 & \text{otherwise.} \end{cases}$$

When $g \in L^q$, $g \geqslant 0$, $\|g\|_q \leqslant 1$, there exists a sequence of step functions g_i such that $g_i \geqslant 0$ and $\|g - g_i\|_q \to 0$ (since $q < \infty$).

As $\varphi_n \in L^p$, one therefore has

$$\int \varphi_n g = \lim_i \int \varphi_n g_i \leqslant \limsup_i \int |f| g_i \leqslant M.$$

By Exercise 6·107, $\|\varphi_n\|_p \leqslant M$, whence $\|f\|_p \leqslant M$. When $p = 1$ note that if one were to have $\|f\|_1 > M$ there would exist a rectangle P such that

$$\int_P |f| = \int |f| \mathbb{1}_P > M,$$

which would contradict (1), because $\|\mathbb{1}_P\|_\infty = 1$.

EXERCISE 6·110: Let $X = \mathbb{R}^n$, $Y = \mathbb{R}^m$, and let f be a positive measurable function on $X \times Y$.

Show that for $p \geqslant 1$:

$$\left\{\int_Y \left(\int_X f(x,y)\mathrm{d}x\right)^p \mathrm{d}y\right\}^{1/p} \leqslant \int_X \left\{\int_Y f(x,y)^p \mathrm{d}y\right\}^{1/p} \mathrm{d}x.$$

(*GENERALISED MINKOWSKI INEQUALITY*)

Δ∇Δ = ∇Δ∇ = Δ∇Δ = ∇Δ∇ = Δ∇Δ

SOLUTION: Set

$$g(y) = \int_X f(x,y)\mathrm{d}x.$$

On writing the function $y \mapsto f(x,y)$ as $f(x,\cdot)$, it is a matter of proving that

$$\|g\|_{L^p(Y)} \leqslant \int_X \|f(x,\cdot)\|_{L^p(Y)} \mathrm{d}x. \qquad (*)$$

Let φ be a positive measurable function on Y. Then

$$\int g(y)\varphi(y)dy = \int dx \int f(x,y)\varphi(y)dy,$$

whence, writing the conjugate of p as q,

$$\int g\varphi \leqslant \|\varphi\|_{L^q(Y)} \int_X \|f(x,\cdot)\|_{L^p(Y)} dx.$$

This inequality implies (*) (cf., Exercise 6·107).

EXERCISE 6·111: For every i, $1 \leqslant i \leqslant n$, let $X_i = \mathbb{R}^{m_i}$, and let f_i be a positive measurable function on $X_1 \times \cdots \times \hat{X}_i \times \cdots \times X_n$ (where the hat indicates the term that must be omitted in the product). Set

$$I_i = \int f_i^{n-1} dx_1 \ldots \widehat{dx_i} \ldots dx_n.$$

Considering f_i as a function on $X_1 \times \cdots \times X_n$ which does not depend upon the i-th variable, show that

$$\int f_1 \cdots f_n dx_1 \ldots dx_n \leqslant (I_1 \cdots I_n)^{1/(n-1)}. \tag{*}$$

Deduce from this that if V is a measurable set of $\mathbb{R}^3$, and if S_1, S_2, S_3 denote the areas of the projections of V onto each of the three coordinate planes, then:

$$\mathrm{vol}(V) \leqslant \sqrt{S_1 S_2 S_3}. \tag{**}$$

ΔVΔ = VΛV = ΔVΔ = VΔV = ΔVΔ

SOLUTION: For $n = 2$ inequality (*) is clear (in fact, equality holds). Assume that this relation has been proved up to order $n - 1$ Then

$$I = \int f_1 \cdots f_n \mathrm{d}x_1 \ldots \mathrm{d}x_n = \int f_1 \mathrm{d}x_2 \ldots \mathrm{d}x_n \int f_2 \cdots f_n \mathrm{d}x_1 .$$

Writing, for $2 \leqslant i \leqslant n$,

$$g_i = \int f_i^{n-1} \mathrm{d}x_1 ,$$

Generalized Hölder's Inequality gives (cf. Exercise 6•106)

$$I \leqslant \int f_1 g_2^{1/(n-1)} \cdots g_n^{1/(n-1)} \mathrm{d}x_2 \ldots \mathrm{d}x_n .$$

Applying Hölder's Inequality again, with $p = n - 1$ and $q = (n-1)\times(n-2)^{-1}$, yields

$$I \leqslant I_1^{1/(n-1)} \left(\int g_2^{1/(n-2)} \cdots g_n^{1/(n-2)} \mathrm{d}x \ldots \mathrm{d}x_n \right)^{(n-2)/(n-1)} .$$

Now, using the induction hypothesis shows that the integral on the right side is majorised by

$$\prod_{i=2}^{n} \left(\int g_i \mathrm{d}x_2 \ldots \widehat{\mathrm{d}x_i} \ldots \mathrm{d}x_n \right)^{1/(n-2)} .$$

Finally, by noticing that for $2 \leqslant i \leqslant n$

$$I_i = \int g_i \mathrm{d}x_2 \ldots \widehat{\mathrm{d}x_i} \ldots \mathrm{d}x_n ,$$

Inequality (*) is obtained.

To obtain (**), it suffices to note that if A_1, A_2, A_3 denote the projections of V onto the coordinate planes, then

$$\mathbb{1}_V(x,y,z) \leqslant \mathbb{1}_{A_1}(x,y)\mathbb{1}_{A_2}(y,z)\mathbb{1}_{A_3}(x,z).$$

EXERCISE 6•112: Show that if $\|f - f_n\|_p \to 0$, then $f_n \to f$ in measure.

Δ∇Δ = ∇Δ∇ = Δ∇Δ = ∇Δ∇ = Δ∇Δ

SOLUTION: If $\alpha > 0$ and $E_n = \{|f - f_n| \geq \alpha\}$, then

$$\alpha^p \text{meas}(E_n) \leq \int_{E_n} |f - f_n|^p \leq \|f - f_n\|_p^p,$$

which shows that $\text{meas}(E_n) \to 0$.

EXERCISE 6·113: Let (f_n) be a sequence of functions in $L^p(X)$, $1 \leq p < \infty$, that converges in measure to a function f.

Show that the following conditions are equivalent:

(a): $f \in L^p(X)$ and $\lim_{n\to\infty} \|f - f_n\|_p = 0$;

(b): The sequence $(|f_n|^p)$ conserves mass and is uniformly integrable (for the definitions of these notions see Exercise 3·58).

ΔVΔ = ∇Δ∇ = Δ∇Λ = ∇Δ∇ = Δ∇Δ

SOLUTION: Assume first that $f \in L^p$ and $f_n \to f$ in L^p. For every measurable set E of X, Minkowski's Inequality gives

$$\left(\int_E |f_n|^p\right)^{1/p} \leq \left(\int_E |f|^p\right)^{1/p} + \|f - f_n\|_p. \tag{1}$$

Let $\varepsilon > 0$. There exists a set B of finite measure, a number $\beta > 0$, and an integer N such that

$$\int_{X-B} |f|^p < \varepsilon 2^{-p},$$

$$\int_E |f|^p < \varepsilon 2^{-p} \qquad \text{if } \text{meas}(E) < \beta,$$

$$\|f - f_n\|_p < \tfrac{1}{2}\varepsilon^{1/p} \text{ if } n > N.$$

By (1), if $n > N$ and $\text{meas}(E) < \beta$:

$$\int_{X-B} |f_n|^p < \varepsilon, \qquad \int_E |f_n|^p < \varepsilon.$$

Let $B_1, \ldots, B_N$ be set of finite measure and $\beta_1, \ldots, \beta_N > 0$ numbers such that

$$\int_{X-B_n} |f_n|^p < \varepsilon \quad \text{if } 1 \leqslant n \leqslant N,$$

$$\int_E |f_n|^p < \varepsilon \qquad \text{if } 1 \leqslant n \leqslant N \text{ and meas}(E) < \beta_n.$$

Setting $A = B \cup B_1 \cup \cdots \cup B_N$ and $\delta = \min(\beta, \beta_1, \ldots, \beta_N)$, then for every integer n,

$$\int_{X-A} |f_n|^p < \varepsilon, \tag{2}$$

$$\int_E |f_n|^p < \varepsilon \quad \text{if meas}(E) < \delta, \tag{3}$$

which proves that the sequence $(|f_n|^p)$ conserves mass and is uniformly integrable.

Now assume that $f_n \to f$ in measure, and that the sequence (f_n) satisfies Condition (ii). For every $\varepsilon > 0$ there exist a set A of finite measure and a number $\delta > 0$ such that the conditions (2) and (3) are satisfied for every integer n. By Fatou's Lemma the function f also satisfies these conditions. Set

$$E_n = \left\{|f - f_n|^p > \frac{\varepsilon}{\text{meas}(A)}\right\} \cap A.$$

There exists an integer N such that $\text{meas}(E_n) < \delta$ for $n > N$, and consequently

$$\|f - f_n\|_p \leqslant \left(\int_{X-A} |f|^p\right)^{1/p} + \left(\int_{X-A} |f_n|^p\right)^{1/p}$$
$$+ \left(\int_{E_n} |f|^p\right)^{1/p} + \left(\int_{E_n} |f_n|^p\right)^{1/p}$$
$$+ \left(\int_{A-E_n} |f - f_n|^p\right)^{1/p}$$
$$\leqslant 5\varepsilon^{1/p}$$

which proves that $f \in L^p$ and $f_n \to f$ in L^p.

REMARK: If $f_n \in L^p(X)$, $f_n \to f$ almost everywhere, and Condition (ii) is satisfied, one again has $f_n \to f$ in L^p. It suffices to replace the sets E_n in the proof above by a set $E \subset A$ such that $\text{meas}(E) < \delta$ and $f_n \to f$ uniformly on $A - E_n$, which is possible by Egoroff's Theorem.

EXERCISE 6·114: Let (f_n) be a sequence of functions in $L^p(X)$, $1 \leqslant p < \infty$.

(a): Show that if $f_n \to f \in L^p(X)$ almost everywhere, and if

$$\lim_{n\to\infty} \|f_n\|_p = \|f\|_p,$$

then $f_n \to f$ in $L^p(X)$.

(b): Show that the conclusion above is still valid if the convergence almost everywhere of f_n to f is replaced by convergence in measure.

Δ∇Δ = ∇Δ∇ = Δ∇Δ = ∇Δ∇ = Δ∇Δ

SOLUTION: (a): Let $\varepsilon > 0$. There exists a set B of finite measure and a number $\delta > 0$ such that

$$\int_{X-B} |f|^p \leqslant \frac{\varepsilon}{2},$$

$$\int_E |f|^p \leqslant \frac{\varepsilon}{2} \quad \text{if meas}(E) < \delta.$$

By Egoroff's Theorem there exists a set $A \subset B$ such that meas$(B - A) < \delta$ and $f_n \to f$ uniformly on A. Therefore, by Fatou's Lemma

$$\int_X |f|^p \leqslant \varepsilon + \int_A |f|^p \leqslant \varepsilon + \liminf_n \int_A |f_n|^p.$$

Since, by hypothesis,

$$\lim_{n\to\infty} \int_X |f_n|^p = \int_X |f|^p,$$

this yields

$$\int_X |f|^p \leqslant \varepsilon + \int_X |f|^p - \limsup_{n\to\infty} \int_{X-A} |f_n|^p,$$

i.e.

$$\limsup_{n\to\infty} \int_{X-A} |f_n|^p \leqslant \varepsilon.$$

Furthermore:

$$\|f - f_n\|_p \leqslant \left(\int_{X-A}|f|^p\right)^{1/p} + \left(\int_{X-A}|f_n|^p\right)^{1/p}$$

$$+ \sup_A|f - f_n| . \text{meas}(A)^{1/p},$$

whence

$$\limsup_{n\to\infty}\|f - f_n\|_p \leqslant 2\varepsilon^{1/p},$$

which proves that $f_n \to f$ in $L^p(X)$.

SOLUTION: (b): By the preceding exercise it suffices to show that the sequence $(|f_n|^p)$ conserves mass and is uniformly integrable. As Fatou's Lemma is valid for convergence in measure, the argument used in part (a) above shows that

$$\int_{X-A}|f^p| \leqslant \varepsilon$$

implies that

$$\limsup_{n\to\infty}\int_{X-A}|f_n|^p \leqslant \varepsilon,$$

which proves that the sequence $(|f_n|^p)$ conserves mass. If this sequence were not uniformly integrable there would exist $\alpha > 0$, a sequence (E_k) of sets of finite measure and some integers $n_1 < n_2 < \cdots < n_k < \cdots$ such that

$$\text{meas}(E_k) \to 0,$$

$$\int_{E_k}|f_{n_k}|^p \geqslant \alpha. \tag{1}$$

By considering a subsequence one would be able to assume additionally that $f_{n_k} \to f$ almost everywhere (cf., Exercise 3·54). By part (a) above f_{n_k} would tend to f in L^p, and consequently the f_{n_k} would be uniformly integrable (cf., the preceding exercise), which contradicts (1).

EXERCISE 6·115: Let f_n be a sequence of integrable functions on a measurable set X of $\mathbb{R}^p$. Assume that meas$(X) = 1$.

(a): Show that if

$$\lim_{n\to\infty}\int_X |\mathrm{Re}(f_n)| = 1,$$

$$\lim_{n\to\infty}\int_X |1 - |f_n|| = 0,$$

then

$$\lim_{n\to\infty}\int_X |\mathrm{Im}(f_n)| = 0.$$

(b): Now assume that

$$\lim_{n\to\infty}\int_X \mathrm{Re}(f_n) = \lim_{n\to\infty}\int_X |f_n| = \lim_{n\to\infty}\int_X |f_n|^{\frac{1}{2}} = 1.$$

Show that

$$\lim_{n}\int_X |1 - f_n| = 0.$$

$$\Delta\nabla\Delta = \nabla\Delta\nabla = \Delta\nabla\Delta = \nabla\Delta\nabla = \Delta\nabla\Delta$$

SOLUTION: (a): Set $f_n = u_n + iv_n$ (u_n, v_n real). And

$$\varepsilon_n = \int_X |1 - |f_n||.$$

Now,

$$|v_n| \leqslant |f_n| \leqslant 1 + |1 - |f_n||,$$

and consequently

$$\int_X |v_n| \leqslant 1 + \varepsilon_n.$$

Therefore the second hypothesis implies

$$\ell = \limsup_{n\to\infty} \int_X |v_n| \leqslant 1.$$

Assume that $0 < \ell \leqslant 1$ and choose a λ such that $0 < \lambda < \ell$. By considering a subsequence, one may assume that

$$\int_X |v_n| \geqslant \lambda$$

for all n. Now let $\alpha, \beta > 0$ be numbers which we will specify how to choose later. Set

$$A_n = (|1 - |f_n|| > \alpha),$$

$$B_n = (|v_n| \leqslant \beta).$$

Then

$$\lim_{n\to\infty} \text{meas}(A_n) = 0.$$

Furthermore

$$\lambda \leqslant \int_{B_n} |v_n| + \int_{X-B_n} (1 + |1 - |f_n||)$$

$$\leqslant \beta\text{meas}(B_n) + 1 - \text{meas}(B_n) + \varepsilon_n.$$

If $0 < \beta < 1$ this would imply

$$\text{meas}(B_n) \leqslant \frac{1 - \lambda + \varepsilon_n}{1 - \beta}.$$

Set $D_n = A_n \cup B_n$. On $X - D_n$ we have simultaneously

$$|f_n| \leqslant 1 + \alpha, \qquad |v_n| > \beta.$$

Hence

$$|u_n| \leqslant \delta = \sqrt{(1 + \alpha)^2 - \beta^2}.$$

Since $|u_n| \leqslant 1 + |1 - |f_n||$ it would follow that

$$\int_X |u_n| \leqslant \delta(1 - \text{meas}(D_n)) + \text{meas}(D_n) + \varepsilon_n$$

$$= \delta + (1 - \delta)\text{meas}(D_n) + \varepsilon_n.$$

If $0 < \delta < 1$, taking into account that

$$\text{meas}(D_n) \leqslant \text{meas}(A_n) + \text{meas}(B_n),$$

one would have

$$1 = \lim_{n\to\infty}\int_X |u_n| \leqslant \delta + (1 - \delta)\frac{1 - \lambda}{1 - \beta} = 1 - (1 - \delta)\frac{\lambda - \beta}{1 - \beta}.$$

One is therefore led to a contradiction if $0 < \beta < \lambda$. It remains to show that α,β can actually be chosen so that $0 < \beta < \lambda$ (which will imply $\gamma < 1$) and $\delta = \sqrt{(1 + \alpha)^2 - \beta^2} < 1$. This is always possible choosing β first and then α so small such that $\delta < 1$.

SOLUTION: (b): With the same notations as in part (a),

$$\lim_{n\to\infty}\int_X u_n = \lim_{n\to\infty}\int_X |f_n| = 1. \tag{1}$$

The inequalities

$$\left|\int_X u_n\right| \leqslant \int_X |u_n| \leqslant \int_X |f_n|$$

imply

$$\lim_{n\to\infty}\int_X |u_n| = 1. \tag{2}$$

Furthermore, the Cauchy-Schwarz Inequality gives

$$\left(\int_X |1 - |f_n||\right)^2 < \int_X (1 - |f_n|^{\frac{1}{2}})^2 \int_X (1 + |f_n|^{\frac{1}{2}})^2$$

With $\varepsilon = \pm 1$,

$$\int_X (1 + \varepsilon|f_n|^{\frac{1}{2}})^2 = 1 + 2\varepsilon\int_X |f_n|^{\frac{1}{2}} + \int_X |f_n|,$$

which, as $n \to \infty$, tends to 4 or 0 according as $\varepsilon = +1$ or -1. From this it follows that

$$\lim_{n\to\infty}\int_X |1 - |f_n|| = 0. \tag{3}$$

Relations (2) and (3) allow us to use part (a), so

$$\lim_{n\to\infty}\int_X |v_n| = 0. \tag{4}$$

But then, from the inequalities:

$$|1 - f_n| \leqslant |1 - |f_n|| + ||f_n| - f_n|$$

$$\leqslant |1 - |f_n|| + |v_n| + |f_n| - u_n ,$$

and from (1),(3) and (4), it follows that

$$\lim_{n\to\infty}\int_X |1 - f_n| = 0.$$

EXERCISE 6·116: Let X be a measurable set of $\mathbb{R}^m$ and f a measurable function on X. As usual we write

$$\|f\|_p = (\int_X |f|^p)^{1/p} \quad \text{if } 0 < p < \infty,$$

$$\|f\|_\infty = \operatorname*{ess\,sup}_X |f|.$$

Assume that f is not equal to zero almost everywhere and set

$$I_f = \{p: \|f\|_p < \infty\}.$$

(a): Show that I_f is an interval. Can one choose f so that I_f is an arbitrary interval in $]0,\infty]$?

(b): If I_f is not empty, show that $\alpha \to \log\|f\|_{1/\alpha}$ is convex on its interval of definition (the convention $1/0 = \infty$ is used). From this deduce that $p \mapsto \|f\|_p$ is continuous on I_f. More precisely, prove that even if r is an endpoint of I_f and $0 < r \leqslant \infty$, then $\|f\|_p \to \|f\|_r$ when $p \in I_f$, $p \to r$, whether $\|f\|_r$ is finite or not.

(c): Show that: $p \mapsto p\log\|f\|_p$ is convex on $I_f - \{\infty\}$.

(d): Deduce from the preceding that if $0 < r < p < s \leqslant \infty$ then

$$\|f\|_p \leqslant \max(\|f\|_r, \|f\|_s).$$

Conclude that $L^r(X) \cap L^s(X) \subset L^p(X)$.

ΔVΔ = VΔV = ΔVΔ = VΔV = ΔVΔ

SOLUTION: One may assume that $f \geqslant 0$. Let $A = (f \geqslant 1)$.

SOLUTION: (a): If $r < p < s$ and $r \in I_f$, $s \in I_f$, when $s < \infty$ one has $f^p \leqslant f^s$ on A and $f^p \leqslant f^r$ on $X - A$, which proves that f^p is integrable, and therefore that $p \in I_f$. If $s = \infty$, $f^p \leqslant f^r$ still holds on $X - A$ and furthermore $f^p \leqslant \|f\|_\infty^p$; since f^r is integrable we have meas$(A) < \infty$ and it again follows that f^p is integrable.

Assume that $X = [0,\infty[$. If $f(x) = e^{-x}$, then $I_f =]0,\infty]$. If $f(x) = n$ when $n \leqslant x \leqslant n + 2^{-n}$ ($n = 0,1,\ldots$) and 0 outside of these intervals $I_f =]0,\infty[$. If $f = 1$, $I_f = \{\infty\}$, and if f is continuous, and $f(x) \sim x^{-1/a}$ $a > 0$ when $x \to \infty$, then $I_f =]-a,\infty]$, but $I_f = [a,\infty]$ if $f(x) \sim x^{-1/a}(\log x)^{-b}$ with $ab > 1$. However, if f is continuous on $]0,\infty[$, $f(x) = 0$ for $x \geqslant 1$, and $f(x) \sim x^{-1/a}$ when $x \to 0$, then $I_f =]0,a[$, but $I_f =]0,a]$ if $f(x) \sim x^{-1/a}(\log x)^{-b}$, $ab > 1$. Lastly, if $f(x) = x$, $I_f = \emptyset$. Considering the sums of two functions of the above type, any integral in $]0,\infty]$ can be obtained for I_f.

SOLUTION: (b): This is a question of proving that if $r \in I_f$, $s \in I_f$

and

$$\frac{1}{p} = \frac{\lambda}{r} + \frac{(1 - \lambda)}{s}, \qquad 0 \leqslant \lambda \leqslant 1,$$

then:

$$\|f\|_p \leqslant \|f\|_r^{\lambda}\|f\|_s^{1-\lambda} \tag{1}$$

If r,s are finite, it suffices to make $h = f^{\lambda p}$, $g = f^{(1-\lambda)p}$ in Hölder's Inequality

$$\int_X hg \leqslant (\int_X h^{r/\lambda p})^{\lambda p/r}(\int_X g^{s/(1-\lambda)p})^{(1-\lambda)p/s}.$$

When $0 < r < s = \infty$, one has $p = r/\lambda > r$, whence

$$\|f\|_p^p \leqslant \|f^{(r/\lambda)-r}\|_\infty \int_X f^r = \|f\|_r^{p\lambda}\|f\|_\infty^{p(1-\lambda)},$$

which shows that (1) is still valid in this case.

By a classical property of convex functions, from this it follows that $\alpha \mapsto \log\|f\|_{1/\alpha}$ is continuous on the interior of its defining interval, and consequently that $p \to \|f\|_p$ is continuous on the interior of I_f. It remains for us to prove that if $p \in I_f$ tends to an endpoint r of I_f, finite or not but different from zero, then $\|f\|_p \to \|f\|_r$ whether $\|f\|_r$ is finite or not. Assume first that r is finite. In this case $f^p \to f^r$, and does so monotonically on each of the sets A and $X - A$. From this it follows that

$$\int_A f^p \to \int_A f^r, \qquad \int_{X-A} f^p \to \int_{X-A} f^r$$

(the Monotonic Convergence Theorem; it will be noticed that as the f^p's are integrable, this Theorem can still be used in the case of decreasing sequences). This proves the result in this case.

If $r = \infty$ and $0 < \gamma < \|f\|_\infty$ then

$$\text{meas}(f > \gamma) > 0 \quad \text{and} \quad \|f\|_p \geqslant \gamma(\text{meas}(f > \gamma))^{1/p},$$

whence

$$\liminf_{p\to\infty}\|f\|_p \geqslant \gamma,$$

and consequently, in view of the arbitrariness of γ,

$$\liminf_{p\to\infty}\|f\|_p \geqslant \|f\|_\infty.$$

On the other hand, if $p > r$, with $r \in I_f$, when $\|f\|_\infty < \infty$ one has

$$\|f\|_p \leqslant \|f\|_\infty^{1-r/p}\|f\|_r^{r/p},$$

whence

$$\limsup_{p\to\infty}\|f\|_p \leqslant \|f\|_\infty.$$

SOLUTION: (c): This is a matter or proving that if $0 < r < s < \infty$ and $0 < \lambda < 1$, then

$$\int_X f^{\lambda r+(1-\lambda)s} \leqslant \left(\int_X f^r\right)^\lambda\left(\int_X f^s\right)^{1-\lambda}$$

It suffices to make $h = f^{\lambda r}$, $g = f^{(1-\lambda)s}$ in Hölder's Inequality:

$$\int_X fg < \left(\int_X f^{1/\lambda}\right)^\lambda\left(\int_X f^{1/(1-\lambda)}\right)^{1-\lambda}.$$

SOLUTION: (d): The inequality is trivial if $\|f\|_r = \infty$ or $\|f\|_s = \infty$. Assume, therefore, that these two numbers are finite. If s is finite the inequality results from the convexity of $\log\|f\|_p$ as a function of $1/p$. If $s = \infty$, for $p < \sigma < \infty$ one has

$$\|f\|_p \leqslant \max(\|f\|_r, \|f\|_\sigma),$$

and by making $\sigma \to \infty$ the inequality sought is obtained because of part (b) above. It is then clear that

$$L^r \cap L^s \subset L^p.$$

EXERCISE 6·117: The notations are the same as in the preceding exercise. It is further assumed that meas$(X) = 1$.

(a): Show that $\|f\|_r \leqslant \|f\|_s$ if $0 < r < s \leqslant \infty$. Can one have $\|f\|_r = \|f\|_s < \infty$?

(b): Assume that $\|f\|_r < \infty$ for one $r > 0$.
Show that:

$$\lim_{p\to 0}\|f\|_p = \exp\left(\int_X \log|f|\right),$$

where by convention $\exp(-\infty) = 0$.

Δ∇Δ = ∇Δ∇ = ΔVΔ = ∇Δ∇ = ΔVΔ

SOLUTION: (a): Let σ be such that $\frac{1}{r} = \frac{1}{s} + \frac{1}{\sigma}$. Then

$$\|fg\|_r \leqslant \|f\|_s \|g\|_\sigma \tag{1}$$

(cf., Exercise 6·106). On making $g = 1$ the desired inequality is obtained. Since (1) is obtained by applying Holder's Inequality to f^r and g^r, equality can be had with $g = 1$ only if f is equal to a constant almost everywhere.

SOLUTION: (b): By Jensen's Formula,

$$\log\|f\|_p = \frac{1}{p}\log\int_X f^p \geqslant \int_X \log f.$$

On the other hand, since $\log u \leqslant u - 1$,

$$\frac{1}{p} \log \int_X f^p \leqslant \frac{1}{p} \left[\int_X f^p - 1\right] = \int_X \frac{1}{p} (f^p - 1).$$

On $A = (f \geqslant 1)$ $p^{-1}(f^p - 1)$ decreases to $\log f$, and on $X - A$ it increases to the same function. Consequently

$$\lim_{p \to 0} \int_X p^{-1}(f^p - 1) = \int_X \log f,$$

which proves that

$$\lim_{p \to 0} \log \|f\|_p = \int_X \log f.$$

NOTE: As in the preceding exercise, we have assumed that $f \geqslant 0$.

EXERCISE 6·118: The notations are the same as in the preceding exercise (in particular, $\text{meas}(X) = 1$).

Find all the functions ϕ on $]0,\infty[$ such that

$$\phi(\lim_{p \to 0} \|f\|_p) = \int_X \phi(f)$$

for every bounded measurable function $f > 0$.

Δ∇Δ = ∇Δ∇ = Δ∇Δ = ∇Δ∇ = Δ∇Δ

SOLUTION: Let $0 < c < 1$, and let $A \subset X$ be such that $\text{meas}(A) = c$. The function f, equal to $x > 0$ on A and to 1 on $X - A$, is such that:

$$\lim_{p \to 0} \|f\|_p = \exp \int_X \log f = \exp(c \log x) = x^c$$

(cf., the preceding exercise). As $\phi(f) = \phi(x)$ on A and $\phi(f) =$

$\phi(1)$ on $X - A$, one must therefore have:

$$\phi(x^c) = c\phi(x) + (1 - c)\phi(1)$$

for $x > 0$ and $0 \leqslant c \leqslant 1$ (the formula is in fact trivial if $c = 0$ or $c = 1$). If we set $\phi(x) = \psi(\log x)$, where ψ is defined on $\mathbb{R}$, the preceding relation becomes

$$\psi(cx) = c\psi(x) + (1 - c)\psi(0), \quad 0 \leqslant c \leqslant 1, \quad x \in \mathbb{R}.$$

From this it follows that there exist three constants a_1, a_2, b such that:

$$\psi(x) = a_1 x + b \quad \text{if } x \geqslant 0,$$

$$\psi(x) = a_2 x + b \quad \text{if } x \leqslant 0.$$

Let us then consider u, v, $0 < u < 1 < v$ such that $uv > 1$ and $B \subset X$ such that $\text{meas}(B) = \frac{1}{2}$. For the function f equal to u on B and to v on $X - B$

$$\lim_{p \to 0} \|f\|_p = \exp(\tfrac{1}{2}\log u + \tfrac{1}{2}\log v) = \sqrt{uv} > 1.$$

Consequently:

$$\phi(\lim_{p \to \infty} \|f\|_p) = \tfrac{1}{2}a_1(\log u + \log v) + b.$$

On the other hand,

$$\int_X \phi(f) = \tfrac{1}{2}(a_2 \log u + b) + \tfrac{1}{2}(a_1 \log v + b).$$

From this it follows that $a_1 = a_2$. Hence ϕ must be of the form

$$\phi(x) = a \log x + b.$$

Conversely, if ϕ is so defined, then

$$\phi(\lim_{p\to 0}\|f\|_p) = a\log(\lim_{p\to 0}\|f\|_p) + b$$

$$= a\int_X \log f + b = \int_X (a\log f + b)$$

$$= \int_X \phi(f).$$

EXERCISE 6·119: For any function ψ increasing on $[1,\infty[$, such that $\psi(1) > 0$ and $\lim_{p\to\infty}\psi(p) = \infty$, show that there exists on $[0,1]$ a measurable function f such that $\lim_{p\to\infty}\|f\|_p = \infty$ and $\|f\|_p \leqslant \psi(p)$ for all p.

ΔVΔ = VΔV = ΔVΔ = VΔV = ΔVΔ

SOLUTION: Let L_ψ be the set of functions f, measurable on $[0,1]$, such that:

$$\|f\|_\psi = \sup_{p\geqslant 1} \frac{\|f\|_p}{\psi(p)} < \infty.$$

It is clear that L_ψ is a vector space and that $f \mapsto \|f\|_\psi$ is a norm on this space. Since $\|f\|_p \leqslant \|f\|_\infty$ (cf., exercise 6·117) it is clear that $L^\infty \subset L_\psi$. Everything reduces to proving that $L^\infty \neq L_\psi$, for then there exists f such that $\|f\|_\infty = \infty$, $\|f\|_\psi \leqslant 1$; by exercise 6·116 one has $\lim_{p\to\infty}\|f\|_p = \infty$, and on the other hand the definition of $\|f\|_\psi$ implies that $\|f\|_p \leqslant \psi(p)$ for all $p \geqslant 1$.

First, let us show that L_ψ provided with the norm $f \mapsto \|f\|_\psi$ is complete. For this, consider a sequence (f_n) of elements of L_ψ such that:

$$\sum_n \|f_n\|_\psi < \infty.$$

For $p \geqslant 1$,

$$\sum_n \|f_n\|_p \leqslant \psi(p) \sum_n \|f_n\|_\psi < \infty.$$

From this it follows that the series:

$$g(x) = \sum_n f_n(x)$$

converges for almost all x. Furthermore, if $s_N = f_1 + \cdots + f_N$,

$$\|g - s_N\|_p \leqslant \sum_{n>N} \|f_n\|_p \leqslant \psi(p) \sum_{n>N} \|f_n\|_\psi,$$

which proves that $g - s_N \in L_\psi$, hence that $g \in L_\psi$, and furthermore

$$\|g - s_N\|_\psi \leqslant \sum_{n>N} \|f_n\|_\psi,$$

and consequently that

$$\lim_{N\to\infty}\|g - s_N\|_\psi = 0.$$

Assume that $L^\infty = L_\psi$. $\psi(1) = 1$ can be assumed without any loss of generality. Since

$$\|f\|_\psi = \sup_{p\geqslant 1} \frac{\|f\|_p}{\psi(p)} \leqslant \|f\|_\infty,$$

the norms $\|\cdot\|_\psi$ and $\|\cdot\|_\infty$ would be equivalent by virtue of a theorem of Banach; that is to say, there would exist a constant M such that for every $f \in L^\infty$

$$\|f\|_\infty \leqslant M\|f\|_\psi.$$

Now this is absurd, as for every $A > \max(M,1)$ there exists p_0 such that $\psi(p) \geqslant A$ for $p > p_0$; then if

$$f = A\mathbb{1}_{[0,A^{-p_0}]}$$

one has

$$\|f\|_\infty = A, \quad \|f\|_p \leqslant \|f\|_\infty = A \leqslant \psi(p) \quad \text{if } p > p_0,$$

and:

$$\|f\|_p \leqslant \|f\|_{p_0} = 1 \leqslant \psi(p) \quad \text{when } 1 \leqslant p \leqslant p_0.$$

EXERCISE 6·120: (a): Let $p \geqslant 1$ and $f \in L^p(\mathbb{R})$. For all $x \in \mathbb{R}$ set

$$F(x) = \int_0^x f(t)dt.$$

Show that $F(x + h) - F(x) = o(|h|^{1-1/p})$ as $h \to 0$, uniformly for $x \in \mathbb{R}$.

(b): Show that if f is integrable on $\mathbb{R}$, is continuously differentiable, and if $f' \in L^p(\mathbb{R})$ for some $p \geqslant 1$, then

$$\lim_{|x|\to\infty} f(x) = 0.$$

SOLUTION: (a): Let $\varepsilon > 0$. There exists $\delta > 0$ such that meas$(E) < \delta$ implies

$$\int_E |f|^p < \varepsilon^p.$$

If $0 < h < \delta$, by Hölder's Inequality, if $p > 1$ one has

$$|F(x + h) - F(x)| = \left|\int_x^{x+h} f\right| \leqslant h^{1-1/p}\left(\int_x^{x+h} |f|^p\right)^{1/p} \leqslant \varepsilon h^{1-1/p}.$$

SOLUTION: (b): By part (a) above, the function f is uniformly continuous, and consequently $f(x) \to 0$ as $|x| \to \infty$ (cf., Exercise 3·37).

EXERCISE 6·121: Let X be a measurable set of $\mathbb{R}^n$ such that $0 <$ meas$(X) < \infty$. Moreover, let f and F be two functions defined on X that are measurable and positive.

Show that if for all $\lambda > 0$

$$\mathrm{meas}(F > \lambda) \leqslant \frac{1}{\lambda}\int_{(F>\lambda)} f,$$

then for all p $(1 < p < \infty)$

$$\left(\int_X F^p\right)^{1/p} \leqslant \frac{p}{p-1}\left(\int_X f^p\right)^{1/p}.$$

ΔVΔ = VΔV = ΔVΔ = VΔV = ΔVΔ

SOLUTION: One can clearly assume that $\int_X f^p < \infty$, as otherwise there would be nothing to prove. Let $F_n = \min(F,n)$; it is clear that $(F_n > \lambda) \subset (F > \lambda)$ and that $(F_n > \lambda) = \emptyset$ if $\lambda \geqslant n$. From this it follows (cf., Exercise 5·98) that

$$\int_X F_n^p = p\int_0^n \lambda^{p-1}\mathrm{meas}(F_n > \lambda)d\lambda$$

$$\leqslant p\int_0^n \lambda^{p-2}d\lambda\int_{(F>\lambda)} f(x)dx$$

$$= p\int_X f(x)dx\int_0^{\min(n,F(x))}\lambda^{p-2}d\lambda = \frac{p}{p-1}\int_X f(x)F_n(x)^{p-1}dx.$$

By Hölder's Inequality, this yields, if $\frac{1}{p} + \frac{1}{q} = 1$

$$\int_X F_n^p < \frac{p}{p-1}\left(\int_X f^p\right)^{1/p}\left(\int_X F_n^{(p-1)q}\right)^{1/q}.$$

Now $(p-1)q = p$, $\int_X F_n^p < \infty$ since $0 \leqslant F_n \leqslant n$ and meas$(X) < \infty$, so

$$\left(\int_X F_n^p\right)^{1/p} \leqslant \frac{p}{p-1}\left(\int_X f^p\right)^{1/p}.$$

The desired result is now obtained by making $n \to \infty$ and using the Monotonic Convergence Theorem.

EXERCISE 6·122: Let $1 < p < \infty$. Show that if f is locally integrable the following conditions are equivalent:

(i): $f \in L^p$;

(ii): There exists a constant M such that for every sequence $P_1, \ldots, P_n$ of disjoint rectangles that are of non-zero measure

$$\sum_{i=1}^{n} \frac{1}{\text{meas}(P_i)^{p-1}}\left|\int_{P_i} f\right|^p \leqslant M. \qquad (*)$$

If these conditions hold, the smallest constant M that can be taken in (*) is equal to $\|f\|_p^p$.

ΔVΔ = VΔV = ΔVΔ = VΔV = ΔVΔ

SOLUTION: If $f \in L^p$ and $\frac{1}{p} + \frac{1}{q} = 1$, then

$$\left|\int_{P_i} f\right|^p \leqslant \text{meas}(P_i)^{p/q}\int_{P_i} |f|^p,$$

and consequently, because $p/q = p - 1$

$$\sum_{i=1}^{n} \frac{1}{\text{meas}(P_i)^{p-1}}\left|\int_{P_i} f\right|^p \leq \sum_{i=1}^{n} \int_{P_i} |f|^p < \|f\|_p^p. \tag{1}$$

Now assume that condition (*) holds. Let $g = \sum_i c_i \mathbb{1}_{P_i}$ be a step function, with $c_i \neq 0$, and the P_i being disjoint non-negligeable rectangles. Then

$$\left|\int fg\right| = \left|\sum_i c_i \int_{P_i} f\right|$$

$$\leq \left(\sum_i \frac{1}{\text{meas}(P_i)^{p-1}}\left|\int_{P_i} f\right|^p\right)^{1/p}\left(\sum_i |c_i|^q \text{meas}(P_i)^{q(p-1)/p}\right)^{1/q}$$

$$\leq M^{1/p}\left(\sum_i |c_i|^q \text{meas}(P_i)\right)^{1/q} = M^{1/p}\|g\|_q.$$

By Exercise 6·109 we therefore have

$$\|f\|_p^p \leq M. \tag{2}$$

Comparing (1) and (2) yields the last part of the problem.

EXERCISE 6·123: Denote by E_n the vector space of step functions on $\mathbb{R}^n$. Let T be a linear mapping from E_n into $L^1_{\text{loc}}(\mathbb{R}^m)$. Assume that for every $f \in E_n$

$$\|Tf\|_{q_0} \leq A_0\|f\|_{p_0}, \qquad \|Tf\|_{q_1} \leq A_1\|f\|_{p_1},$$

where $1 \leq p_i, q_i \leq \infty$, $A_0 > 0$, $A_1 > 0$.

Prove that if

$$\frac{1}{p} = \frac{1-t}{p_0} + \frac{t}{p_1}, \qquad \frac{1}{q} = \frac{1-t}{q_0} + \frac{t}{q_1}, \qquad 0 \leq t \leq 1,$$

for all $f \in E_n$

$$\|Tf\|_q \leqslant A_0^{1-t} A_1^t \|f\|_p. \qquad \text{(RIESZ-THORIN THEOREM)}$$

(Set $\alpha_i = 1/p_i$, $\alpha = 1/p$, $\beta_i = 1/q_i$, $\beta = 1/q$ and, for all $z \in C$, $\alpha(z) = (1 - z)\alpha_0 + z\alpha_1$, $\beta(z) = (1 - z)\beta_0 + z\beta_1$. Let $f \in E_n$ and $g \in E_n$ be such that $\|f\|_p = \|g\|_{q'} = 1$, where $\frac{1}{q} + \frac{1}{q'} = 1$. The function f may then be written:

$$f = \sum_{k=1}^{u} r_k e^{i\vartheta_k} 1\!\!1_{A_k},$$

where $r_k > 0$, $\vartheta_k \in \mathbb{R}$, and the A_k are disjoint rectangles of $\mathbb{R}^n$; similarly

$$g = \sum_{\ell=1}^{v} \rho_\ell e^{i\varphi_\ell} 1\!\!1_{B_\ell},$$

with analogous conditions. If $\alpha > 0$ and $\beta < 1$ then set

$$f_z = \sum_k r_k^{\alpha(z)/\alpha} e^{i\vartheta_k} 1\!\!1_{A_k},$$

$$g_z = \sum_\ell \rho_\ell^{(1-\beta(z))/(1-\beta)} e^{i\varphi_\ell} 1\!\!1_{B_\ell},$$

$$F(z) = \frac{1}{A_0^{1-z} A_1^z} \int_{\mathbb{R}^m} T(f_z) g_z.$$

Show that F is an entire function, bounded on the strip $0 \leqslant \text{Re}(z) \leqslant 1$, and that for all $y \in \mathbb{R}$

$$|F(iy)| \leqslant 1, \qquad |F(1 + iy)| \leqslant 1.$$

By considering for $\varepsilon > 0$ the functions $G_\varepsilon(z) = e^{\varepsilon z^2} F(z)$ to which one can apply the maximum principle, deduce that $|F(z)| \leqslant 1$ if $0 \leqslant \mathrm{Re}(z) \leqslant 1$. Next examine the case where $\alpha(1-\beta) = 0$).

Use the Riesz-Thorin Theorem to prove that if $f \in L^p$, $g \in L^q$, $\frac{1}{r} = \frac{1}{p} + \frac{1}{q} - 1 \geqslant 0$, then f and g are convolvable, and $\|f*g\|_r \leqslant \|f\|_p\|g\|_q$ (cf., Exercise 6·106).

$$\Delta\nabla\Delta = \nabla\Delta\nabla = \Delta\nabla\Delta = \nabla\Delta\nabla = \Delta\nabla\Delta$$

SOLUTION: Note first of all that

$$\alpha(t) = \alpha, \qquad \beta(t) = \beta, \tag{1}$$

and that if $y \in \mathbb{R}$,

$$\begin{aligned} \mathrm{Re}\alpha(iy) &= \alpha_0, \qquad \mathrm{Re}\beta(iy) = \beta_0, \\ \mathrm{Re}\alpha(1+iy) &= \alpha_1, \qquad \mathrm{Re}\beta(1+iy) = \beta_1. \end{aligned} \tag{2}$$

Also notice that if z belongs to the strip $0 \leqslant \mathrm{Re}(z) \leqslant 1$ then so do $\alpha(z), \beta(z), 1-z$, and that if $a > 0$ then $\min(1,a) \leqslant |a^z| \leqslant \max(1,a)$. Since

$$F(z) = \frac{1}{A_0^{1-z}A_1^z} \sum_{k,\ell} r_k^{\alpha(z)/\alpha} \rho_\ell^{(1-\beta(z))/(1-\beta)} e^{i(\vartheta_k+\varphi_\ell)} \int_{B_\ell} T(\amalg_{A_k}),$$

it follows that F is an entire function bounded by a constant M for $0 \leqslant \mathrm{Re}(z) \leqslant 1$. Hence by (2), upon setting $q_i' = 1/(1-\beta_i)$, $q' = 1/(1-\beta)$, one has

$$\|f_{iy}\|_{p_0}^{p_0} = \|f_{1+iy}\|_{p_1}^{p_1} = \sum_k r_k^{1/\alpha}\mathrm{meas}(A_k) = \|f\|_p^p = 1,$$

$$\|g_{iy}\|_{q_0'}^{q_0'} = \|g_{1+iy}\|_{q_1'}^{q_1'} = \sum_\ell \rho_\ell^{1/(1-\beta)}\mathrm{meas}(B_\ell) = \|g\|_{q'}^{q'} = 1,$$

that is to say:

$$\|f_{iy}\|_{p_0} = \|f_{1+iy}\|_{p_1} = \|g_{iy}\|_{q_0'} = \|g_{1+iy}\|_{q_1'} = 1$$

(if one of the numbers p_0, p_1, q_0' or q_1' is equal to ∞, these formulae are still true; for example, if $p_0 = \infty$ then $\mathrm{Re}\,\alpha(iy) = 0$ and $\|f_{iy}\|_\infty = \max_k |r_k^{\alpha(iy)/\alpha}| = 1$).

Hölder's Inequality then gives

$$|F(iy)| \leqslant \frac{1}{A_0}\|T(f_{iy})\|_{q_0}\|g_{iy}\|_{q_0'} \leqslant 1,$$

$$|F(1+iy)| \leqslant \frac{1}{A_1}\|T(f_{1+iy})\|_{q_1}\|g_{1+iy}\|_{q_1'} \leqslant 1.$$

If $\varepsilon > 0$ and $G(z) = e^{\varepsilon z^2}F(z)$, if $0 \leqslant x \leqslant 1$, $y \in \mathbb{R}$, then

$$|G(x+iy)| = e^{\varepsilon(x^2-y^2)}|F(z)| \leqslant e^{\varepsilon(1-y^2)}M,$$

$$|G(iy)| \leqslant e^{\varepsilon},$$

$$|G(1+iy)| \leqslant e^{\varepsilon}.$$

The first of these inequalities shows that there exists T such that:

$$|G(x+iy)| \leqslant e^{\varepsilon} \quad \text{if } 0 \leqslant x \leqslant 1,\ |y| \geqslant T.$$

On the perimeter of the rectangle with vertices $\pm iT, 1 \pm iT$ the modulus of G is bounded by e^{ε}, by the Maximum Principle this also holds over the whole of this rectangle, and consequently over the whole strip $0 \leqslant \mathrm{Re}(z) \leqslant 1$. In particular

$$|G(t)| = |e^{\varepsilon t^2}F(t)| \leqslant e^{\varepsilon}.$$

Making $\varepsilon \to 0$, because $f_t = f$, $g_t = g$ by Equation (1), yields

$$|F(t)| = \frac{1}{A_0^{1-t}A_1^t} \left|\int T(f)g\right| \leqslant 1.$$

By exercise 6·109 this proves that:

$$\|T(f)\|_q \leqslant A_0^{1-t}A_1^t$$

when $f \in E_n$ and $\|f\|_p = 1$. By homogeneity, for $f \in E_n$

$$\|T(f)\|_q \leqslant A_0^{1-t}A_1^t\|f\|_p.$$

Now examine the cases where $\alpha(1 - \beta) = 0$. In the first place, if $\alpha = 0$ one has, for example, $\alpha_0 = 0$; if $\alpha_1 > 0$ then $t = 0$, and there is nothing to prove; if $\alpha_0 = \alpha_1 = 0$ then

$$\|Tf\|_{q_0} < A_0\|f\|_\infty, \qquad \|Tf\|_{q_1} \leqslant A_1\|f\|_\infty.$$

As the function $\beta \mapsto \log\|f\|_{1/\beta}$ is convex on the interval with end-points β_0, β_1 (cf., Exercise 6·116),

$$\|Tf\|_q \leqslant (A_0\|f\|_\infty)^{1-t}(A_1\|f\|_\infty)^t = A_0^{1-t}A_1^t\|f\|_\infty.$$

Finally, when $\beta = 1$ one can restrict oneself to examining the case where $\beta_0 = \beta_1 = 1$ and $\alpha > 0$; the function:

$$F(z) = \frac{1}{A_0^{1-z}A_1^z}\int_{\mathbb{R}^m} T(f_z)g$$

is again entire and bounded on the strip $0 \leqslant \text{Re}(z) \leqslant 1$, and

$$|F(iy)| \leqslant \frac{1}{A_0}\|T(f_{iy})\|_1\|g\|_\infty \leqslant 1,$$

$$|F(1 + iy)| \leqslant \frac{1}{A_1} \|T(f_{1+iy})\|_1 \|g\|_\infty < 1,$$

and everything finishes as above.

Let $g \in L^q$, f be a step function, and set

$$T_g(f) = f * g.$$

If $\frac{1}{q} + \frac{1}{q'} = 1$ then

$$\|T_g(f)\|_\infty \leqslant \|g\|_q \|f\|_{q'},$$

$$\|T_g(f)\|_q \leqslant \|g\|_q \|f\|_1.$$

Then if

$$\frac{1}{r} = \frac{t}{q}, \qquad \frac{1}{p} = \frac{1 - t}{q'} + t, \qquad 0 \leqslant t \leqslant 1, \tag{3}$$

we have

$$\|T_g(f)\|_r = \|f * g\|_r \leqslant \|f\|_p \|g\|_q. \tag{4}$$

The relations (3) are equivalent to $\frac{1}{r} = \frac{1}{p} + \frac{1}{q} - 1 \geqslant 0$. Now assume that $f \in L^p$, $g \in L^q$, $f \geqslant 0$, $g \geqslant 0$, $1 \leqslant p < \infty$. There exists a sequence (f_n) of positive step functions such that

$$f_n \to f \text{ almost everywhere,}$$

$$\|f - f_n\|_p \to 0.$$

As the space L^r is complete one deduces from (4) the existence of a function $h \in L^r$ such that $f_n * g \to h$ in L^r. Also

$$\|h\|_r \leqslant \|f\|_p \|g\|_q.$$

By considering a subsequence, it can be assumed that

$$f_n * g \mapsto h \quad \text{almost everywhere.}$$

By Fatou's Lemma, for almost all x,

$$\int f(y)g(x - y)dy \leqslant \liminf \int f_n(y)g(x - y)dy = h(x) < \infty.$$

From this it follows that f and g are convolvable and that $f*g \leqslant h$ almost everywhere, whence

$$\|f*g\|_r \leqslant \|h\|_r \leqslant \|f\|_p \|g\|_q.$$

In the general case it is noted that f and g are convolvable if and only if $|f|$ and $|g|$ are, and in this case $|f*g| \leqslant |f|*|g|$. Finally, if $p = \infty$, one necessarily has $q = 1$, $r = \infty$, and in this case the result is classical.

EXERCISE 6·124: Let $X \subset \mathbb{R}^n$, $Y \subset \mathbb{R}^n$ be measurable sets such that $\text{meas}(X) > 0$, $\text{meas}(Y) > 0$. If f is a function on X and $\lambda > 0$, set

$$f_\lambda(x) = \begin{cases} f(x) & \text{if } |f(x)| \leqslant \lambda, \\ \lambda f(x)|f(x)|^{-1} & \text{if } |f(x)| \geqslant \lambda, \end{cases}$$

$$f^\lambda = f - f_\lambda.$$

Furthermore, let $1 \leqslant p_0 < p_1 \leqslant \infty$ and E be a vector subspace of $L^{p_0}(X) \cap L^{p_1}(X)$ such that $f \in E$ implies that $f_\lambda \in E$ for $\lambda > 0$. Finally, let T be a mapping of E into the set of measurable functions on Y such that

$$|T(f + g)| < |T(f)| + |T(g)| \tag{1}$$

for any $f \in E$, $g \in E$.

(a): Assume that $p_1 < \infty$ and that there exist numbers $A_0 > 0$, $A_1 > 0$ such that for all $f \in E$ and all $\lambda > 0$

$$\text{meas}(|Tf| > \lambda) \leqslant \left(\frac{A_i \|f\|_{p_i}}{\lambda}\right)^{p_i}, \quad i = 0,1. \tag{2}$$

Show that for all $p, p_0 < p < p_1$, there exists a constant A_p, that depends only upon p_0, p_1, p, A_0, A_1, such that

$$\|Tf\|_p \leqslant A_p \|f\|_p, \qquad f \in E. \tag{3}$$

(Use Exercise 5·98 and the decomposition $f = f^\lambda + f_\lambda$, and evaluate the quantities $\text{meas}(|f^\lambda| > t)$, $\text{meas}(|f_\lambda| > t)$).

(b): Show that (3) still holds when $p_1 = \infty$, if the inequality in (2) for $i = 1$ is replaced by

$$\|Tf\|_\infty \leqslant A_\infty \|f\|_\infty, \qquad f \in E. \tag{2'}$$

(c): Assume in addition that $\text{meas}(Y) < \infty$ and that if $f \in E$, $\lambda > 0$, and

$$\tilde{f}_\lambda(x) = \begin{cases} f(x) & \text{if } |f(x)| \leqslant \lambda, \\ 0 & \text{otherwise,} \end{cases}$$

then $\tilde{f}_\lambda \in E$. Set $\tilde{f}^\lambda = f - \tilde{f}_\lambda$.

Show that if $1 = p_0 < p_1 \leqslant \infty$ and (2) is satisfied (and (2') if $p_1 = \infty$), then if $0 < p < 1$ there exists a constant A_p such that

$$\|Tf\|_p \leqslant A_p \|f\|_1, \qquad f \in E. \tag{4}$$

(Evaluate the integrals:

$$\int_0^a p\lambda^{p-1}\text{meas}(|Tf| > \lambda)d\lambda, \qquad \int_a^\infty p\lambda^{p-1}\text{meas}(|Tf| > \lambda)d\lambda,$$

then make $a = \|f\|_1$).

(d): The hypotheses are as for Part (c), and assume further than meas$(X) < \infty$.

Show that there exist constants B, C such that

$$\|Tf\|_1 \leqslant B + C\int_X |f| \log^+|f|. \tag{5}$$

(Use the decomposition $f = \tilde{f}^{\sqrt{\lambda}} + \tilde{f}_{\sqrt{\lambda}}$, then evaluate the integrals:

$$\int_0^1 \text{meas}(|Tf| > \lambda)d\lambda, \qquad \int_1^\infty \text{meas}(|Tf| > \lambda)d\lambda.$$

Also note that there exists a constant M such that

$$u \leqslant M(1 + u\log^+ u)$$

for all real u).

NOTE: Formula (3) is a weakened form of a more general theorem of Marcinkiewicz (cf., for example, R.E. Edwards: *Fourier Series, Vol. II*, (Holt, Rinehart and Winston, Inc.), pp. 157 et seq.).

Δ∇Δ = ∇Δ∇ = ΔVΔ = ∇Δ∇ = Δ∇Δ

SOLUTION: (a): Let

$$\varphi(\lambda) = \text{meas}(|f| > \lambda), \qquad \psi(\lambda) = \text{meas}(|Tf| > \lambda).$$

Since $|Tf| \leqslant |Tf^\lambda| + |Tf_\lambda|$ by (1), it is clear that

$$(|Tf| > \lambda) \subset (|Tf^\lambda| > \tfrac{1}{2}\lambda) \cup (|Tf_\lambda| > \tfrac{1}{2}\lambda),$$

and consequently:

$$\psi(\lambda) \leqslant \text{meas}(|Tf^\lambda| > \tfrac{1}{2}\lambda) + \text{meas}(|Tf_\lambda| > \tfrac{1}{2}\lambda) \tag{6}$$

Furthermore, by (2)

$$\left.\begin{aligned} \text{meas}(|Tf^\lambda| > \tfrac{1}{2}\lambda) &\leqslant \left(\frac{2A_0\|f^\lambda\|_{p_0}}{\lambda}\right)^{p_0}, \\ \text{meas}(|Tf_\lambda| > \tfrac{1}{2}\lambda) &\leqslant \left(\frac{2A_1\|f_\lambda\|_{p_1}}{\lambda}\right)^{p_1}. \end{aligned}\right\} \tag{7}$$

From (6),(7) and Exercise 5·98 it follows that

$$\begin{aligned} \|Tf\|_p^p &= p\int_0^\infty \lambda^{p-1}\psi(\lambda)\mathrm{d}\lambda \\ &\leqslant p(2A_0)^{p_0}\int_0^\infty \lambda^{p-p_0-1}\mathrm{d}\lambda\int_X |f^\lambda(x)|^{p_0}\mathrm{d}x \\ &\quad + p(2A_1)^{p_1}\int_0^\infty \lambda^{p-p_1-1}\mathrm{d}\lambda\int_X |f_\lambda(x)|^{p_1}\mathrm{d}x. \end{aligned} \tag{8}$$

Now note that

$$(|f_\lambda| > t) = \begin{cases} (|f| > t) & \text{if } 0 < t < \lambda, \\ \varnothing & \text{if } t \geqslant \lambda, \end{cases}$$

and that $|f^\lambda(x)| > t > 0$ if and only if $|f(x)| > \lambda$, and

$$|f(x)|(1 - \lambda|f(x)|^{-1}) > t,$$

or in other words

$$(|f^\lambda| > t) = (|f| > t + \lambda).$$

Thus

$$\operatorname{meas}(|f^\lambda| > t) = \varphi(t + \lambda),$$

$$\operatorname{meas}(|f_\lambda| > t) = \begin{cases} \varphi(t) & \text{if } 0 < t < \lambda, \\ 0 & \text{if } t \geqslant \lambda. \end{cases} \tag{9}$$

Using Exercise 5·98 again to evaluate the integrals $\int |f^\lambda|^{p_0}$ and $\int |f_\lambda|^{p_1}$, it follows from (8) and (9) that

$$\|Tf\|_p^p \leqslant pp_0(2A_0)^{p_0}\int_0^\infty \lambda^{p-p_0-1}\,d\lambda\int_0^\infty t^{p_0-1}\varphi(t+\lambda)dt$$

$$+\, pp_1(2A_1)^{p_1}\int_0^\infty \lambda^{p-p_1-1}\,d\lambda\int_0^\lambda t^{p_1-1}\varphi(t)dt. \tag{10}$$

Since $p - p_0 - 1 > -1$ and $t - \lambda \leqslant t$,

$$\int_0^\infty \lambda^{p-p_0-1}\,d\lambda\int_0^\infty t^{p_0-1}\varphi(t+\lambda)dt \leqslant \int_0^\infty \lambda^{p-p_0-1}\,d\lambda\int_\lambda^\infty t^{p_0-1}\varphi(t)dt$$

$$= \int_0^\infty t^{p_0-1}\varphi(t)dt\int_0^t \lambda^{p-p_0-1}\,d\lambda$$

$$= \frac{1}{p-p_0}\int_0^\infty t^{p-1}\varphi(t)dt$$

$$= \frac{1}{p(p-p_0)}\,\|f\|_p^p. \tag{11}$$

Similarly, because $p - p_1 - 1 < -1$,

$$\int_0^\infty \lambda^{p-p_1-1}\,d\lambda\int_0^\lambda t^{p_1-1}\varphi(t)dt = \int_0^\infty t^{p_1-1}\varphi(t)dt\int_t^\infty \lambda^{p-p_1-1}\,d\lambda =$$

$$= \frac{1}{p_1 - p}\int_0^\infty t^{p-1}\varphi(t)dt$$

$$= \frac{1}{p(p_1 - p)}\,\|f\|_p^p. \tag{12}$$

(10),(11) and (12) yield

$$\|Tf\|_p^p \leqslant \left(\frac{p_0(2A_0)^{p_0}}{p - p_0} + \frac{p_1(2A_1)^{p_1}}{p_1 - p}\right)\|f\|_p^p,$$

which proves (3).

SOLUTION: (b): When $p_1 = \infty$ and (2') holds, write

$$\text{meas}(|Tf| > \lambda) \leqslant \text{meas}(|Tf^{\lambda/2A_\infty}| > \tfrac{1}{2}\lambda) + \text{meas}(|Tf_{\lambda/2A_\infty}| > \tfrac{1}{2}\lambda).$$

Now the set $(|Tf_{\frac{1}{2}\lambda A_\infty}| > \lambda/2)$ has measure zero, since almost every-

$$|Tf_{\lambda/2A_\infty}| \leqslant A_\infty\|f_{\lambda/2A_\infty}\|_\infty \leqslant A_\infty\frac{\lambda}{2A_\infty} = \frac{\lambda}{2}..$$

From this it follows that

$$\|Tf\|_p^p \leqslant p\int_0^\infty \lambda^{p-1}\text{meas}(|Tf^{\lambda/2A_\infty}| > \tfrac{1}{2}\lambda)d\lambda$$

$$\leqslant p_0 p(2A_0)^{p_0}\int_0^\infty \lambda^{p-p_0-1}d\lambda\int_0^\infty t^{p_0-1}\text{meas}(|f^{\lambda/2A_\infty}| > t)dt$$

$$= p_0 p(2A_0)^{p_0}\int_0^\infty \lambda^{p-p_0-1}d\lambda\int_0^\infty t^{p_0-1}\varphi\left(t + \frac{\lambda}{2A_\infty}\right)dt \leqslant$$

(Contd)

(Contd) $$\leqslant p_0 p(2A_0)^{p_0}\int_0^\infty \lambda^{p-p_0-1}\,d\lambda\int_{\lambda/2A_\infty}^\infty t^{p_0-1}\varphi(t)dt$$

$$= p_0 p(2A_0)^{p_0}\int_0^\infty t^{p_0-1}\varphi(t)dt\int_0^{2A_\infty t}\lambda^{p-p_0-1}\,d\lambda$$

$$= \frac{p_0 p(2A_0)^{p_0}(2A_\infty)^{p-p_0}}{p-p_0}\int_0^\infty t^{p-1}\varphi(t)dt$$

$$= \frac{p_0(2A_0)^{p_0}(2A_\infty)^{p-p_0}}{p-p_0}\,\|f\|_p^p.$$

SOLUTION: (c): It can be assumed that $a = \|f\|_1 > 0$. It can also be assumed that $1 = p_0 < p_1 < \infty$; indeed, if $p_1 = \infty$, and if $p_0 < p_1' < \infty$, by part (b)

$$\text{meas}(|Tf| > \lambda) \leqslant \frac{1}{\lambda^{p_1'}}\int_{|Tf|>\lambda}|Tf|^{p_1'}$$

$$\leqslant \frac{\|Tf\|_{p_1'}^{p_1'}}{\lambda^{p_1'}} \leqslant \left(\frac{A_{p_1'}\|f\|_{p_1'}}{\lambda}\right)^{p_1'}.$$

This being so,

$$\int_0^a p\lambda^{p-1}\psi(\lambda)d\lambda \leqslant \text{meas}(Y)\int_0^a p\lambda^{p-1}d\lambda = a^p\text{meas}(Y). \tag{13}$$

On the other hand

$$\psi(\lambda) \leqslant \text{meas}(|T\tilde{f}^\lambda| > \tfrac{1}{2}\lambda) + \text{meas}(|T\tilde{f}_\lambda| > \tfrac{1}{2}\lambda) \leqslant$$

$$\leqslant \frac{2A_0\|f^\lambda\|_1}{\lambda} + \left(\frac{2A_1\|\tilde{f}_\lambda\|_{p_1}}{\lambda}\right)^{p_1},$$

whence:

$$\int_a^\infty p\lambda^{p-1}\psi(\lambda)d\lambda \leqslant 2pA_0\int_a^\infty \lambda^{p-2}d\lambda\int_{|f(x)|>\lambda}|f(x)|dx$$

$$+ p(2A_1)^{p_1}\int_a^\infty \lambda^{p-p_1-1}d\lambda\int_{|f(x)|\leqslant\lambda}|f(x)|^{p_1}dx$$

In the last integral $|f(x)|^{p_1} \leqslant \lambda^{p_1-1}|f(x)|$, and consequently

$$\int_a^\infty p\lambda^{p-1}\psi(\lambda)d\lambda < 2pA_0\int_X|f(x)|dx\int_{a<\lambda<|f(x)|}\lambda^{p-2}d\lambda$$

$$+ p(2A_1)^{p_1}\|f\|_1\int_a^\infty\lambda^{p-2}d\lambda$$

$$= \frac{2pA_0}{1-p}\int_X|f(x)|(a^{p-1} - |f(x)|^{p-1})_+dx$$

$$+ \frac{p(2A_1)^{p_1}}{1-p}a^{p-1}\|f\|_1.$$

Since $(a^{p-1} - |f(x)|^{p-1})_+ \leqslant a^{p-1}$, and since $a = \|f\|_1$, this yields

$$\int_a^\infty p\lambda^{p-1}\psi(\lambda)d\lambda \leqslant \frac{p[2A_0 + (2A_1)^{p_1}]}{1-p}\|f\|_1^p, \tag{14}$$

and, taking (13),(14) into account,

$$\|Tf\|_p^p \leqslant \left\{\text{meas}(Y) + \frac{p[2A_0 + (2A_1)^{p_1}]}{1-p}\right\}\|f\|_1^p.$$

SOLUTION: (d): On one hand

$$\int_0^1 \psi(\lambda)\,d\lambda \leqslant \text{meas}(Y), \tag{15}$$

and on the other

$$\psi(\lambda) \leqslant \text{meas}(|T\tilde{f}^{\sqrt{\lambda}}| > \tfrac{1}{2}\lambda) + \text{meas}(|T\tilde{f}_{\sqrt{\lambda}}| > \tfrac{1}{2}\lambda)$$

$$\leqslant \frac{2A_0\|\tilde{f}^{\sqrt{\lambda}}\|_1}{\lambda} + \left(\frac{2A_1\|\tilde{f}_{\sqrt{\lambda}}\|_{p_1}}{\lambda}\right)^{p_1},$$

so

$$\int_1^\infty \psi(\lambda)\,d\lambda \leqslant 2A_0\int_1^\infty \frac{d\lambda}{\lambda}\int_{|f(x)|>\sqrt{\lambda}} |f(x)|\,dx$$

$$+ (2A_1)^{p_1}\int_1^\infty \frac{d\lambda}{\lambda^{p_1}}\int_{|f(x)|\leqslant\sqrt{\lambda}} |f(x)|^{p_1}\,dx.$$

In the last integral $|f(x)|^{p_1} \leqslant \lambda^{(p_1-1)/2}|f(x)|$, whence

$$\int_1^\infty \psi(\lambda)\,d\lambda \leqslant 2A_0\int_X |f(x)|\,dx\int_{1<\lambda<|f(x)|^2} \lambda^{-1}\,d\lambda$$

$$+ (2A_1)^{p_1}\|f\|_1\int_1^\infty \lambda^{-(p_1+1)/2}\,d\lambda$$

$$= 2A_0\int_X |f(x)|\log^+|f(x)|^2\,dx + \frac{2(2A_1)^{p_1}}{p_1 - 1}\|f\|_1. \tag{16}$$

If M is such that $u \leqslant M(1 + u\log^+ u)$, then

$$\|f\|_1 < M(\text{meas}(X) + \int_X |f|\log^+|f|). \tag{17}$$

Furthermore, since $\log^+ u^2 - 2\log^+ u$ it follows from (15),(16) and (17) that

$$\|Tf\|_1 \leqslant B + C\int_X |f|\log^+|f|,$$

whence:

$$B = \text{meas}(Y) + \frac{2M(2A_1)^{p_1}}{p_1 - 1}\,\text{meas}(X),$$

$$C = 4A_0 + \frac{2M(2A_1)^{p_1}}{p_1 - 1}.$$

EXERCISE 6·125: Set

$$u = \begin{cases} 1 & \text{if } x \leqslant e, \\ \log x & \text{if } x \geqslant e, \end{cases}$$

(a): If f and g are two positive measurable functions such that

$$\int f^2 u(f) < \infty, \qquad \int \frac{g^2}{u(g)} < \infty,$$

show that fg is integrable.

(b): Assume that f is as above and that (g_n) is a sequence of positive measurable functions such that:

$$\lim_n \int \frac{g_n^2}{u(g_n)} = 0.$$

Show that

$$\lim_n \int f g_n = 0.$$

(c): Now assume that g is as in part (a), and that

$$\lim_n \int f_n^2 u(f_n) = 0.$$

Show that

$$\lim_n \int f_n g = 0.$$

Δ∇Δ = ∇Δ∇ = Δ∇Δ = ∇Δ∇ = Δ∇Δ

SOLUTION: (a): Let us show that for $x \geqslant 0$, $y \geqslant 0$,

$$xy \leqslant 2x^2 u(x) + \frac{y^2}{u(y)}, \tag{1}$$

which will show that fg is integrable if $f^2 u(f)$ and $g^2/u(g)$ are. If $x \leqslant y/u(y)$, then $xy \leqslant y^2/u(y)$, and (1) holds. Assume, therefore, that $y \leqslant xu(y)$. If $y \leqslant e$ then $y \leqslant x$, and consequently $xy \leqslant x^2 \leqslant x^2 u(x)$; if $y \geqslant e$, then noticing that $\log u \leqslant u/e$, one obtains

$$\sqrt{y} \leqslant \frac{2}{e} \frac{y}{\log y} \leqslant \frac{2}{e} x \leqslant x,$$

and consequently $\log y \leqslant 2\log x$, whence:

$$xy \leqslant x^2 \log y \leqslant 2x^2 \log x \leqslant 2x^2 u(x).$$

SOLUTION: (b): Let $0 < r < 1$. By replacing x by rf and y by g_n in (1), integrating and noting that $u(rf) \leq u(f)$, one obtains

$$\int fg_n \leq 2r\int f^2u(f) + \frac{1}{r}\int\frac{g_n^2}{u(g_n)} ,$$

so

$$\limsup_n \int fg_n \leq 2r\int f^2u(f),$$

and on making $r \to 0$

$$\lim_n \int fg_n = 0.$$

SOLUTION: (c): Now replace x by f_n and y by $r^{-1}g$ in (1). This yields

$$\int f_n g \leq 2r\int f_n^2u(f_n) + \frac{1}{r}\int\frac{g^2}{u(r^{-1}g)} ,$$

so

$$\limsup_n \int f_n g \leq \frac{1}{r}\int\frac{g^2}{u(r^{-1}g)} . \tag{2}$$

If $r > 1$

$$\int\frac{g^2}{u(r^{-1}g)} = \int_{g\leq re} \frac{g^2}{u(g)}\,u(g) + \int_{g>re} \frac{g^2}{u(g)}\,\frac{\log g}{\log g - \log r}$$

$$\leq \log(re)\int\frac{g^2}{u(g)} ,$$

and consequently:

$$\lim_{r\to\infty} \frac{1}{r}\int \frac{g^2}{u(r^{-1}g)} = 0,$$

which proves, because of (2), that

$$\lim_n \int f_n g = 0.$$

CHAPTER 7

The Space L^2

EXERCISE 7·126: Let X be a set of $\mathbb{R}^p$ such that meas$(X) = 1$ and $A_1,\ldots,A_n$ a finite sequence of measurable sets of X such that:

$$\text{meas}(A_i) \geqslant c > 0 \quad \text{for all } i.$$

Show that if

$$nc \geqslant 1, \qquad \varepsilon \geqslant \frac{c(1-c)}{n-1},$$

there exist two indices i,j $(i < j)$ such that

$$\text{meas}(A_i \cap A_j) \geqslant c^2 - \varepsilon.$$

Δ∇Δ = ∇Δ∇ = ΔVΔ = ∇Δ∇ = ΔVΔ

SOLUTION: Applying the Cauchy-Schwarz inequality to the function

$$\sum_{i=1}^{n} \mathbb{1}_{A_i},$$

and setting

$$\lambda = \sum_{i=1}^{n} \text{meas}(A_i)$$

yields

$$\lambda^2 \leqslant \lambda + 2 \sum_{i<j} \text{meas}(A_i \cap A_j).$$

If one had

$$\text{meas}(A_i \cap A_j) < c^2 - \varepsilon$$

for every pair $i \neq j$, taking into account that $\lambda^2 - \lambda$ is increasing for $\lambda \geqslant 1$, and that $\lambda \geqslant nc \geqslant 1$, one would have

$$n^2c^2 - nc \leqslant \lambda^2 - \lambda < 2 \cdot \frac{n(n-1)}{2}(c^2 - \varepsilon)$$

so

$$\varepsilon < \frac{c(1-c)}{n-1},$$

contrary to the hypothesis.

EXERCISE 7·127: Let f be a piecewise continuously differentiable function on the interval $[0,a]$.

Show that if $f(0) = 0$, then

$$\int_0^a |f'(t)f(t)|\,dt \leqslant \frac{a}{2}\int_0^a |f'(t)|^2 dt.$$

When does equality hold?

Δ∇Δ = ∇Δ∇ = Δ∇Δ = ∇Δ∇ = Δ∇Δ

SOLUTION: Let

$$u(t) = \int_0^t |f'(s)|\,ds.$$

Now,

$$|f(t)| \leqslant u(t) \quad \text{and} \quad u'(t) = |f'(t)|.$$

Consequently,

$$\int_0^a |f'(t)f(t)|\,dt \leqslant \int_0^a u'(t)u(t)\,dt = \tfrac{1}{2}u^2(a).$$

Furthermore,

$$u^2(a) = \left(\int_0^a |f'(t)|\,dt\right)^2$$

$$\leqslant \int_0^a dt \int_0^a |f'(t)|^2\,dt = a\int_0^a |f'(t)|^2\,dt,$$

whence the result.

In order to have equality it is necessary first of all that $|f| = u$, that is to say

$$\left|\int_0^t f'(s)\,ds\right| = \int_0^t |f'(s)|\,ds,$$

which implies that $f' = e^{i\alpha}|f'|$, where α is a real constant. Next, in order that the Cauchy-Schwarz inequality become an equality, it is necessary that $|f'|$ be constant. Taking into account that $f(0) = 0$, it follows that $f(t) = \lambda t$, $\lambda \in \mathbb{C}$.

EXERCISE 7·128: Let X be a measurable set of $\mathbb{R}^p$ such that $\text{meas}(X) < \infty$ and (f_n) a sequence of measurable functions on X such that

$$|f_n| \leqslant M, \qquad \int_X |f_n|^2 = 1 \quad \text{for all } n.$$

(a): Show that if a sequence of complex numbers (a_n) is such that the series $\sum_n a_n f_n(x)$ converges for almost all $x \in X$, then $\lim_{n\to\infty} a_n = 0$.

(b): Show that there exists a sequence (b_n) of complex numbers such that

$$\lim_{n\to\infty} b_n = 0 \quad \text{and} \quad \sum_n |b_n f_n(x)|^2 = \infty$$

for all x's belonging to a subset of X having strictly positive measure.

ΔVΔ — VΔV — ΔVΔ — VΔV — ΔVΔ

SOLUTION: We shall assume that $\text{meas}(X) = 1$.

SOLUTION (a): If the series $\sum a_n f_n(x)$ converges almost everywhere, in particular one has $a_n f_n(x) \to 0$ almost everywhere. Let $\delta > 0$ be such that $1 - M^2\delta > 0$. By Egoroff's theorem there exists a set A of X such that

$$\text{meas}(X - A) < \delta \quad \text{and} \quad \varepsilon_n = \sup_A |a_n f_n(x)| \to 0.$$

Then

$$|a_n|^2 = \int_X |a_n f_n|^2 = \int_A |a_n f_n|^2 + \int_{X-A} |a_n f_n|^2 \leqslant \qquad \text{(Contd)}$$

(Contd) $\leqslant \varepsilon_n^2 + M^2|a_n|^2\delta,$

so

$$|a_n|^2 \leqslant \frac{\varepsilon_n^2}{1 - M^2\delta},$$

which proves that $a_n \to 0$.

SOLUTION (b): Choose a sequence (b_n) such that $b_n \to 0$ and $\sum |b_n|^2 = \infty$ (for example, $b_n = 1/\sqrt{n}$). Assume that $\sum |b_n f_n(x)|^2 < \infty$ almost everywhere. Choosing δ as in part (a), there exists a set A such that $\text{meas}(X - A) < \delta$, and

$$\sum_{n \geqslant n_0} |b_n f_n(x)|^2 \leqslant 1$$

for all $x \in A$ whenever n_0 is large enough (Egoroff's Theorem). Now, for every n.

$$1 = \int_A |f_n|^2 + \int_{X-A} |f_n|^2$$

$$\leqslant \int_A |f_n|^2 + M^2\delta,$$

so

$$1 \geqslant \text{meas}(A) \geqslant \sum_{n \geqslant n_0} |b_n|^2 \int_A |f_n|^2 \geqslant (1 - M^2\delta) \sum_{n \geqslant n_0} |b_n|^2,$$

which is absurd.

EXERCISE 7·129: Let $\mathcal{U}$ be an open set of $\mathbb{R}^p$ and $H = L^2(\mathcal{U})$. Recall that the Gram determinant associated with elements $f_1,\ldots,$ f_n of the Hilbert space $L^2(\mathcal{U})$ is defined by

$$G(f_1,\ldots,f_n) = \det\|(f_i|f_j)\|_{1\leqslant i,j\leqslant n}.$$

Prove that:

$$G(f_1,\ldots,f_n) = \frac{1}{n!}\int_{\mathcal{U}^n} |\det\|f_i(x_j)\|_{1\leqslant i,j\leqslant n}|^2 dx_1 dx_2 \ldots dx_n.$$

ΔVΔ = VΔV = ΔVΔ = VΔV = ΔVΔ

SOLUTION: We have

$$|\det\|f_i(x_j)\||^2 = \sum_{\tau,\sigma} \varepsilon(\tau\sigma) \prod_{i=1}^{n} f_{\tau(i)}(x_i)\overline{f_{\sigma(i)}(x_i)},$$

where τ and σ run over the permutation group of $\{1,2,\ldots,n\}$ and ε denotes the signature. On setting

$$a_{ij} = (f_i|f_j) = \int_{\mathcal{U}} f_i\bar{f}_j,$$

the right side of the formula to be proved becomes:

$$\frac{1}{n!}\sum_{\tau,\sigma} \varepsilon(\tau\sigma)\prod_{i=1}^{n} a_{\tau(i),\sigma(i)} = \frac{1}{n!}\sum_{\tau,\sigma} \varepsilon(\tau\sigma^2)\prod_{i=1}^{n} a_{\tau\sigma(i),\sigma(i)}$$

$$= \frac{1}{n!}\sum_{\sigma}\sum_{\tau} \varepsilon(\tau)\prod_{i=1}^{n} a_{\tau(i),i}$$

$$= \det\|a_{ij}\|.$$

EXERCISE 7·130: Let K be a function defined on $\mathbb{R}^2$ that is measurable and such that

$$A^2 = \iint_{\mathbb{R}^2} |K(x,y)|^2 dx dy < \infty.$$

For every function $f \in L^2(\mathbb{R})$ the function Tf is defined by

$$Tf(x) = \int K(x,y) f(x) dy. \tag{*}$$

Show that this defines a linear transformation of $L^2(\mathbb{R})$ into itself, and that

$$\|Tf\|_2 \leqslant A \|f\|_2. \tag{**}$$

Show next that if T' is defined analogously with kernel K', then $T'T$ has kernel K'', where

$$K''(x,y) = \int K'(x,z) K(z,y) dz.$$

$$\Delta\nabla\Delta = \nabla\Delta\nabla = \Delta\nabla\Delta = \nabla\Delta\nabla = \Delta\nabla\Delta$$

SOLUTION: By Fubini's theorem, for almost all x the function $y \mapsto K(x,y)$ is square summable, and consequently $Tf(x)$ is defined. By the Cauchy-Schwarz inequality

$$|Tf(x)|^2 \leqslant \|f\|_2 \int |K(x,y)|^2 dy.$$

Integrating with respect to x yields (**). Nevertheless, it remains to prove that Tf is measurable. Let E be a subset of $\mathbb{R}$ with finite measure. Then

$$\int dx \int |1\!\!1_E(x) K(x,y) f(y)| dy \leqslant \|f\|_2 \int_E dx \left(\int |K(x,y)|^2 dy\right)^{\frac{1}{2}} < \infty,$$

because by Fubini's Theorem the function

$$x \mapsto \left(\int |K(x,y)|^2 dy\right)^{\frac{1}{2}}$$

is square integrable. The same Theorem implies that the function

$$x \mapsto 1\!\!1_E(x)Tf(x) = \int 1\!\!1_E(x)K(x,y)f(y)dy$$

is measurable. From this it follows that Tf is measurable.

The last formula results from the following calculations:

$$T'Tf(x) = \int K'(x,z)dz \int K(z,y)f(y)dy$$

$$= \int f(y)dy \int K'(x,z)K(z,y)dz,$$

and

$$\int |K''(x,y)|^2 dx dy \leq \int dx dy \left(\int |K'(x,z)|^2 dz\right)\left(\int |K(z,y)|^2 dz\right)$$

$$= \int |K'(x,z)|^2 dx dz \int |K(z,y)|^2 dz dy.$$

EXERCISE 7·131: If $f \in L^2(0,1)$, for $0 \leq x \leq 1$ set

$$Wf(x) = f(x) + 2\int_0^x e^{x-t} f(t)dt. \qquad (*)$$

Also let

$$\varphi_1(x) = e^{-x}, \qquad \varphi_2(x) = e^{x}.$$

Show that (*) defines a linear mapping W of $L^2(0,1)$ into itself, such that:

(a): $W\varphi_1 = \varphi_2$;

(b): If $(f|\varphi_1) = 0$, then $(Wf|\varphi_2) = 0$ and $\|Wf\|_2 = \|f\|_2$;

(c): If $(g|\varphi_2) = 0$ there exists $f \in L^2(0,1)$ such that:

$(f|\varphi_1) = 0$ and $g = Wf$.

Deduce from this that W is a bijection.

Δ∇Δ = ∇Δ∇ = Δ∇Δ = ∇Δ∇ = Δ∇Δ

SOLUTION (a): We have

$$W\varphi_1(x) = e^{-x} + 2\int_0^x e^{x-2t}dt$$

$$= e^{-x} - [e^{x-2t}]_{t=0}^{t=x} = e^x = \varphi_2(x).$$

SOLUTION (b): Set $W = 1 + U$, and determine U^*:

$$(U^*f|h) = (f|Uh) = 2\int_0^1 f(x)dx\int_0^x e^{x-t}\overline{h(t)}dt$$

$$= 2\int_0^1 \overline{h(t)}dt\int_t^1 e^{x-t}f(x)dx.$$

Hence

$$U^*f(x) = 2\int_x^1 e^{t-x}f(t)dt.$$

Now calculate U^*Uf when

$$(f|\varphi_1) = \int_0^1 e^{-t} f(t)dt = 0: \qquad (**)$$

$$U^*Uf(x) = 4\int_x^1 e^{t-x}dt\int_0^t e^{t-u}f(u)du$$

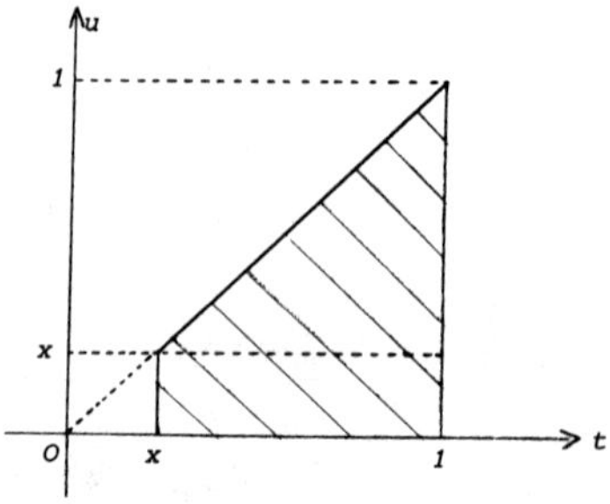

$$= 4\int_0^x e^{-(u+x)}f(u)du$$

$$\times \int_x^1 e^{2t}dt$$

$$+ 4\int_x^1 e^{-(u+x)}f(u)du\int_u^1 e^{2t}dt$$

$$= 2e^{2-x}\int_0^1 e^{-u}f(u)du$$

$$- 2\int_0^x e^{x-u}f(u)du - 2\int_x^1 e^{u-x}f(u)du$$

$$= - Uf(x) - U^*f(x).$$

Thus, whenever $(f|\varphi_1) = 0$,

$$W^*Wf = (1 + U + U^* + U^*U)f = f.$$

From this it follows that

$$(Wf|\varphi_2) = (Wf, W\varphi_1) = (W^*Wf|\varphi_1) = (f|\varphi_1) = 0,$$

and

$$\|Wf\|_2^2 = (Wf|Wf) = (W^*Wf|f) = (f|f) = \|f\|_2^2.$$

SOLUTION (c): Now assume that $(g|\varphi_2) = 0$. Then

$$UU^*g(x) = 4\int_0^x e^{x-t}dt \times \int_t^1 e^{u-t} g(u)du$$

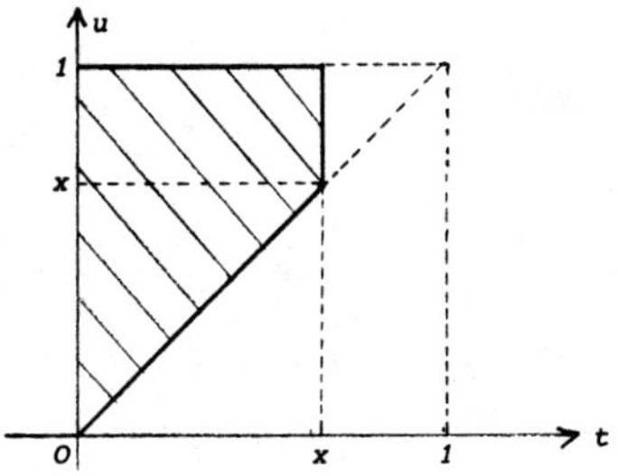

$$= 4\int_0^x e^{x+u} g(u)du \times \int_0^u e^{-2t}dt$$

$$+ 4\int_x^1 e^{x+u} g(u)du \int_0^x e^{-2t}dt$$

$$= 2e^x\int_0^1 e^u g(u)du - 2\int_0^x e^{x-u}g(u)du - 2\int_x^1 e^{u-x}g(u)du$$

$$= - Ug(x) - U^*g(x).$$

Consequently $WW^*g = g$ when $(g|\varphi_2) = 0$. Setting $f = W^*g$, one has $g = Wf$, and furthermore

$$(f|\varphi_1) = (W^*g|\varphi_1) = (g,W\varphi_1) = (g|\varphi_2) = 0.$$

Thus W is an isometry of $\{\varphi_1\}^\perp$ onto $\{\varphi_2\}^\perp$. From this it certainly follows that W is bijective. In fact, if

$$(f|\varphi_1) = 0 \quad \text{and} \quad W(f + \lambda\varphi_1) = Wf + \lambda\varphi_2 = 0,$$

then

$$Wf = 0 \quad \text{and} \quad \lambda = 0,$$

and consequently

$$f + \lambda\varphi_1 = 0$$

because

$$\|f\|_2 = \|Wf\|_2 = 0.$$

Note that if $g \in L^2(0,1)$ then

$$g = g - \frac{(g|\varphi_2)}{(\varphi_2|\varphi_2)}\varphi_2 + \frac{(g|\varphi_2)}{(\varphi_2|\varphi_2)}\varphi_2.$$

Setting

$$f = W^*\left(g - \frac{(g|\varphi_2)}{(\varphi_2|\varphi_2)}\varphi_2\right) + \frac{(g|\varphi_2)}{(\varphi_2|\varphi_2)}\varphi_1,$$

we have $Wf = g$. Notice also that $W^*\varphi_2 = \lambda\varphi_1$ and

$$\lambda(\varphi_1|\varphi_1) = (W^*\varphi_2|\varphi_1) = (\varphi_2|W\varphi_1) = (\varphi_2|\varphi_2),$$

so that

$$f = W^*g - \left(\frac{1}{(\varphi_1|\varphi_1)} - \frac{1}{(\varphi_2|\varphi_2)}\right)(g|\varphi_2)\varphi_1.$$

Now,

$$(\varphi_1|\varphi_1) = \frac{e^2 - 1}{2e^2}, \qquad (\varphi_2|\varphi_2) = \frac{e^2 - 1}{2},$$

whence

$$f = W^*g - 2(g|\varphi_2)\varphi_1.$$

This relation may be further written as

$$f(x) = g(x) - 2\int_0^x e^{t-x} g(t)\,dt,$$

which is the "inversion formula" of:

$$g(x) = f(x) + 2\int_0^x e^{x-t} f(t)\,dt.$$

EXERCISE 7·132: Let f be a continuously differentiable function on $\mathbb{R}$ such that

$$\int_{-\infty}^{+\infty} x^2 |f(x)|^2 dx < \infty, \qquad \int_{-\infty}^{+\infty} |f'(x)|^2 dx < \infty.$$

(a): Show that for $x \geqslant 0$

$$x|f(x)|^2 \leqslant 4\left(\int_x^\infty t^2 |f(t)|^2 dt\right)^{\frac{1}{2}} \left(\int_x^\infty |f'(t)|^2 dt\right)^{\frac{1}{2}}.$$

(b): From this deduce that

$$\int_{-\infty}^{+\infty} |f(x)|^2 dx \leqslant 2\left(\int_{-}^{+} x^2 |f(x)|^2 dx\right)^{\frac{1}{2}} \left(\int_{-\infty}^{+\infty} |f'(x)|^2 dx\right)^{\frac{1}{2}}.$$

(This inequality is Heisenberg's Uncertainty Principle in Quantum Mechanics).

Δ∇Δ = ∇Δ∇ = Δ∇Δ = ∇Δ∇ = Δ∇Δ

SOLUTION (a): If $x \geqslant 0$ let $f_0 = f(x)$, and

$$\alpha^2 = \int_x^\infty t^2 |f(t)|^2 dt, \qquad \beta^2 = \int_x^\infty |f'(t)|^2 dt.$$

For all $t \geqslant 0$,

$$f(x+t) - f(x) = \int_x^{x+t} f'(u)\mathrm{d}u,$$

so, by the Cauchy-Schwarz inequality,

$$|f(x+t) - f(x)| \leqslant \beta\sqrt{t}.$$

Assume that $0 \leqslant h \leqslant f_0^2/\beta^2$, so that for $0 \leqslant t \leqslant h$

$$|f(x+t)| \geqslant f_0 - \beta\sqrt{h} \geqslant 0.$$

Then it is clear that

$$\alpha \geqslant x\sqrt{h}(f_0 - \beta\sqrt{h})$$

Choose the value of $\sqrt{h}$ that maximizes the right side of this inequality, i.e. $\sqrt{h} = f_0/2\beta$. This yields

$$xf_0^2 \leqslant 4\alpha\beta,$$

which is the desired inequality.

SOLUTION (b): From the first part (since an analogous inequality holds when $x \leqslant 0$) it follows that

$$\lim_{|x|\to\infty} x|f(x)|^2 = 0.$$

But then, by integrating by parts

$$\int_{-\infty}^{+\infty} |f(x)|^2\mathrm{d}x = -\int_{-\infty}^{+\infty} x(f'(x)\overline{f(x)} + f(x)\overline{f'(x)})\mathrm{d}x,$$

whence, using the Cauchy-Schwarz inequality,

$$\int_{-\infty}^{+\infty} |f(x)|^2 dx \leqslant 2\int_{-\infty}^{+\infty} x|f||f'|$$

$$\leqslant 2\left(\int_{-\infty}^{+\infty} x^2|f|^2\right)^{\frac{1}{2}}\left(\int_{-\infty}^{+\infty} |f'|^2\right)^{\frac{1}{2}}.$$

EXERCISE 7·133: (I): Let X be a measurable set of $\mathbb{R}^p$ such that $0 < \text{meas}(X) < \infty$, and $(f_n)_{n\geqslant 1}$ an orthonormal sequence of $L^2(X)$. For every integer $N \geqslant 1$ set

$$K_N(x,y) = \sum_{n=1}^{N} f_n(x)\overline{f_n(y)},$$

$$L_N(x) = \int_X |K_N(x,y)|\,dy.$$

(a): Assume that all the functions f_n are real and that there exists a constant A such that $|L_N(x)| \leqslant A$ for every integer $N \geqslant 1$ and all $x \in X$.

Show that if $(a_n)_{n\geqslant 1}$ is a sequence of complex numbers such that $\sum |a_n|^2 < \infty$, and that if

$$s_N(x) = \sum_{n=1}^{N} a_n f_n(x),$$

then for almost all $x \in X$

$$\sup_N |s_N(x)| < \infty.$$

(Assume first that the a_n's are real and introduce a measurable mapping ν of X into $\{1,2,\ldots,N\}$ such that

$$s_N^*(x) = \max_{n \leqslant N} s_n(x) = s_{\nu(x)}(x),$$

and then show that

$$\int_X s_N^*(x)dx$$

is bounded by a constant).

(b): Take the same hypotheses and notations as in part (a). Show that if

$$f = \sum_{n=1}^{\infty} a_n f_n$$

in the sense of convergence in $L^2(X)$, then for almost all $x \in X$

$$f(x) = \sum_{n=1}^{\infty} a_n f_n(x).$$

(Introduce an increasing sequence $(\omega_n)_{n\geqslant 1}$ of numbers greater than zero, such that $\omega_n \to \infty$ and $\sum \omega_n^2 |a_n|^2 < \infty$, and a sequence of integers $N_1 < N_2 < \cdots < N_k < \cdots$ such that $s_{N_k}(x) \to f(x)$ almost everywhere; next bound $|s_N(x) - s_{N_k}(x)|$ when $N_k \leqslant N < N_{k+1}$, using part (a) applied to $\omega_n a_n$).

(c): Assume that $X = [0,1]$, and that for every function f continuous on X

$$\lim_N \sum_{n=1}^{N} (f|f_n)f_n(x) = f(x)$$

uniformly for $x \in X$.

Show that there exists a constant A such that $|L_N(x)| \leqslant A$ for all N and all x.

(II): We call the *HAAR SYSTEM* the sequence of functions $(f_n)_{n\geqslant 0}$ defined on $X = [0,1]$ in the following way: $f_0 = 1$, and if $n \geqslant 1$ and $n = 2^m + k$, $0 \leqslant k < 2^m$, $0 \leqslant x < 1$,

$$f_n(x) = \begin{cases} 2^{m/2} & \text{if } \dfrac{2k}{2^{m+1}} \leqslant x < \dfrac{2k+1}{2^{m+1}}, \\[2ex] -2^{m/2} & \text{if } \dfrac{2k+1}{2^{m+1}} \leqslant x < \dfrac{2k+2}{2^{m+1}}, \\[2ex] 0 & \text{otherwise} \end{cases}$$

Assume in addition that

$$f_n(1) = f_n(1-0).$$

(a): Show that $(f_n)_{n\geqslant 0}$ is an orthonormal system in $L^2(X)$.

(b): Show that for all $n \geqslant 0$ there exists a sequence:

$$0 = x_0^n < x_1^n < \cdots < x_{n+1}^n = 1$$

which possesses the following property: In order that a function f belongs to the vector subspace V_n generated by $f_0, f_1, \ldots, f_n$, it is necessary and sufficient that f is constant on each of the intervals $[x_i^n, x_{i+1}^n[$, $0 \leqslant i \leqslant n$, and that $f(1) = f(1-0)$.

From this deduce that if $f \in L^2(X)$, its orthogonal projection on V_n is, on each interval $[x_i^n, x_{i+1}^n[$, equal to:

$$\frac{1}{x_{i+1}^n - x_i^n}\int_{x_i^n}^{x_{i+1}^n} f(x)\,dx.$$

(c): From this deduce that if f is continuous on X then

$$f(x) = \sum_{n=0}^{\infty} (f|f_n)f_n(x) \tag{1}$$

uniformly on X. In particular, $(f_n)_{n \geq 0}$ is an orthonormal basis of $L^2(X)$. Using parts (I)(c) and (I)(b), deduce from this that for every function $f \in L^2(X)$ the relation (1) is satisfied for almost all $x \in X$.

(III): The *RADEMACHER FUNCTIONS* are the functions defined on X by the formulae

$$r_n(x) = \operatorname{sgn}(\sin 2^{n+1}\pi x), \qquad n \geq 0.$$

(a): Show that if $0 \leq n_1 < n_2 < \cdots < n_p$, then

$$\int_0^1 r_{n_1}(x) r_{n_2}(x) \cdots r_{n_p}(x)\,dx = 0.$$

In particular, $(r_n)_{n \geq 0}$ is an orthonormal system in $L^2(X)$.

(b): Let $(a_n)_{n \geq 0}$ be a sequence of real numbers such that $\sum a_n^2 < \infty$ and $f = \sum a_n r_n$ in $L^2(X)$.

Show that for almost all $x \in X$

$$f(x) = \sum_{n=0}^{\infty} a_n r_n(x).$$

(Express the Rademacher functions in terms of the Haar functions, and use part (II)(c)).

(c): With a_n and f as above, show that for all $p > 0$

$$A_p \|f\|_2 \leq \|f\|_p \leq B_p \|f\|_2,$$

where

$$A_p = \min(1, 2^{-(2-p)/p}), \qquad B_p = \left(1 + \frac{p}{2}\right)^{\frac{1}{2}}.$$

(For the second inequality assume first that $p = 2k$, k an integer; for the first, if $0 < p < 2$ write $2 = \lambda p + 4(1 - \lambda)$, $0 < \lambda < 1$).

Deduce from this that for every real number μ

$$\int_0^1 \exp(\mu|f(x)|^2)dx < \infty.$$

(d): Show that the functions 1 and $r_{n_1} r_{n_2} \cdots r_{n_p}$ $(0 \leqslant n_1 < n_2 < \cdots < n_p)$ form an orthonormal basis of $L^2(X)$ (the *WALSCH BASIS*).

$$\Delta\nabla\Delta = \nabla\Delta\nabla = \Delta\nabla\Delta = \nabla\Delta\nabla = \Delta\nabla\Delta$$

SOLUTION (I):(a): One can clearly reduce to the case where the a_n's are real. Let $f = \sum a_n f_n$ in $L^2(X)$. Then $a_n = (f|f_n)$, so

$$s_N(x) = \int_X K_N(x,y)f(y)dy.$$

For $0 \leqslant n \leqslant N$ let

$$A_n = \{x : s_p(x) < s_n(x) \text{ if } 0 \leqslant p < n \text{ and } s_p(x) \leqslant s_n(x) \text{ if } n \leqslant p \leqslant N\}.$$

The sets A_n are measurable, mutually disjoint, and have union X. Setting $\nu(x) = n$ when $x \in A_n$, one has

$$s_N^*(x) = \max_{n \leqslant N} s_n(x) = s_{\nu(x)}(x)$$

$$= \int_X K_{\nu(x)}(x,y)f(y)dy.$$

Note that

$$K_{\nu(x)}(x,y) = \sum_{n=0}^{N} 1\!\!1_{A_n}(x) K_n(x,y),$$

and that consequently this function is square integrable on $X \times X$. This remark allows us to justify the various interchanges in the orders of the integration carried out in the calculations which follow:

$$\int_X s_N^*(x)dx = \int_X f(y)dy \int_X K_{\nu(x)}(x,y)dx$$

$$\leqslant \|f\|_2 \{\int_X dy (\int_X K_{\nu(x)}(x,y)dx)^2\}^{\frac{1}{2}}.$$

Now,

$$\int_X dy (\int_X K_{\nu(x)}(x,y)dx)^2 = \iint_{X\times X} d\alpha d\beta \int_X K_{\nu(\alpha)}(\alpha,y) K_{\nu(\beta)}(\beta,y)dy,$$

and furthermore,

$$\int_X K_{\nu(\alpha)}(\alpha,y) K_{\nu(\beta)}(\beta,y)dy = \sum_{r=0}^{\nu(\alpha)} \sum_{s=0}^{\nu(\beta)} f_r(\alpha) f_s(\beta) \int_X f_r(y) f_s(y)dy.$$

Taking into account the orthonormality of the f_n's, this yields

$$\int_X dy (\int_X K_{\nu(x)}(x,y)dx)^2 = \iint_{\nu(\alpha)\leqslant\nu(\beta)} K_{\nu(\alpha)}(\alpha,\beta)d\alpha d\beta$$

$$+ \iint_{\nu(\beta)<\nu(\alpha)} K_{\nu(\beta)}(\alpha,\beta)d\alpha d\beta.$$

The first integral of the right side is bounded in modulus by

$$\int_X d\alpha \int_{\omega(\alpha)\leqslant\omega(\beta)} |K_{\nu(\alpha)}(\alpha,\beta)| d\beta \leqslant \int_X L_{\nu(\alpha)}(\alpha)d\alpha \leqslant A\text{meas}(X).$$

Since $K_{\nu(\beta)}(\alpha,\beta) = K_{\nu(\beta)}(\beta,\alpha)$ the second integral is bounded by the same quantity, so

$$\int_X s_N^*(x)dx \leqslant \|f\|_2\sqrt{2A\text{meas}(X)}. \tag{2}$$

The functions s_N^* are integrable, as the s_N are, and the maximum of a finite number of integrable functions is also; further, $s_N^* \leqslant s_{N+1}^*$ and $s_N^*(x) \to \sup_n s_n(x)$. The Lebesgue Monotonic Convergence Theorem and inequality (2) show that $\sup_n s_n(x)$ is integrable, and consequently that $\sup_n s_n(x) < \infty$ for almost all x. Replacing f by $-f$ one deduces from this that $\sup_n(-s_n(x)) < \infty$, that is to say $\inf_n s_n(x) > -\infty$, and therefore that $\sup_n|s_n(x)| < \infty$ almost everywhere.

SOLUTION (I):(b): Here again it may be assumed that the a_n's are real. Let $n_1 < n_2 < \cdots$ be a sequence of integers such that

$$\sum_{n\geqslant n_k} a_n^2 \leqslant k^{-4}.$$

Set $n_0 = 0$, $\omega_n = 1$ if $0 \leqslant n < n_1$, and $\omega_n = k$ if $n_k \leqslant n < n_{k+1}$ and $k \geqslant 1$. Then

$$\sum_{n=0}^{\infty} \omega_n^2 a_n^2 = \sum_{n<n_1} a_n^2 + \sum_{k=1}^{\infty} k^2 \sum_{n_k\leqslant n<n_{k+1}} a_n^2$$

$$\leqslant \sum_{n<n_1} a_n^2 + \sum_{k=1}^{\infty} k^{-2} < \infty.$$

Moreover, it is clear that the sequence (ω_n) is increasing and that $\omega_n \to \infty$. Since $s_N \to f$ in $L^2(X)$ there exists a subsequence s_{N_k} such that

$$\lim_k s_{N_k}(x) = f(x) \quad \text{almost everywhere.} \tag{3}$$

By part (I)(a), for almost all x

$$M_x = \sup_N \left| \sum_{n=0}^{N} \omega_n a_n f_n(x) \right| < \infty. \tag{4}$$

If x is such that (3) and (4) are satisfied, and if $N_k \leqslant N < N_{k+1}$, Abel's bound gives:

$$|s_N(x) - s_{N_k}(x)| = \left| \sum_{N_k < n \leqslant N} \omega_n^{-1} \omega_n a_n f_n(x) \right|$$

$$\leqslant 2M_x \omega_{N_k}^{-1},$$

which shows that $s_N(x) \to f(x)$.

SOLUTION (I):(c): Let $C(X)$ be the space of continuous functions on X, with the norm of uniform convergence. If $f \in C(X)$ let us set:

$$s_N(f;x) = \sum_{n=1}^{N} (f|f_n)f_n(x) = \int_0^1 K_N(x,y)f(y)\mathrm{d}y.$$

For fixed N and x the mapping $f \mapsto s_N(f;x)$ is linear; denote it by $s_N(\cdot;x)$. This map is continuous and its norm is

$$\|s_N(\cdot;x)\| = \sup\left\{ \left| \int_0^1 K_N(x,y)f(y)\mathrm{d}y \right| : f \in C(X), \|f\|_\infty \leqslant 1 \right\}$$

$$= \int_0^1 |K_N(x,y)|\mathrm{d}y = L_N(x).$$

(This flows from exercise 6·108 on observing that one can replace the space of step functions by the space of continuous functions).

Assume that

$$\sup_{N,x} L_N(x) = \infty.$$

Then there would exist a sequence (x_k) of points in X and a sequence (N_k) of integers such that

$$L_{N_k}(x_k) = \|s_{N_k}(\cdot;x_k)\| \to \infty.$$

By the Banach-Steinhaus theorem there would exist $f \in C(I)$ such that

$$\sup_k |s_{N_k}(f;x_k)| = \infty.$$

But this is impossible, because by hypothesis $s_N(f;x) \to f(x)$ uniformly in x.

= = = = = = =

SOLUTION (II):(a): It is clear that

$$\int_0^1 f_n^2 = 1, \qquad n \geq 0.$$

Furthermore, if $0 \leq n_1 < n_2$ it can be seen without difficulty that $f_{n_1} f_{n_2} = \varepsilon f_{n_2}$, with $\varepsilon = 0, 1$, or -1, and consequently

$$\int_0^1 f_{n_1} f_{n_2} = \varepsilon \int_0^1 f_{n_2} = 0.$$

SOLUTION (II):(b): The result is obvious if $n = 0$. When $n = 2^m + k$, $0 \leq k < 2^m$, let us consider the end points x_i^n $(0 \leq i \leq n + 1)$ of the intervals obtained by putting end to end, starting at 0, $2k + 2$ intervals of length $2^{-(m+1)}$, then $2^m - k - 1$ intervals of length 2^{-m} (observe that $(2k + 2)2^{-(m+1)} + (2^m - k - 1)2^{-m} = 1$). Let $\phi_1^n, \phi_2^n, \ldots, \phi_{n+1}^n$ be the characteristic functions of the intervals $[0, x_1^n[, [x_1^n, x_2^n[, \ldots, [x_n^n, x_{n+1}^n]$. It is clear that each function f_i, $0 \leqslant i \leqslant n$, is a linear combination of the ϕ_j^n, $1 \leqslant j \leqslant n + 1$; since each of these systems of functions is linearly independent and they have the same cardinality, they generate the same vector subspace V_n, which proves the first part of the question.

If $f \in L^2(X)$, let

$$h = \sum_{j=1}^{n+1} \lambda_j \phi_j^n$$

be its orthogonal projection onto V_n. Then

$$\int_0^1 (f - h)\phi_i^n = 0, \qquad 1 \leqslant i \leqslant n + 1,$$

and $\phi_i^n \phi_j^n = 0$ if $i \neq j$, while $\phi_i^n \phi_i^n = \phi_i^n$, so

$$\lambda_i = \frac{1}{x_i^n - x_{i-1}^n} \int_{x_{i+1}^n}^{x_i^n} f(x)dx.$$

SOLUTION (II):(c): If f is continuous on X:

$$s_N(x) = \sum_{n=0}^{N} (f|f_n)f_n(x)$$

is its orthogonal projection onto V_n, and consequently if $\phi_i^N(x) = 1$, $2^m \leqslant N < 2^{m+1}$, and

$$\varepsilon_m = \sup\{|f(x) - f(y)| : |x - y| \leqslant 2^{-m}\},$$

then

$$|f(x) - s_N(x)| = \left|\frac{1}{x_i^N - x_{i-1}^N}\int_{x_{i-1}^N}^{x_i^N} (f(x) - f(y))dy\right| \leqslant \varepsilon_m,$$

which proves that in this case

$$f(x) = \sum_{n=0}^{\infty} (f|f_n)f_n(x)$$

uniformly in x.

As the continuous functions are dense in $L^2(X)$, and as every continuous function is the uniform limit, and hence also the limit in $L^2(X)$, of its orthogonal projections onto V_n, it follows that the f_n form an orthonormal basis of $L^2(X)$. Finally, by parts (I)(c) and (I)(b) formula (1) holds almost everywhere whenever $f \in L^2(X)$

REMARK: This last result can be proved independently of (I) by using part (II)(b) and Lebesgue's differentiation theorem (cf., Chapter 9). It will be noted that as the linear mapping

$$f \mapsto \sum_{i=0}^{n} (f|f_i)f_i$$

is continuous from $L^1(X)$ into V_n, and as $L^2(X)$ is dense in $L^1(X)$,

$$\sum_{i=0}^{n} (f|f_i)f_i = \sum_{i=1}^{n+1} \lambda_i(f)\phi_i^n,$$

with

$$\lambda_i(f) = \frac{1}{x_i^n - x_{i-1}^n}\int_{x_{i-1}^n}^{x_i^n} f(x)dx$$

still holds for $f \in L^1(X)$. This and Lebesgue's theorem on differentiation show that in this case (1) is still true almost everywhere.

SOLUTION (III):(a): Observe that r_n can be considered as a function defined on $\mathbb{R}$ with period 2^{-n}. If $n_1 < n_2 < \cdots < n_p$ the function $r_{n_1} r_{n_2} \cdots r_{n_p}$ has period 2^{-n_1}, and consequently

$$\int_0^1 r_{n_1} r_{n_2} \cdots r_{n_p} = 2^{n_1} \int_0^{2^{-n_1}} r_{n_1} r_{n_2} \cdots r_{n_p}$$

$$= 2^{n_1} \left(\int_0^{2^{-(n_1+1)}} r_{n_2} \cdots r_{n_p} - \int_{2^{-(n_1+1)}}^{2^{-n_1}} r_{n_2} \cdots r_{n_p} \right).$$

Since $r_{n_2} \cdots r_{n_p}$ has period 2^{-n_2}, and since $2^{-(n_1+1)} = 2^{n_2-n_1-1} 2^{-n_2}$, the term in parentheses equals zero. In particular, r_{n_1} and r_{n_2} are orthogonal if $n_1 \neq n_2$, and as $r_n^2 = 1$ almost everywhere the system (r_n) is certainly orthonormal.

SOLUTION (III):(b): It is clear that

$$r_n = 2^{n/2} \sum_{2^n \leq i < 2^{n+1}} f_i.$$

It follows that

$$\sum_{n=0}^{N} a_n r_n$$

is the orthogonal projection of f onto $V_{2^{N+1}-1}$; by part (II)(c) for almost all x

$$\lim_{N\to\infty} \sum_{n=0}^{N} a_n r_n(x) = f(x).$$

SOLUTION (III):(c): If $k \geq 1$ is an integer,

$$\left|\sum_{n=0}^{N} a_n r_n\right|^{2k} = \sum_{\beta_0+\cdots+\beta_n = 2k} \frac{(2k)!}{\beta_0!\cdots\beta_N!} a_0^{\beta_0}\cdots a_N^{\beta_N} r_0^{\beta_0}\cdots r_N^{\beta_N}.$$

Taking into account that $r_n^2 = 1$ and part (III)(a), this yields

$$\int_0^1 \left|\sum_{n=0}^{N} a_n r_n\right|^{2k} = \sum_{\beta_0+\cdots+\beta_N=k} \frac{(2k)!}{(2\beta_0)!\cdots(2\beta_N)!} a_0^{2\beta_0}\cdots a_N^{2\beta_N}.$$

Now note that

$$\left(\sum_{n=0}^{N} a_n^2\right)^k = \sum_{\beta_0+\cdots+\beta_N=k} \frac{k!}{\beta_0!\cdots\beta_N!} a_0^{2\beta_0}\cdots a_N^{2\beta_N},$$

and that for every integer $n \geq 0$, $2^n \leq (2n)!/n! \leq 2^n n^n$, so that

$$\frac{(2k)!}{k!}\,\frac{\beta_0!\cdots\beta_N!}{(2\beta_0)!\cdots(2\beta_N)!} \leq \frac{2^k k^k}{2^{\beta_0+\cdots+\beta_N}} = k^k.$$

It follows that

$$\int_0^1 \left|\sum_{n=0}^{N} a_n r_n\right|^{2k} \leq k^k\left(\sum_{i=0}^{N} a_n^2\right)^k \leq k^k \|f\|_2^{2k}.$$

By the preceding part, the function under the summation sign tends almost everywhere to f^{2k} as $N \to \infty$. Fatou's Lemma then gives

$$\|f\|_{2k} \leqslant k^{\frac{1}{2}}\|f\|_2. \tag{5}$$

When $p > 0$ consider the integer $k \geqslant 1$ such that $2k - 2 < p \leqslant 2k$; by exercise 6·117(a)

$$\|f\|_p \leqslant \|f\|_{2k} \leqslant k^{\frac{1}{2}}\|f\|_2 \leqslant \left(1 + \frac{p}{2}\right)^{\frac{1}{2}}\|f\|_2,$$

which proves the first inequality.

If $p \geqslant 2$ then $\|f\|_2 \leqslant \|f\|_p$ (Exercise 6·117(a)); when $0 < p < 2$ there exists $0 < \lambda < 1$ such that $2 = \lambda p + 4(1 - \lambda)$, and then (cf., Exercise 6·116(c))

$$\|f\|_2^2 \leqslant \|f\|_p^{\lambda p}\|f\|_4^{4(1-\lambda)}.$$

But by (5)

$$\|f\|_4 \leqslant 2^{\frac{1}{2}}\|f\|_2,$$

so

$$2^{-2(1-\lambda)}\|f\|_2^{4\lambda-2} \leqslant \|f\|_p^{\lambda p}.$$

Since

$$4\lambda - 2 = \lambda p \quad \text{and} \quad \frac{1 - \lambda}{\lambda} = \frac{2 - p}{2},$$

this yields

$$2^{-(2-p)/p}\|f\|_2 \leqslant \|f\|_p,$$

which proves the second inequality.

Finally, to prove that

$$\int_0^1 \exp(\mu f^2) < \infty,$$

we can restrict ourselves to the case where $\mu > 0$. Let N be such that

$$\sum_{n \geq N} a_n^2 < \frac{1}{2e\mu}, \tag{6}$$

and let

$$h = \sum_{n<N} a_n r_n, \qquad g = \sum_{n \geq N} a_n r_n.$$

Since $\exp(\mu f^2) \leqslant \exp(2\mu h^2)\exp(2\mu g^2)$, and since h is bounded, it suffices to prove that $\exp(2\mu g^2)$ is integrable. Now by (5),

$$\int_0^1 \exp(2\mu g^2) = \sum_{n=0}^{\infty} \frac{(2\mu)^n}{n!}\int_0^1 g^{2n} \leqslant \sum_{n=0}^{\infty} (2\mu\|g\|_2^2)^n \frac{n^n}{n!}.$$

D'Alembert's Criterion shows that the last series converges if $2e\mu\|g\|_2^2 < 1$, which is assured by (6).

SOLUTION (III):(d): By part (III)(a) the Walsch system is orthonormal. For $n \geqslant 0$ and $0 \leqslant k < 2^n$ let $\phi_{n,k}$ be the characteristic function of $[k2^{-n},(k+1)2^{-n}]$. It is easily verified that

$$\tfrac{1}{2}(1 + r_n)\phi_{n,k} = \phi_{n+1,2k}, \qquad \tfrac{1}{2}(1 - r_n)\phi_{n,k} = \phi_{n+1,2k+1}.$$

Taking into account that $r_n^2 = 1$, it follows from this that the vector subspace generated by the Walsch system contains all the functions $\phi_{n,k}$, and that it is consequently dense in $L^2(X)$.

EXERCISE 7·134: Let (α_n) be a sequence of distinct numbers strictly greater than $-\frac{1}{2}$.

Show that the function t^{α_n} form a total sequence in $L^2(0,1)$ if and only if one of the following three conditions is satisfied:

(a): There exists a subsequence of the sequence (α_n) that tends towards a finite limit that is also greater than $-\frac{1}{2}$;

(b): $\lim \alpha_n = -\frac{1}{2}$ and $\sum |\alpha_n + \frac{1}{2}| = \infty$;

(c): $\lim \alpha_n = \infty$ and $\sum \alpha_n^{-1} = \infty$.

From this deduce *MUNTZ'S THEOREM*: If $\alpha_n > 0$ and $\lim \alpha_n = \infty$, the sequence of functions formed by 1 and the t^{α_n} is total in $C(0,1)$ if and only if $\sum \alpha_n^{-1} = \infty$.

Δ∇Δ = ∇Δ∇ = ΔVΔ = ∇Δ∇ = Δ∇Δ

SOLUTION: First of all we are going to calculate the distance in $L^2(0,1)$ from t^p to the vector subspace generated by $t^{\alpha_1},\ldots,t^{\alpha_n}$ ($p \geqslant 0$ an integer). Recall that if $x_1,\ldots,x_n$ are elements of a Hilbert space and if $G(x_1,\ldots,x_n) = \det\|(x_i|x_j)\|$, then $x_1,\ldots,x_n$ are linearly independent if and only if $G(x_1,\ldots,x_n) \neq 0$, and in this case the distance from a point x to the vector subspace generated by $x_1,\ldots,x_n$ is given by

$$d = \sqrt{\frac{G(x,x_1,\ldots,x_n)}{G(x_1,\ldots,x_n)}} \,. \tag{*}$$

(The proof of this result is given at the end of the exercise). In the present case,

$$G(t^{\alpha_1},\ldots,t^{\alpha_n}) = \det\left\|\frac{1}{\alpha_i + \alpha_j + 1}\right\| .$$

Now,

$$\det\left\|\frac{1}{a_i + b_j}\right\| = \frac{\prod_{i<j} (a_i - a_j)(b_i - b_j)}{\prod_{i,j} (a_i + b_j)} .$$

(This is called the *CAUCHY DETERMINANT FORMULA*, and is proved at the end of the exercise). Setting $a_i = b_i = \alpha_i + \frac{1}{2}$, this yields

$$G(t^{\alpha_1},\ldots,t^{\alpha_n}) = \frac{\prod_{i<j} (\alpha_i - \alpha_j)^2}{\prod_{i,j} (\alpha_i + \alpha_j + 1)} .$$

Equation (*) then gives for the distance from t^p to the vector subspace generated by $t^{\alpha_1},\ldots,t^{\alpha_n}$:

$$d = \frac{1}{\sqrt{2p+1}}\cdot\frac{|p - \alpha_1|\cdots|p - \alpha_n|}{(p + \alpha_1 + 1)\cdots(p + \alpha_n + 1)} .$$

By Weierstrass's Theorem and the property that $C(0,1)$ is dense in $L^2(0,1)$, the sequence (t^{α_n}) will be total in $L^2(0,1)$ if and only if for every integer $p \geqslant 0$

$$\liminf_{n\to\infty} \frac{|p - \alpha_1|\cdots|p - \alpha_n|}{(p + \alpha_1 + 1)\cdots(p + \alpha_n + 1)} = 0. \qquad (**)$$

Assume first that (α_n) has an accumulation point ℓ, $-\frac{1}{2} < \ell < \infty$. By considering, if need be, a subsequence, it can be assumed that $-\frac{1}{2} < a \leqslant \alpha_n \leqslant b < \infty$ for all n. But then considering the range of the function $\alpha \to |p - \alpha|/(p + \alpha + 1)$, $\alpha > -\frac{1}{2}$, shows that there exists λ such that $0 \leqslant \lambda < 1$ and

$$\frac{|p - \alpha_n|}{p + \alpha_n + 1} \leqslant \lambda \quad \text{for all } n,$$

and consequently (**) is satisfied. Next, assume that $\alpha_n = -\frac{1}{2} + \varepsilon_n$, with $\varepsilon_n > 0$ and $\varepsilon_n \to 0$. Then whenever n is large enough for $\alpha_n < p$ to be true,

$$\frac{|p - \alpha_n|}{p + \alpha_n + 1} = 1 - \frac{2\varepsilon_n}{p + \varepsilon_n + \frac{1}{2}} .$$

Since:

$$\frac{2\varepsilon_n}{p + \varepsilon_n + \frac{1}{2}} \sim \frac{2\varepsilon_n}{p + \frac{1}{2}} ,$$

it follows that accordings as $\sum \varepsilon_n$ is finite or not, the limit of

$$\frac{|p - \alpha_1|\cdots|p - \alpha_n|}{(p + \alpha_1 + 1)\cdots(p + \alpha_n + 1)}$$

is strictly positive or zero (at least if p is different from all the α_n's, which is the case when p is large enough). Consequently, in this case the sequence (t^{α_n}) is total in $L^2(0,1)$ if and only if

$$\sum \varepsilon_n = \sum (\alpha_n + \tfrac{1}{2}) = \infty.$$

Lastly, if $\alpha_n \to \infty$, then for large enough

$$\frac{|p - \alpha_n|}{p + \alpha_n + 1} = 1 - \frac{2p + 1}{p + \alpha_n + 1} ,$$

$$\frac{2p + 1}{p + \alpha_n + 1} \sim \frac{2p + 1}{\alpha_n} .$$

From this it follows that the sequence (t^{α_n}) is total in $L^2(0,1)$ if and only if $\sum \alpha_n^{-1} = \infty$. This accomplishes the proof of the first part of the exercise.

Assume that $\alpha_n > 0$ and $\alpha_n \to \infty$. It is clear that the sequence (t^{α_n}) together with 1 can be total in $C(0,1)$ only if it is total in $L^2(0,1)$, hence if $\sum \alpha_n^{-1} = \infty$. It remains to prove the converse. One can assume that $\alpha_n > 1$ for all n. Let P be a polynomial. Applying what has gone before to the sequence $(\alpha_n - 1)$, for all $\varepsilon > 0$ one can determine numbers c_n, only a finite number of which are non-zero, such that

$$\int_0^1 |P'(t) - \sum c_n t^{\alpha_n - 1}|^2 dt < \varepsilon^2.$$

Since for $0 \leqslant x \leqslant 1$

$$P(x) - P(0) - \sum c_n \alpha_n^{-1} x^{\alpha_n} = \int_0^x (P'(t) - \sum c_n t^{\alpha_n - 1}) dt,$$

the Cauchy-Schwarz inequality gives

$$|P(x) - P(0) - \sum c_n \alpha_n^{-1} x^{\alpha_n}| < \varepsilon, \qquad 0 \leqslant x \leqslant 1.$$

As the polynomials are dense in $C(0,1)$, Müntz's Theorem is proved.

If $x_1,\ldots,x_n$ are linearly independent elements of a Hilbert space H, then for all $x \in H$ there exist complex numbers $\lambda_1,\ldots,\lambda_n$ and a vector $z \in H$ that are uniquely determined by the conditions

$$\begin{cases} x = z + \sum \lambda_i x_i, \\ (z|x_i) = 0, \quad 1 \leqslant i \leqslant n. \end{cases}$$

If d denotes the distance from x to the subspace generated by $x_1,\ldots,x_n$ then $d^2 = (z|z) = (x|z)$, so that $\lambda_1,\ldots,\lambda_n,d^2$ are the unique solution of the linear system

$$\begin{cases} \sum_i \lambda_i(x_i|x_j) = (x|x_j), \quad 1 \leqslant j \leqslant n, \\ \\ d^2 + \sum_i \lambda_i(x|x_i) = (x|x). \end{cases}$$

That this system has a solution implies that $G(x_1,\ldots,x_n) \neq 0$; (*) then follows from Cramer's rule.

═ ═ ═ ═ ═ ═ ═

Let us now prove the Cauchy Determinant Formula. It is clear that

$$\det\left\|\frac{1}{a_i + b_j}\right\|_{1\leqslant i,j\leqslant n} = \frac{R}{\prod_{i,j}(a_i + b_i)},$$

where R is a polynomial of degree $n^2 - n$ homogeneous in the a_i and the b_i. If for $i < j$ $a_i = a_j$ (*resp.* $b_i = b_j$) the determinant has two identical rows (*resp.* columns) and consequently is zero. The polynomial R is therefore divisible by

$$\prod_{i<j}(a_i - a_j)(b_i - b_j),$$

and since this polynomial is of degree $n^2 - n$,

$$\det\left\|\frac{1}{a_i + b_j}\right\|_{1\leqslant i,j\leqslant n} = C_n \frac{\prod_{i<j}(a_i - a_j)(b_i - b_j)}{\prod_{i,j}(a_i + a_j)},$$

where C_n is a constant. Finally, it is verified without difficulty that

$$\lim_{b_1\to\infty}\left(\lim_{a_1\to\infty} a_1 \det\left\|\frac{1}{a_i+b_j}\right\|_{1\leqslant i,j\leqslant n}\right) = \det\left\|\frac{1}{a_i+b_j}\right\|_{2\leqslant i,j\leqslant n},$$

and that

$$\lim_{b_1\to\infty}\left(\lim_{a_1\to\infty} a_1 \frac{\prod_{1\leqslant i<j\leqslant n}(a_i-a_j)(b_i-b_j)}{\prod_{1\leqslant i,j\leqslant n}(a_i+b_j)}\right)$$

$$= \frac{\prod_{2\leqslant i<j\leqslant n}(a_i-a_j)(b_i-b_j)}{\prod_{2\leqslant i,j\leqslant n}(a_i+b_j)},$$

so that $C_n = C_{n-1}$. Since $C_1 = 1$ the formula is proved.

EXERCISE 7·135: Let f be a holomorphic function in the disc $U = \{z : |z| < 1\}$.

Show that if f is injective and if

$$f(z) = \sum_{n\geqslant 0} c_n z^n \quad \text{for} \quad |z| < 1,$$

the area of $f(U)$ is equal to

$$\pi \sum_{n=0}^{\infty} n|c_n|^2.$$

Δ∇Δ = ∇Δ∇ = ΔVΔ = ∇Δ∇ = ΔVΔ

SOLUTION: the Jacobian of f is equal to $|f'|^2$, so that:

$$\operatorname{meas}(f(\mathcal{U})) = \int_0^1 r\mathrm{d}r \int_0^{2\pi} |f'(re^{i\vartheta})|^2 \mathrm{d}\vartheta.$$

Furthermore, if $0 \leqslant r < 1$,

$$f'(re^{i\vartheta}) = \sum_{n=1}^{\infty} nr^{n-1} c_n e^{i(n-1)\vartheta}$$

Plancherel's Formula gives

$$\int_0^{2\pi} |f'(re^{i\vartheta})|^2 \mathrm{d}\vartheta = 2\pi \sum_{n=1}^{\infty} n^2 r^{2n-2} |c_n|^2,$$

so

$$\operatorname{meas}(f(\mathcal{U})) = 2\pi \sum_{n=1}^{\infty} n^2 |c_n|^2 \int_0^1 r^{2n-1} \mathrm{d}r$$

$$= \pi \sum_{n=1}^{\infty} n |c_n|^2.$$

EXERCISE 7·136: Let f be a continuous function on the disc $|z| \leqslant 1$ and holomorphic in its interior. Assume that

$$f(z) = 1 + \sum_{n=1}^{\infty} a_n z^n \quad \text{for } |z| < 1.$$

Show that if

$$\frac{1}{2\pi} \int_0^{2\pi} |f(e^{i\vartheta})| \mathrm{d}\vartheta < 1 + \tfrac{1}{4}|a_1|^2,$$

then the function f possesses at least one root in the disc $|z| < 1$.

Δ∇Δ = ∇Δ∇ = ΔΔ∇ = ∇Δ∇ = ΔΔ∇

SOLUTION: If f were never zero in the disc $|z| < 1$ there would exist a function g, holomorphic in this disc, such that $g(0) = 1$ and $f = g^2$. Let

$$g(z) = 1 + \frac{a_1}{2} z + \sum_{n=2}^{\infty} b_n z^n$$

be the expansion of g as an entire series in the disc $|z| < 1$. For $0 \leqslant r < 1$ Plancherel's Formula yields:

$$1 + \tfrac{1}{4}|a_1|^2 r^2 + \sum_{n=2}^{\infty} |b_n|^2 r^{2n} = \frac{1}{2\pi}\int_0^{2\pi} |g(re^{i\vartheta})|^2 d\vartheta$$

$$= \frac{1}{2\pi}\int_0^{2\pi} |f(re^{i\vartheta})| d\vartheta.$$

Since

$$\lim_{r \to 1^-} f(re^{i\vartheta}) = f(e^{i\vartheta})$$

uniformly in ϑ, it follows that

$$\frac{1}{2\pi}\int_0^{2\pi} |f(e^{i\vartheta})| d\vartheta \geqslant 1 + \tfrac{1}{4}|a_1|^2.$$

EXERCISE 7·137: Let U be an open set of $\mathbb{C}$ and $H(U)$ the set of holomorphic functions in U.

Prove that $H(U) \cap L^2(U)$ is closed in $L^2(U)$.

ΔΔ∇ = ∇Δ∇ = ΔΔ∇ = ∇Δ∇ = ΔΔ∇

SOLUTION: Let $f_n \in H(U) \cap L^2(U)$ and let $f_n \to f$ in $L^2(U)$. If D is a closed disc with centre z_0 and radius $r > 0$, and is contained in U, there exist r_1, r_2 such that $r < r_1 < r_2$, and the closed disc with centre z_0 and of radius r_2 is also contained in U. For all $z \in D$ and all ρ, $r_1 \leqslant \rho \leqslant r_2$, we have (Cauchy's Formula)

$$f_n(z) = \frac{1}{2\pi i}\int_{|\xi - z_0| = \rho} \frac{f_n(\xi)}{\xi - z}\, d\xi$$

$$= \frac{\rho}{2\pi}\int_0^{2\pi} f_n(z_0 + \rho e^{i\vartheta}) \frac{e^{i\vartheta}}{z_0 + \rho e^{i\vartheta} - z}\, d\vartheta,$$

and consequently

$$f_n(z) = \frac{1}{2\pi(r_2 - r_1)}\int_{r_1}^{r_2} \rho\, d\rho \int_0^{2\pi} f_n(z_0 + \rho e^{i\vartheta}) \frac{e^{i\vartheta}}{z_0 + \rho e^{i\vartheta} - z}\, d\vartheta$$

$$= \frac{1}{2\pi(r_2 - r_1)}\iint_{r_1 \leqslant |\xi - z_0| \leqslant r_2} f_n(\xi) \frac{\xi - z_0}{|\xi - z_0|(\xi - z)}\, du\, dv,$$

where $\xi = u + iv$.

The function equal to $(\xi - z_0)|\xi - z_0|^{-1}(\xi - z)^{-1}$ if $r_1 \leqslant |\xi - z_0| \leqslant r_2$ and to zero otherwise belongs to $L^2(U)$, and thus

$$\lim_n f_n(z) = \frac{1}{2\pi(r_2 - r_1)}\iint_{r_1 \leqslant |\xi - z_0| \leqslant r_2} f(\xi) \frac{\xi - z_0}{|\xi - z_0|(\xi - z)}\, du\, dv.$$

Since, for $z \in D$ and $r_1 \leqslant |\xi - z_0| \leqslant r_2$,

$$\left|\frac{\xi - z_0}{|\xi - z_0|(\xi - z)}\right| \leqslant \frac{1}{r_1 - r},$$

the function

$$g(z) = \lim_n f_n(z)$$

is holomorphic in $\mathring{D}$, and consequently in U. Clearly $g = f$ almost everywhere in U, which proves the result.

CHAPTER 8
Convolution Products and Fourier Transforms

EXERCISE 8·138: Let A be a measurable set of $\mathbb{R}$ with positive measure.

Show that $A - A$ is a neighbourhood of zero. Deduce from this that if A is an additive subgroup of $\mathbb{R}$ that is measurable and is distinct from $\mathbb{R}$, then $\text{meas}(A) = 0$.

$$\Delta\nabla\Delta = \nabla\Delta\nabla = \Delta\nabla\Delta = \nabla\Delta\nabla = \Delta\nabla\Delta$$

SOLUTION: First assume that A is bounded, and let φ be its characteristic function. Since $\varphi \in L^1 \cap L^\infty$ the function $\varphi * \check{\varphi}$ is continuous. Furthermore,

$$\varphi * \check{\varphi}(0) = \text{meas}(A) > 0.$$

Hence there exists a neighbourhood V of zero on which $\varphi * \check{\varphi} > 0$; but then $V \subset A - A$, which proves the property in this case. In the general case there exists a measurable and bounded set B such that $B \subset A$, $\text{meas}(B) > 0$. Then $A - A \supset B - B$ is a neighbourhood of zero by what has preceded.

If A is an additive measurable subgroup of $\mathbb{R}$, and if $\text{meas}(A) > 0$, then $A = A - A$ is a neighbourhood of zero, which implies that $A = \mathbb{R}$.

EXERCISE 8·139: Let f be a measurable function on $\mathbb{R}$ such that for all the elements a of an everywhere dense set A of $\mathbb{R}$

$$f(x + a) = f(x) \quad \text{for almost all } x.$$

Show that there exists a constant c such that $f(x) = c$ for almost all x.

Δ∇Δ = ∇Δ∇ = ΔVΔ = ∇Δ∇ = Δ∇Δ

SOLUTION: First assume that f is bounded. Let (φ_i) be a compact approximate identity in L^1. The functions $f_i = \varphi_i * f$ are continuous and $f_i(x + a) = f_i(x)$ for all i, all $a \in A$, and all $x \in \mathbb{R}$. From this it follows, because A is dense in $\mathbb{R}$, that each of the functions f_i is equal to a constant c_i. By extracting a suitable subsequence from the sequence (φ_i), one can assume that $f_i(x) \to f(x)$ for almost all x. From this it follows that $c_i \to c$ and that $f(x) = c$ for almost all x.

In the general case, consider the functions f_n equal to f where $|f| \leqslant n$ and to zero otherwise. It is clear that the functions f_n still satisfy the condition of the problem. Therefore there exist constants c_n such that $f_n(x) = c_n$ almost everywhere. If for one n $c_n \neq 0$, then $f(x) = c_n$ almost everywhere. Otherwise, for all n, $f_n = 0$ almost everywhere, and consequently $f = 0$ almost everywhere.

EXERCISE 8·140: Let $f \in L^\infty(\mathbb{R})$ and $f_t(x) = f(x - t)$.

Show that if

$$\lim_{t \to 0} \|f_t - f\|_\infty = 0,$$

then there exists a function g that is uniformly continuous on $\mathbb{R}$ such that $f = g$ almost everywhere.

$$\Delta\nabla\Delta = \nabla\Delta\nabla = \Delta\nabla\Delta = \nabla\Delta\nabla = \Delta\nabla\Delta$$

SOLUTION: If (φ_n) is an approximate identity such that $\varphi_n \geqslant 0$, $\int \varphi_n = 1$, and $\varphi_n(x) = 0$ when $|x| \geqslant \varepsilon_n$ (with $\varepsilon_n \to 0$), then

$$\|\varphi_n * f - f\|_\infty \leqslant \sup_{|y| \leqslant \varepsilon_n} \|f_y - f\|_\infty. \tag{*}$$

In fact, if $\psi \in L^1$ and $\|\psi\|_1 = 1$ then

$$\left|\int (f*\varphi_n - f)\psi\right| \leqslant \int |f(x - y) - f(x)|\varphi_n(y)|\psi(x)|\,dx dy$$

$$= \int_{|y|\leqslant\varepsilon_n} \varphi_n(y)dy \int |f_y(x) - f(x)|\,|\psi(x)|\,dx$$

$$\leqslant \sup_{|y|\leqslant\varepsilon_n} \|f_y - f\|_\infty.$$

It follows from (*) that $\varphi_n * f \to f$ in L^∞. But the $\varphi_n * f$ are uniformly continuous, and since for such functions the "sup" and the "ess sup" coincide, the $\varphi_n * f$ form a Cauchy sequence in UC^∞, and consequently $\varphi_n * f \to g$ in UC^∞. Clearly $f = g$ almost everywhere.

EXERCISE 8·141: Let I be an open interval of $\mathbb{R}$, and let $f \in L^1_{loc}(I)$ be such that for some integer n

$$\int_I f\varphi = 0$$

for every function $\varphi \in D^\infty(I)$ satisfying the conditions

$$\int_I \varphi(x)dx = \int_I x\varphi(x)dx = \cdots = \int_I x^n\varphi(x)dx = 0.$$

Show that f is a polynomial of degree at most n.

SOLUTION: More generally, we shall prove the following result: If $g_1,\dots,g_n$ are elements of $L^1_{\text{loc}}(I)$ and if $\int_I f\varphi = 0$ for every function $\varphi \in \mathcal{D}^\infty(I)$ satisfying

$$\int_I g_i\varphi = 0, \qquad 1 \leqslant i \leqslant n, \tag{*}$$

then f is equal (almost everywhere) to a linear combination of the g_i.

Let $E = L^1_{\text{loc}}(I)$, $F = \mathcal{D}^\infty(I)$, and denote by G the subspace of E generated by the g_i. Let G_0 be the subspace of F formed by the $\varphi \in F$ such that (*) holds (and consequently $\int g\varphi = 0$ for all $g \in G$). Let u be the canonical linear mapping of F onto F/G_0. There exists a linear mapping v of F into G^*, the dual of G, defined by

$$\langle g, v(\varphi)\rangle = \int g\varphi, \qquad g \in G, \qquad \varphi \in F.$$

The kernel of v is exactly G_0. Hence there exists an *injective* linear mapping ω of F/G_0 into G^* such that:

$$v = \omega \circ u.$$

Since G is finite-dimensional, $(G^*)^*$ is canonically identified with G, and ${}^t\omega$ is identified with a *surjective* linear mapping of G onto $(F/G_0)^*$. Now, the linear form defined on F by

$$\varphi \mapsto \int f\varphi$$

is zero on G_0 by hypothesis. Hence it defines a linear form ϕ on F/G_0, that is to say, an element of $(F/G_0)^*$, such that for all

$\varphi \in F$:

$$\int f\varphi = \langle \phi, u(\varphi) \rangle .$$

Let $g \in G$ be such that ${}^t\omega(g) = \phi$. Then for all $\varphi \in F$

$$\int f\varphi = \langle {}^t\omega(g), u(\varphi) \rangle = \langle g, (\omega_0 u)(\varphi) \rangle$$
$$= \langle g, v(\varphi) \rangle = \int g\varphi .$$

Therefore, by the fundamental lemma of the Calculus of Variations, $f = g$ almost everywhere.

= = = = = = =

For readers not so familiar with duality and quotient vector spaces, here is another proof.

Firstly, it can be assumed that the g_i are linearly independent in $L^1_{loc}(I)$. Consider the linear mapping of $\mathcal{D}^\infty(I)$ into $\mathbb{C}^n$ given by

$$\varphi \mapsto (\lambda_i = \int g_i \varphi)_{1 \leqslant i \leqslant n} .$$

This mapping is surjective, otherwise there would exist complex numbers $a_1, \ldots, a_n$, *not all zero*, such that for all $\varphi \in \mathcal{D}^\infty(I)$

$$0 = \sum_i a_i \int g_i \varphi = \int (\sum_i a_i g_i)\varphi .$$

By the fundamental lemma of the Calculus of Variations, this would imply that $\sum a_i g_i = 0$ (almost everywhere), contrary to the hypothesis made earlier about the linear independence of the g_i. From this it follows that for all $j = 1, 2, \ldots, n$ one can find $\varphi_j \in \mathcal{D}^\infty(I)$ such that

$$\int g_i \varphi_j = \delta_{ij}, \qquad 1 \leqslant i \leqslant n.$$

Then let

$$C_j = \int f\varphi_j, \qquad 1 \leqslant j \leqslant n,$$

$$g = \sum_j C_j \varphi_j,$$

and let us show that for all $\varphi \in \mathcal{D}^\infty(I)$

$$\int f\varphi = \int g\varphi,$$

which will prove the property. Note that if

$$\varphi_0(x) = \varphi(x) - \sum_j \varphi_j(x) \int g_j \varphi,$$

then for all $i = 1,2,\ldots,n$

$$\int g_i \varphi_0 = 0,$$

and consequently

$$\int f\varphi_0 = 0.$$

From this it follows that

$$\int f\varphi = \sum_j C_j \int g_j \varphi = \int g\varphi.$$

EXERCISE 8·142: Let f be a locally integrable complex function on $[0,\infty[$.

(a): Show that for all $\alpha > 0$ the integral

$$f_\alpha(x) = \frac{1}{\Gamma(\alpha)}\int_0^x (x - t)^{\alpha-1} f(t)\mathrm{d}t$$

exists for almost all $x \geqslant 0$, and that f_α is locally integrable.

(b): Show that $(f_\alpha)_\beta = f_{\alpha+\beta}$.

(c): Show that if $f \in L^p_{\text{loc}}$ $(p > 1)$ then f_α is continuous whenever $\alpha > 1/p$.

ΔVΔ = VΔV = ΔVΔ = VΔV = ΔVΔ

SOLUTION (a): For every real number $\alpha > 0$

$$\int_0^a \mathrm{d}x \int_0^x (x - t)^{\alpha-1}|f(t)|\mathrm{d}t = \int_0^a |f(t)|\mathrm{d}t \int_t^a (x - t)^{\alpha-1}\mathrm{d}x$$

$$= \frac{1}{\alpha}\int_0^a |f(t)|(a - t)^\alpha \mathrm{d}t < \infty,$$

which proves that

$$\int_0^x (x - t)^{\alpha-1}|f(t)|\mathrm{d}t < \infty$$

for almost all x, and that f_α (which is therefore defined almost everywhere) belongs to L^1_{loc}.

SOLUTION (b):

$$(f_\alpha)_\beta(x) = \frac{1}{\Gamma(\alpha)\Gamma(\beta)}\int_0^x (x - t)^{\beta-1}\mathrm{d}t\int_0^t (t - u)^{\alpha-1} f(u)\mathrm{d}u$$

$$= \frac{1}{\Gamma(\alpha)\Gamma(\beta)}\int_0^x f(u)\mathrm{d}u\int_u^x (x - t)^{\beta-1}(t - u)^{\alpha-1}\mathrm{d}t.$$

On setting $t = u + (x - u)\tau$ this yields

$$(f_\alpha)_\beta(x) = \frac{1}{\Gamma(\alpha)\Gamma(\beta)}\int_0^x (x - u)^{\alpha+\beta-1} f(u)du \int_0^1 \tau^{\alpha-1}(1 - \tau)^{\beta-1} d\tau$$

$$= \frac{1}{\Gamma(\alpha + \beta)}\int_0^x (x - u)^{\alpha+\beta-1} f(u)du$$

$$= f_{\alpha+\beta}(x).$$

The premutation of the orders of integration is valid for almost all x, because, by part (a), the integral obtained by replacing $f(u)$ by $|f(u)|$ in the penultimate line is finite for almost all x.

SOLUTION (c): Let $a > 0$. If $f^a = \mathbb{1}_{[0,a]}\cdot f$, and

$$g^a(x) = \frac{x^{\alpha-1}}{\Gamma(\alpha)}, \qquad 0 < x \leqslant a,$$

and $g^a(x) = 0$ if $x > a$, it is clear that for $0 \leqslant x \leqslant a$

$$f_\alpha(x) = g^a * f^a(x).$$

Now, $f^a \in L^p$, and if $\frac{1}{p} + \frac{1}{q} = 1$ then $g^a \in L^p$ if $q(\alpha - 1) > -1$, that is to say, if $\alpha > 1 - \frac{1}{q} = \frac{1}{p}$. In this case it is known that $g^a * f^a$ is continuous.

EXERCISE 8·143: Prove that there is a sequence of functions $f_n \in L^2(\mathbb{R}^p)$ such that $\|f_n\|_2 = 1$ and $f_n * \tilde{f}_n \to 1$ uniformly on every compact set of $\mathbb{R}^p$.

ΔVΔ = VΔV = ΔVΔ = VΔV = ΔVΔ

SOLUTION: First assume that $p = 1$, and let

$$g_n = \sqrt{n}\,\mathbb{1}_{[-1/2n,1/2n]}.$$

Then $\|g_n\|_2 = 1$ and $\bar{g}_n = g_n$. If $f_n = \hat{g}_n$ is the Fourier transform of g_n, then

$$\|f_n\|_2 = \|g_n\|_2 = 1,$$

and

$$f_n * \tilde{f}_n = \hat{g}_n * \hat{\bar{g}}_n = \widehat{g_n \bar{g}_n} = \widehat{g_n^2} = \sqrt{n}\hat{g}_n = \sqrt{n} f_n.$$

Now, since $g_n \in L^1(\mathbb{R})$:

$$f_n(x) = \sqrt{n}\int_{-1/2n}^{1/2n} e^{-2\pi ixy}\,dy = \frac{1}{\sqrt{n}}\left(\frac{\sin\frac{\pi x}{n}}{\frac{\pi x}{n}}\right),$$

and it is clear that $\sqrt{n} f_n(x) \to 1$ uniformly on every compact set of $\mathbb{R}$. When $p > 1$ it suffices to consider the functions

$$f_n(x_1)f_n(x_2)\cdots f_n(x_p).$$

REMARK: To determine f_n one may draw some insight from the property that $\widehat{f_n * \tilde{f}_n} = |\hat{f}_n|^2$ must tend, in the sense of distributions, to the Dirac measure at the origin, whence the idea of taking an approximate identity for $|\hat{f}_n|^2$. It is not difficult to verify that $\|f_n\|_2 = 1$, for

$$\frac{1}{n}\int_{-\infty}^{+\infty}\left(\frac{\sin\frac{\pi x}{n}}{\frac{\pi x}{n}}\right)^2 dx = \frac{1}{\pi}\int_{-\infty}^{+\infty}\left(\frac{\sin x}{x}\right)^2 dx = 1.$$

The direct verification of the relation $f_n * f_n = \sqrt{n} f_n$ is a little more fastidious.

EXERCISE 8·144: Let V be a connected open set of $\mathbb{R}^p$ and let $f \in L^1_{\text{loc}}(V)$.

Show that the following conditions are equivalent:

(i): There exists a constant c such that $f(x) = c$ for almost all $x \in V$;

(ii): For every compact set $K \subset V$, as $h \to 0$

$$\int_K |f(x + h) - f(x)|\,dx = o(|h|).$$

$$\Delta\nabla\Delta = \nabla\Delta\nabla = \Delta\nabla\Delta = \nabla\Delta\nabla = \Delta\nabla\Delta$$

SOLUTION: First assume that f can be extended to an integrable function on $\mathbb{R}^p$. Let (φ_n) be an approximate identity formed by continuous functions such that $\varphi_n(x) = 0$ if $|x| \geqslant 1/n$, and let $f_n = f * \varphi_n$. If $x \in V$,

$$0 < \delta < \frac{1}{n} \qquad \text{and} \qquad K = B(x, \delta) \subset V,$$

then

$$\begin{aligned} |f_n(x + h) - f_n(x)| & \\ &\leqslant \int_{|y| \leqslant 1/n} |f(x + h - y) - f(x - y)|\,|\varphi_n(y)|\,dy \\ &\leqslant \|\varphi_n\|_\infty \int_K |f(y + h) - f(y)|\,dy = o(|h|). \end{aligned}$$

Thus, at every point of V the derivative of f_n is zero. Hence f_n is equal on V to a constant c_n. By choosing a suitable subsequence it can be assumed that $f_n(x) \to f(x)$ for almost all x. Consequently c_n tends to a limit c, and $f(x) = c$ for almost all $x \in V$.

In the general case, consider an increasing sequence of relatively compact *connected* open sets (V_n), the union of which is V, and which are such that $\bar{V}_n \subset V$. The function equal to f on V_n and to zero elsewhere is integrable and satisfies Condition (ii) of the problem for every compact set $K \subset V_n$ (for $K + h \subset V_n$ as soon as h is small enough). Hence there exists a constant c_n such that $f(x) = c_n$ almost everywhere on V_n. It follows immediately that all the c_n's are equal to the same constant c, and that $f(x) = c$ almost everywhere on V.

REMARK: It may be interesting to indicate how one proves that every *connected* open set V of $\mathbb{R}^p$ is the union of a sequence of connected open sets V_n such that $\bar{V}_n \subset V_{n+1}$, with the $\bar{V}_n$ being compact.

Let (W_n) be a sequence of relatively compact open sets with union V and such that $\bar{W}_n \subset W_{n+1}$. Let us show that for every compact set K such that $K \subset V$ there exists an integer n such that K is contained in a connected component of W_n. There certainly exists an integer m such that $K \subset W_m$. One then has $K \subset O_1 \cup \cdots \cup O_p$, where the O_i are connected components of W_m. Let us consider the connected compact set obtained by taking the union of $\bar{O}_1, \ldots, \bar{O}_p$ and of $(p - 1)$ polygonal lines contained in V and joining a point of O_i to a point of O_{i+1} $(1 \leqslant i \leqslant p - 1)$. This compact set is contained in a W_n, and as it is connected it is in fact contained in a connected component of W_n. This proves the assetion made above. From this it follows that it is possible to construct a sequence of integers $1 = n_0 < n_1 < \cdots$ and a sequence of open sets V_k such that:

$$\begin{cases} V_k \text{ is a connected component of } W_{n_k}, \\ \bar{W}_{n_k} \subset V_{k+1}. \end{cases}$$

It is clear that the V_k satisfy the required conditions.

EXERCISE 8·145: Let V be an open set of $\mathbb{R}$ and $f \in L^1_{loc}(V)$. Show that the following two conditions are equivalent:

(i): f is equal almost everywhere to a function g that has bounded variation on every compact interval contained in V;

(ii): For every compact interval $[a,b]$ contained in V,

$$\int_a^b |f(x + h) - f(x)|\,dx = O\,|h| \qquad \text{as } h \to 0.$$

Δ∇Δ = ∇Δ∇ = Δ∇Δ = ∇Δ∇= Δ∇Δ

SOLUTION: It may be assumed that V is an interval. If f is increasing on V, then for $0 < h \leqslant 1$

$$\int_a^b |f(x + h) - f(x)|\,dx = \int_b^{b+h} f(x)\,dx - \int_a^{a+h} f(x)\,dx \leqslant 2Mh,$$

where M denotes the upper bound of $|f|$ on $[a,b + 1]$. The case where h is negative is treated analogously. As every function with bounded variation is a linear combination of increasing functions, the implication (i) => (ii) is demonstrated.

In order to prove (ii) => (i) one can clearly assume that f is real. Let $[a,b] \subset V$. There exists $\delta > 0$ such that

$$[a - 2\delta, b + 2\delta] \subset V.$$

By replacing f by zero outside this interval, it can be assumed that f is integrable and that there exists a constant M such that

$$\int_{a-\delta}^{b+\delta} |f(x + h) - f(x)|\,dx \leqslant M|h| \qquad \text{if } |h| < \delta.$$

Then let (φ_i) be an approximate identity formed by continuously differentiable functions whose supports are contained in $]-\delta,\delta[$ and are such that $\|\varphi_i\|_1 = 1$. Extracting, if necessary, a subsequence, it may be assumed that

$$f_i(x) = (f*\varphi_i)(x) \to f(x) \quad \text{if } x \notin E,$$

E a set of measure zero. Lastly, by moving a slightly to the left it can be assumed that we still have $[a - 2\delta, b + 2\delta] \subset V$, and, further, that $a \notin E$. If $|h| < \delta$ then

$$\int_a^b \left|\frac{f_i(x+h) - f_i(x)}{h}\right| dx$$

$$\leqslant \frac{1}{|h|}\int_a^b dx \int_{-\delta}^{\delta} |f(x+h-y) - f(x-y)|\,|\varphi_i(y)|\,dy$$

$$\leqslant \frac{1}{|h|}\int_{-\delta}^{\delta} |\varphi_i(y)|\,dy \int_{a-\delta}^{b+\delta} |f(x+h) - f(x)|\,dx$$

$$\leqslant M.$$

The functions f_i are continuously differentiable, and consequently, by Fatou's Lemma

$$V(f_i;a,b) = \int_a^b |f_i'(x)|\,dx \leqslant M.$$

If $a = x_0 < x_1 < \cdots < x_n \leqslant b$, therefore for all i

$$\sum_{k=0}^{n-1} |f_i(x_{k+1}) - f_i(x_k)| \leqslant M.$$

Furthermore, if $x_k \notin E$ for $k = 0,\ldots,n$, then passing to the limit when $i \to \infty$ yields

$$\sum_{k=0}^{n-1} |f(x_{k+1}) - f(x_k)| \leqslant M.$$

Now consider the set of sequences $\Delta = \{a = x_0 < x_1 < \cdots < x_n \leqslant b\}$ such that $x_k \notin E$, and let us agree to write $\Delta \leqslant x$ if $x_n \leqslant x$. Set

$$P(\Delta) = \sum_{k=0}^{n-1} (f(x_{k+1}) - f(x_k))_+,$$

$$N(\Delta) = \sum_{k=0}^{n-1} (f(x_{k+1}) - f(x_k))_-.$$

By the preceding,

$$P(\Delta) + N(\Delta) \leqslant M.$$

Finally, when $a \leqslant x \leqslant b$ let

$$P(x) = \sup_{\Delta \leqslant x} P(\Delta),$$

$$N(x) = \sup_{\Delta \leqslant x} N(\Delta).$$

The numbers $P(x)$ and $N(x)$ are finite and increase with x. When $x \notin E$ we need only consider the sequences Δ for which $x_n = x$. In this case

$$P(\Delta) - N(\Delta) = f(x) - f(a),$$

so

$$P(x) - N(x) = f(x) - f(a).$$

Thus the function f coincides almost everywhere on $[a,b]$ with the function $P - N + f(a)$, which has bounded variation.

If $[a_n,b_n]$ is an increasing sequence of compact intervals the union of which is V, there exist functions g_n of bounded variation which coincide almost everywhere with f on $[a_n,b_n]$. One can assume that g_n is right continuous on $[a_n,b_n]$. It immediately follows from this that g_{n+1} coincides with g_n on $[a_n,b_n]$. Setting $g(x) = g_n(x)$ if $a_n \leqslant x \leqslant b_n$ defines a function on V with bounded variation on every compact interval contained in V, and which is equal to f almost everywhere.

EXERCISE 8·146: Let f be an integrable continuous function of bounded variation on $\mathbb{R}$, and $\hat{f}$ its Fourier transform.

Show that

$$\sum_{n=-\infty}^{+\infty} f(2\pi n) = \frac{1}{2\pi} \sum_{n=-\infty}^{+\infty} \hat{f}(n). \qquad \textit{(POISSON'S FORMULA)}$$

(Prove that the function

$$f^{\natural}(x) = \sum_{n=-\infty}^{+\infty} f(x + 2\pi n)$$

is continuous, has period 2π, and has bounded variation on $[0,2\pi]$, and then use exercise 3·42).

$$\Delta\nabla\Delta = \nabla\Delta\nabla = \Delta\nabla\Delta = \nabla\Delta\nabla = \Delta\nabla\Delta$$

SOLUTION: Let

$$\begin{cases} V(x) = V(f;-\infty,x), \\ \lim_{x\to\infty} V(x) = V < \infty. \end{cases}$$

Proceeding as in exercise 3·42 it is seen that $f^{\natural}(x)$ is defined for almost all x, is integrable on x, and that

$$\int_0^{2\pi} f^{\natural}(x)\mathrm{d}x = \int_{-\infty}^{+\infty} f(x)\mathrm{d}x. \qquad (1)$$

Let x_0 be a point where $f(x_0)$ is defined. If $x_0 \leqslant x \leqslant x_0 + 2\pi$ then

$$|f(x + 2\pi n) - f(x_0 + 2\pi n)| \leqslant V(x + 2\pi n) - V(x_0 + 2\pi n)$$

$$\leqslant V(x_0 + 2\pi(n + 1)) - V(x_0 + 2\pi n),$$

and furthermore

$$\sum_{n=-\infty}^{+\infty} (V(x_0 + 2\pi(n + 1)) - V(x_0 + 2\pi n) = V < \infty.$$

This shows that $f^{\natural}$ is defined and continuous on $[x_0, x_0 + 2\pi]$, and consequently on the whole of $\mathbb{R}$.

If $0 = x_0 < x_1 < \cdots < x_k = 2\pi$ then

$$\sum_{i=0}^{k-1} |f^{\natural}(x_{i+1}) - f^{\natural}(x_i)|$$

$$= \sum_{i=0}^{k-1} \left| \sum_{n=-\infty}^{+\infty} f(x_{i+1} + 2\pi n) - f(x_i + 2\pi n) \right|$$

$$\leqslant \sum_{i=0}^{k-1} \sum_{n=-\infty}^{+\infty} |f(x_{i+1} + 2\pi n) - f(x_i + 2\pi n)|$$

$$\leqslant V.$$

Applying (1) to the function $e^{-inx} f(x)$ yields

$$\frac{1}{2\pi}\int_0^{2\pi} e^{-inx} f^{\natural}(x)dx = \frac{1}{2\pi}\int_{-\infty}^{+\infty} e^{-inx} f(x)dx = \frac{1}{2\pi} \hat{f}(n). \tag{2}$$

By the Jordan-Dirichlet Theorem and (2),

$$\sum_{n=-\infty}^{+\infty} f(2\pi n) = f^{\natural}(0) = \sum_{n=-\infty}^{+\infty} \frac{1}{2\pi}\int_0^{2\pi} e^{-inx} f^{\natural}(x) = \frac{1}{2\pi}\sum_{n=-\infty}^{+\infty} \hat{f}(n).$$

EXERCISE 8·147: (a): Let $I =]a,b[$ and f,g be locally integrable functions on I.

Show that the following properties are equivalent:

(i): After modification of f on a set of measure zero

$$f(y) - f(x) = \int_x^y g(t)\mathrm{d}t, \qquad a < x \leqslant y < b;$$

(ii): For every function $\varphi \in \mathcal{D}^{\infty}(I)$:

$$\int_a^b f(x)\varphi'(x)\mathrm{d}x + \int_a^b g(x)\varphi(x)\mathrm{d}x = 0.$$

Next, show that if the pair (f,g_1) has the same properties, then $g = g_1$ almost everywhere. We then make the convention $g = Df$. If g is continuous it will be noted that $Df = f'$.

(b): Let

$$H^2(I) = \{f : f \in L^2(I), Df \in L^2(I)\}.$$

Prove that the scalar product

$$(f|g) = \int_a^b (f(x)\overline{g(x)} + Df(x).\overline{Dg(x)})\mathrm{d}x$$

defines a Hilbert space structure on $H^2(I)$.

(c): If $f \in H^2(\mathbb{R})$ show that

$\widehat{Df}(y) = iy\hat{f}(y)$ almost everywhere.

(d): Let $b = \infty$ and $f \in H^2(I)$.
Show that $\lim_{x\to\infty} f(x) = 0$.

(e): Assume that a,b are finite. By (i) every function of $H^2(I)$ can be extended by continuity to a and b.
Show that if $a \leqslant c \leqslant b$ there exists $\psi_c \in H^2(I)$ such that

$$f(c) = \int_a^b f(x)\psi_c(x)dx + \int_a^b Df(x)D\psi_c(x)dx, \qquad f \in H^2(I).$$

Determine ψ_c explicitly.

(f): Again assume that a,b are finite and set

$$H_0^2(I) = \{f : f \in H^2(I), f(a) = f(b) = 0\}.$$

Show that $\mathcal{D}^\infty(I)$ is dense in $H_0^2(I)$, and that for $a < c < b$ there exists $\vartheta_c \in H_0^2(I)$ such that

$$f(c) = \int_a^b Df(x).D\vartheta_c(x)dx, \qquad f \in H_0^2(I).$$

(g): Let f be an indefinitely differentiable function on $\mathbb{R}^2 - \{0\}$, zero for $|x| \geqslant 1$, and equal to $\log(\log(1/|x|))$ for $0 < |x| \leqslant \frac{1}{2}$.
Show that $f, \partial f/\partial x_1, \partial f/\partial x_2 \in L^2(\mathbb{R}^2)$, and that for $\varphi \in \mathcal{D}^\infty(\mathbb{R}^2)$

$$\int_{\mathbb{R}^2}\left(f\frac{\partial\varphi}{\partial x_i} + \frac{\partial f}{\partial x_i}\varphi\right) = 0, \qquad i = 1,2.$$

What conclusion can be drawn from this?

Δ∇Δ = ∇Δ∇ = Δ∇Δ = ∇Δ∇ = Δ∇Δ

SOLUTION (a): If f, g satisfy (i) and $\varphi \in \mathcal{D}^{\infty}(I)$, let α, β be such that $a < \alpha < \beta < b$ and $\varphi(x) = 0$ if $x \notin [\alpha, \beta]$. Then

$$\int_a^b g(x)\varphi(x)dx = \int_\alpha^\beta g(x)dx\int_\alpha^x \varphi'(t)dt = \int_\alpha^\beta \varphi'(t)dt\int_t^\beta g(x)dx$$

$$= f(\beta)\int_\alpha^\beta \varphi'(t)dt - \int_\alpha^\beta f(t)\varphi'(t)dt$$

$$= -\int_a^b f(t)\varphi'(t)dt.$$

Thus (i) => (ii).

Now assume that (ii) is satisfied. Let $a < x_0 < b$, and

$$f_1(x) = \int_{x_0}^x g(t)dt, \qquad a < x < b.$$

By what has just been proved,

$$\int_a^b f_1\varphi' + \int_a^b g\varphi = 0, \qquad \varphi \in \mathcal{D}^{\infty}(I).$$

Comparing this with (ii) yields

$$\int_a^b (f - f_1)\varphi' = 0, \qquad \varphi \in \mathcal{D}^{\infty}(I).$$

When φ runs over $\mathcal{D}^{\infty}(I)$ φ' runs over the set of functions $\psi \in \mathcal{D}^{\infty}(I)$ such that $\int \psi = 0$. By exercise 8·141 there exists a constant C such that $f - f_1 = C$ almost everywhere, which proves (ii) => (i).

Lastly, if the pairs $(f,g), (f,g_1)$ satisfy (ii), by substraction one obtains

$$\int_a^b (g - g_1)\varphi = 0, \qquad \varphi \in \mathcal{D}^\infty(I),$$

whence $g = g_1$ almost everywhere.

SOLUTION (b): The Formula defines a positive alternating form on $H^2(I)$, and $(f|f) = 0$ implies that $\|f\|_{L^2(I)}$, whence $f = 0$. It remains to prove that $H^2(I)$ is complete in the norm associated with this scalar product. If (f_i) is a Cauchy sequence in $H^2(I)$, (f_i) and (Df_i) are Cauchy sequences in $L^2(I)$. Hence there exist $f, g \in L^2(I)$ such that $f_i \to f$, $Df_i \to g$ in $L^2(I)$. If $\varphi \in \mathcal{D}^\infty(I)$, then

$$\int_a^b f_i \cdot \varphi' + \int_a^b Df_i \cdot \varphi = 0,$$

and on passing to the limit:

$$\int_a^b f\varphi' + \int_a^b g\varphi = 0.$$

Consequently $g = Df$ and $f_i \to f$ in $H^2(I)$.

SOLUTION (c): If $\varphi \in \mathcal{D}^\infty(\mathbb{R})$, then by Plancherel's Formula and the relation $\widehat{\varphi'}(y) = iy\hat{\varphi}(y)$,

$$\int_{-\infty}^{+\infty} \widehat{Df}(y)\overline{\hat{\varphi}(y)}dy = \int_{-\infty}^{+\infty} Df(x)\overline{\varphi(x)}dx = -\int_{-\infty}^{+\infty} f(x)\overline{\varphi'(x)}dx$$

$$= -\int_{-\infty}^{+\infty} \hat{f}(y)\overline{iy\hat{\varphi}(y)}dy = \int_{-\infty}^{+\infty} iy\hat{f}(y)\overline{\hat{\varphi}(y)}dy.$$

As the $\varphi \in \mathcal{D}^\infty(\mathbb{R})$ are dense in $L^2(\mathbb{R})$, and as the Fourier transformation is an isometry of $L^2(\mathbb{R})$ onto itself, the $\hat{\varphi}$ are dense in

$L^2(\mathbb{R})$. Hence $\widehat{Df}(y) = iy\hat{f}(y)$ almost everywhere.

SOLUTION (d): *First Proof:* Note that if $A = \|Df\|_{L^2(I)}$ and $a < x < y$, then

$$|f(y) - f(x)| \leqslant A\sqrt{y - x}.$$

If $f(x)$ were not to tend to zero as $x \to \infty$ there would exist $\varepsilon > 0$ and a sequence of points x_n such that

$$x_{n+1} > x_n + \frac{\varepsilon^2}{4A^2}, \qquad |f(x_n)| \geqslant \varepsilon.$$

Consequently, for $x_n \leqslant x \leqslant x_n + \varepsilon^2/4A^2$

$$|f(x)| \geqslant |f(x_n)| - |f(x) - f(x_n)| \geqslant \varepsilon - A\frac{\varepsilon}{2A} = \frac{\varepsilon}{2}.$$

But then

$$\int_a^\infty |f(x)|^2 dx \geqslant \sum_n \int_{x_n}^{x_n + c^2/4A^2} |f(x)|^2 dx \geqslant \sum_n \frac{\varepsilon^2}{4}\frac{\varepsilon^2}{4A^2} = \infty.$$

Second Proof: We can reduce to the case where $I = \mathbb{R}$ by the fol- artifice: if $a > -\infty$, f is extended by continuity at a (cf. (i) of part (a)); let $\alpha < a$, and let ϑ be a continuous function with support contained in $[\alpha,a]$ and such that

$$\int_\alpha^a \vartheta(t)dt = f(a).$$

By extending Df to $[-\infty,a]$ by setting $Df(t) = \vartheta(t)$ for $t \leqslant a$, and setting

$$f(x) = f(a) + \int_a^x Df(t)dt, \qquad x \in \mathbb{R},$$

an extension of f is obtained that belongs to $H^2(\mathbb{R})$. Now, for $f \in H^2(\mathbb{R})$ the functions $\hat{f}(y)$ and $iy\hat{f}(y)$ belong to $L^2(\mathbb{R})$ by part (c). From this it follows that $(1 + y^2)^{\frac{1}{2}}\hat{f}(y) \in L^2(\mathbb{R})$; since $(1 + y^2)^{-\frac{1}{2}} \in L^2(\mathbb{R})$ we have $\hat{f} \in L^1(\mathbb{R})$. But then f is the inverse Fourier transform of $\hat{f}$, and therefore tends to zero at infinity.

SOLUTION (e): Let us show that $f \to f(c)$ is a continuous linear functional on $H^2(I)$. To do that, write

$$f(c) = f(x) + \int_x^c Df(t)dt, \qquad a \leq x \leq b,$$

so

$$|f(c)| \leq (b - a)^{\frac{1}{2}}\|Df\|_{L^2(I)} + |f(x)|,$$

or

$$|f(c)|^2 \leq 2(b - a)\|Df\|^2_{L^2(I)} + 2|f(x)|^2. \tag{*}$$

Integrating with respect to x from a to b yields

$$|f(c)|^2 \leq 2(b - a)\|Df\|^2_{L^2(I)} + \frac{2}{b - a}\|f\|^2_{L^2(I)}, \tag{**}$$

which proves our assertion.

The existence and uniqueness of ψ_c results from this. In particular, if $\varphi \in \mathcal{D}^\infty(I)$,

$$\int_a^c \varphi'(x)dx = \int_a^b \varphi(x)\psi_c(x)dx + \int_a^b \varphi'(x)D\psi_c(x)dx,$$

which can again be written, by introducing the characteristic function χ_c of the interval $[a,c]$,

$$\int_a^b \varphi'(x)\{D\psi_c - \chi_c(x)\}dx + \int_a^b \varphi(x)\psi_c(x)dx = 0.$$

Comparison with (ii) of part (a) shows that this implies

$$D\psi_c - \chi_c \in H^2(I) \qquad \text{and} \qquad D(D\psi_c - \chi_c) = \psi_c.$$

Since ψ_c is continuous $D\psi_c - \chi_c$ is continuously differentiable and its derivative is ψ_c. In other words,

$$\begin{cases} (1) & \psi_c \text{ is continuous;} \\ (2) & \psi_c'' = \psi \quad \text{on } [a,b] - \{c\}; \\ (3) & \psi_c'(c - 0) - \psi_c'(c + 0) = 1. \end{cases}$$

Furthermore, if $a < \gamma < c$ and if $f(x) = (\gamma - x)_+$, then $Df = -\chi_\gamma$, and consequently,

$$0 = \int_a^\gamma (\gamma - x)\psi_c(x)dx - \int_a^\gamma D\psi_c(x)dx,$$

i.e.,

$$\psi_c(\gamma) - \psi_c(a) = \int_a^\gamma (\gamma - x)\psi_c(x)dx.$$

If M is the maximum of $|\psi_c|$ on I, it follows from this that

$$|\psi_c(\gamma) - \psi_c(a)| \leqslant M\int_a^\gamma (\gamma - x)dx \leqslant \tfrac{1}{2}M(\gamma - a)^2,$$

and consequently $\psi'(a) = 0$. Proceeding analogously for the other end point, to conditions (1),(2),(3) can be added

(4) $\psi_c'(a) = \psi_c'(b) = 0$

(this is in the case where $a < c < b$). From (2) and (4) one obtains

$$\psi(c) = \begin{cases} A\cosh(x - a), & a < x < c, \\ B\cosh(b - x), & c < x \leqslant b. \end{cases}$$

Conditions (1) and (2) imply the system

$$\begin{cases} A\cosh(c - a) - B\cosh(b - c) = 0, \\ A\sinh(c - a) + B\sinh(b - c) = 1. \end{cases}$$

From this it is not difficult to deduce that

$$\psi_c(x) = \begin{cases} \dfrac{\cosh(b - c)\cosh(x - a)}{\sinh(b - a)}, & a \leqslant x \leqslant c, \\[2ex] \dfrac{\cosh(c - a)\cosh(b - x)}{\sinh(b - a)}, & c \leqslant x \leqslant b. \end{cases}$$

This expression for ψ_c is still valid when $c = a$ or $c = b$. (If $c = a$, for example, then (3) disappears and (4) reduces to $\psi_c'(b) = 0$).

SOLUTION (f): It is clear that $\mathcal{D}^\infty(I) \subset H_0^2(I)$. Furthermore, $H_0^2(I)$ is the orthogonal complement of $\{\psi_a, \psi_b\}$. To prove that $\mathcal{D}^\infty(I)$ is dense in $H_0^2(I)$ it suffices to prove that every $f \in H^2(I)$ orthogonal to $\mathcal{D}^\infty(I)$ is a linear combination of ψ_a and ψ_b. Now, the condition:

$$\int_a^b f(x)\varphi(x)\mathrm{d}x + \int_a^b Df(x)\varphi'(x)\mathrm{d}x = 0, \qquad \varphi \in \mathcal{D}^\infty(I),$$

implies that $Df \in H^2(I)$ and $D(Df) = f$; but then as f and Df are continuous this latter condition can be written $f'' = f$; the orthogonal complement of $\mathcal{D}^\infty(I)$ is thus two-dimensional; because it contains ψ_a and ψ_b the result is proved.

The mapping $f \to Df$ is a continuous *bijection* of $H_0^2(I)$ onto the orthogonal complement of 1 in $L^2(I)$. By a theorem of Banach this mapping is bicontinuous, which means that $f \to \|Df\|_{L^2(I)}$ is, on $H_0^2(I)$, a norm equivalent to that induced by the norm of $H^2(I)$. The existence (and uniqueness) of ϑ_c follows from this.

The orthogonal projection of χ_c onto the orthogonal of 1 in $L^2(I)$ is:

$$\chi_c(x) - \frac{1}{b-a}\int_a^b \chi_c(t)\mathrm{d}t = \begin{cases} \dfrac{b-c}{b-a}, & a \leqslant x \leqslant c, \\[2ex] -\dfrac{c-a}{b-a}, & c \leqslant x \leqslant b, \end{cases}$$

and consequently

$$\vartheta_c(x) = \begin{cases} \dfrac{(b-c)(x-a)}{(b-a)}, & a \leqslant x \leqslant c, \\[2ex] \dfrac{(c-a)(b-x)}{(b-a)}, & c \leqslant x \leqslant b, \end{cases}$$

REMARK: One can avoid using Banach's theorem by showing that

$$f \to \left(|f(a)|^2 + \int_a^b |Df(t)|^2 \mathrm{d}t\right)^{\frac{1}{2}}$$

is a norm on $H^2(I)$ equivalent to the one introduced in part (b) above. One of the two necessary bounds follows from (**) proved in part (e); the second results from (*), this time integrated with respect to c, setting $x = a$.

SOLUTION (g): Set

$$x = (x_1, x_2), \qquad |x| = r = (x_1^2 + x_2^2)^{\frac{1}{2}}.$$

Then

$$\int_{|x|<\frac{1}{2}} |f(x)|^2 dx = 2\pi\int_0^{\frac{1}{2}} r(\log\log\frac{1}{r})^2 dr < \infty.$$

Moreover, if $0 < r < \frac{1}{2}$,

$$\left|\frac{\partial f}{\partial x_i}\right| = \left|\frac{x_i}{r} f'(r)\right| \leqslant |f'(r)| = \frac{1}{r|\log r|},$$

whence

$$\int_{|x|\leqslant\frac{1}{2}} \left|\frac{\partial f}{\partial x_i}(x)\right|^2 dx$$

$$\leqslant 2\pi\int_0^{\frac{1}{2}} \frac{dr}{r|\log r|^2} \qquad \text{(BERTRAND'S INTEGRAL!)}$$

Finally, if $\varphi \in \mathcal{D}^\infty(\mathbb{R}^2)$, and if

$$\omega = f\varphi dx_2 \quad \text{on } \mathbb{R}^2 - \{0\},$$

then

$$d\omega = \left(\frac{\partial f}{\partial x_1}\varphi + f\frac{\partial\varphi}{\partial x_1}\right) dx_1 \wedge dx_2,$$

so (by Stokes' theorem)

$$\int_{0<\varepsilon\leqslant|x|\leqslant R} \left(\frac{\partial f}{\partial x_1}\varphi + f\frac{\partial\varphi}{\partial x_1}\right) dx = \int_{|x|=R} f\varphi dx_2 - \int_{|x|=\varepsilon} f\varphi dx_2.$$

If $R > 1$ this equation can be written as

$$\int_{|x|\geqslant\varepsilon} \left(\frac{\partial f}{\partial x_1}\varphi + f\frac{\partial\varphi}{\partial x_1}\right) dx$$

$$= -\varepsilon\log\log\frac{1}{\varepsilon}\int_0^{2\pi} \varphi(\varepsilon\cos\vartheta, \varepsilon\sin\vartheta)\cos\vartheta d\vartheta.$$

Making $\varepsilon \to 0$, this yields

$$\int_{\mathbb{R}^2}\left(\frac{\partial f}{\partial x_1}\varphi + f\frac{\partial \varphi}{\partial x_1}\right)dx = 0.$$

An identical relation is obtained by replacing x_1 with x_2.

This shows that the space $H^2(\mathbb{R}^2)$, analogous to $H^2(\mathbb{R})$ in two dimensions, is not made up only of continuous functions. This property can be linked with the second proof of part (d) and the fact $(1 + |x|^2)^{-\frac{1}{2}}$ does not belong to $L^2(\mathbb{R}^2)$.

EXERCISE 8·148: All the functions considered are measurable and have period 2π. $\int f$ means $\int_0^{2\pi} f(x)dx$, and $f \in L^1$ if $\int|f| < \infty$.

(a): Assume that

$$\int|f|\log(1 + |f|) < \infty. \tag{*}$$

Show that $f \in L^1$, and that if g is a function such that $\exp(\lambda^{-1}|g|) \in L^1$ for a real number $\lambda > 1$, then the convolution product $f*g$ is defined everywhere, and

$$|f*g| \leqslant \lambda\log\lambda\int|f| + \lambda\int|f|\log(1 + |f|) + \int(e^{\lambda^{-1}|g|} - 1).$$

(Use the inequality

$$ab \leqslant a\log(1 + a) + e^b - 1,$$

valid if $a \geqslant 0$ and $b \geqslant 0$).

(b): Set

$$G(x) = \log\left|\frac{1}{\cos x}\right|,$$

and $f_n(x) = f(x)$ if $|f(x)| \geqslant n$, $f_n(x) = 0$ otherwise.

Show that $f*G$ is defined everywhere and that there exists a continuous function h_n such that

$$f*G = h_n + f_n*G.$$

Conclude from this that $f*G$ is continuous.

Δ∇Δ = ∇Δ∇ = ΔVΔ = ∇Δ∇ = ΔVΔ

SOLUTION (a): Noting that $\log 3 \geqslant 1$ gives

$$\int|f| = \int_{|f|\leqslant 2}|f| + \int_{|f|>2}|f| \leqslant 4\pi + \int|f|\log(1 + |f|),$$

which proves that $f \in L^1$. Furthermore, if $a, b \geqslant 0$ then

$$ab \leqslant a\log(1 + a) + e^b - 1. \qquad \text{(YOUNG'S INEQUALITY)}$$

Consequently, if $\lambda > 1$

$$\int|f(x)g(y - x)|dx < \lambda\int|f|\log(1 + \lambda|f|) + \int(e^{\lambda^{-1}|g|} - 1).$$

Now

$$\log(1 + \lambda|f|) \leqslant \log\lambda + \log(1 + |f|),$$

so

$$|f*g| \leqslant \lambda\log\lambda\int|f| + \lambda\int|f|\log(1 + |f|) + \int(e^{\lambda^{-1}|g|} - 1).$$

SOLUTION (b): If x_0 is a root of $\cos x$, then as $x \to x_0$

$$G(x) \sim \log\left|\frac{1}{x - x_0}\right| ,$$

and consequently

$$e^{\lambda^{-1}G(x)} \sim |x - x_0|^{-1/\lambda}$$

This shows that $G \in L^1$ and $\exp(\lambda^{-1}G) \in L^1$ if $\lambda > 1$. The function $f*g$ is therefore defined everywhere, as is f_n*G, moreover, since f_n also satisfies (*). As for $h_n = (f - f_n)*G$, this is also a function defined everywhere, since $f - f_n$ is bounded. More precisely, h_n is continuous, as $L^\infty * L^1 \subset C$. For $\lambda > 1$ and every integer n

$$\|f*G - h_n\|_\infty$$

$$\leqslant \lambda\log\lambda\int|f_n| + \lambda\int|f_n|\log(1 + |f_n|) + \int(e^{\lambda^{-1}G} - 1).$$

Since $|f_1| \geqslant |f_2| \geqslant \cdots$ and $|f_n| \to 0$, this yields

$$\limsup_{n\to\infty}\|f*G - h_n\|_\infty \leqslant \int(e^{\lambda^{-1}G} - 1).$$

Making $\lambda \to \infty$ shows that $h_n \to f*G$ uniformly, and therefore that $f*G$ is continuous.

EXERCISE 8·149: Prove that the algebra $L^1(\mathbb{R}^n)$ does not have a unit element.

Δ∇Δ = ∇Δ∇ = Δ∇Δ = ∇Δ∇ = Δ∇Δ

FIRST SOLUTION: If $f \in L^1$ were such that $f*g = g$ for all $g \in L^1$, then taking the Fourier transform gives $\hat{f}\hat{g} = \hat{g}$. Now, for all $x \in \mathbb{R}^n$ there exists $g \in L^1$ such that $\hat{g}(x) \neq 0$, and consequently $\hat{f}(x) \equiv 1$, which is impossible, because $\hat{f}$ must tend to zero at infinity.

SECOND SOLUTION: There would exist $\alpha > 0$ such that

$$\int_{|x|\leqslant\alpha} |f(x)|\,dx \leqslant 1.$$

Let φ be the characteristic function of the ball with centre zero and radius $\alpha/2$. Then if $|x| \leqslant \alpha/2$ one would have

$$f*\varphi(x) = \int_{|y|\leqslant\alpha} f(x-y)\,dy \leqslant \int_{|y|\leqslant\alpha} |f(y)|\,dy < 1,$$

a contradiction.

EXERCISE 8·150: Prove that the algebra $L^1(\mathbb{R}^n)$ possesses divisors of zero.

Δ∇Δ = ∇Δ∇ = Δ∇Δ = ∇Δ∇ = Δ∇Δ

SOLUTION: *First Proof:* Let V_1 and V_2 be two non-empty and disjoint bounded open sets of $\mathbb{R}^n$. Let φ_1,φ_2 be non-zero indefinitely differentiable functions compactly supported in V_1,V_2 respectively. If $f_1 = \bar{F}(\varphi_1)$, $f_2 = \bar{F}(\varphi_2)$, then $f_1 \in L^1(\mathbb{R}^n)$, $f_2 \in L^1(\mathbb{R}^n)$, $f_1 \neq 0$, $f_2 \neq 0$, and $f_1*f_2 = 0$, since $F(f_1*f_2) = \varphi_1\varphi_2 = 0$.

Second Proof: We shall assume that $n = 1$. If $a \in \mathbb{R}$, denote by τ_a the operator of translation by a, that is to say, that $\tau_a f(x) = f(x-a)$. It is known that τ_a is an operator on L^1 with norm 1, and that

$$\tau_a \circ \tau_b = \tau_{a+b}, \qquad \tau_a(f*g) = \tau_a f*g.$$

For every summable sequence $\alpha = (\alpha_n)_{n\in\mathbb{Z}}$:

$$\tau_\alpha = \sum_{n=-\infty}^{+\infty} \alpha_n \tau_n \tag{1}$$

is a continuous operator of L^1 (the series (1) converges normally in $L(L^1,L^1)$). If $\beta = (\beta_n)_{n\in\mathbb{Z}}$ is another summable sequence, then

$$\tau_\alpha \circ \tau_\beta = \tau_\gamma, \qquad \gamma_n = \sum_{p+q=n} \alpha_p \beta_q. \tag{2}$$

Now assume that we have determined α, β so that

$$\tau_\alpha \circ \tau_\beta = 0, \quad \alpha \neq 0, \quad \beta \neq 0. \tag{3}$$

If $\varphi \in L^1$, then

$$(\tau_\alpha \varphi * \tau_\beta \varphi) = (\tau_\alpha \circ \tau_\beta)(\varphi * \varphi) = 0.$$

It will be possible to ensure that $\tau_\alpha \varphi \neq 0$, $\tau_\beta \varphi \neq 0$ by taking for φ a function zero outside $]0,1[$ and such that $\|\varphi\|_1 > 0$; then the functions $\tau_n \varphi$ will have disjoint supports, so that, for example,

$$\|\tau_\alpha \varphi\|_1 = \|\varphi\|_1 \sum_n |\alpha_n| > 0.$$

In order to determine α, β so that (3) is satisfied, note that if

$$u(x) = \sum_n \alpha_n e^{inx}, \qquad v(x) = \sum_n \beta_n e^{inx},$$

then:

$$uv = 0, \tag{4}$$

and conversely (4) implies (3). Consider the function $|\sin x|$; its Fourier coefficients are

$$c_n = \frac{1}{2\pi}\int_{-\pi}^{\pi} e^{-inx} |\sin x|\, dx = \frac{1}{\pi}\int_0^{\pi} \cos nx \sin x\, dx.$$

It can be shown that

$$c_{2n+1} = 0, \qquad c_{2n} = \frac{2}{\pi}\frac{1}{1-4n^2}.$$

The function $|\sin x|$ being continuous and piecewise continuously differentiable, the elementary theory of Fourier series ensures that

$$|\sin x| = \sum_{n=-\infty}^{+\infty} c_n e^{inx}.$$

But then

$$\begin{aligned}(\sin x)_+ &= \frac{1}{4i}(e^{ix} - e^{-ix}) + \tfrac{1}{2}\sum_{n=-\infty}^{+\infty} c_n e^{inx},\\ -(\sin x)_- &= \frac{1}{4i}(e^{ix} - e^{-ix}) - \tfrac{1}{2}\sum_{n=-\infty}^{+\infty} c_n e^{inx},\end{aligned} \tag{5}$$

so that the sequences α and β defined by

$$\begin{cases} \alpha_1 = \beta_1 = \dfrac{1}{4i}, \\ \alpha_{-1} = \beta_{-1} = -\dfrac{1}{4i}, \\ \alpha_{2n} = -\beta_{2n} = \dfrac{1}{\pi}\dfrac{1}{1-4n^2}, \\ \alpha_{2n+1} = \beta_{2n+1} = 0 \quad \text{if } n \neq 0 \text{ and } -1, \end{cases}$$

satisfy (3). One can, if one wants, obtain real sequences on replacing x by $x + \pi/2$ in (5), which gives

$$\begin{cases} \alpha_1 = \beta_1 = \alpha_{-1} = \beta_{-1}, \\ \alpha_{2n} = -\beta_{2n} = \dfrac{1}{\pi}\dfrac{(-1)^n}{1-4n^2}, \qquad \alpha_{2n+1} = \beta_{2n+1} = 0 \quad \text{if } n \neq 0 \text{ and } -1. \end{cases}$$

EXERCISE 8·151: Let L_+ be the set of functions locally integrable on $\mathbb{R}$ and zero on $]-\infty,0[$. Two functions of L_+ that are equal almost everywhere are identified.

(a): Show that if $f,g \in L_+$, then they are convolvable, and $f*g \in L_+$.

(b): Show that if $f \in L_+$ and if $f*f = 0$ almost everywhere on $[0,2a]$, $a > 0$, then for every integer $n \geqslant 1$

$$\left|\int_{-a}^{a} e^{nx} f(a - x)\mathrm{d}x\right|^2 \leqslant \iint_{\substack{u>-a\\ v>-a\\ u+v>0}} |f(a - u)f(a - v)|\,\mathrm{d}u\mathrm{d}v.$$

Using Exercise 3·72, deduce from this that $f = 0$ almost everywhere on $[0,a]$.

(c): Show that if $f,g \in L_+$ and $f*g = 0$, then, setting $f_1(x) = xf(x)$, $g_1(x) = xg(x)$,

$$f*g_1 + f_1*g = 0,$$

and consequently $f*g_1 = 0$.

(d): Conclude from the preceding that the algebra L_+ does not possess divisors of zero (*TITCHMARSH'S THEOREM*).

$\Delta\nabla\Delta = \nabla\Delta\nabla = \Delta\nabla\Delta = \nabla\Delta\nabla = \Delta\nabla\Delta$

SOLUTION (a): Let $M > 0$ and let $f_M(x) = f(x)$ if $x \leqslant M$, $f_M(x) = 0$ if $x > M$, g_M being define analogously.

Then if $x \leqslant M$,

$$\int_{-\infty}^{+\infty} |f(y)g(x - y)|\mathrm{d}y = \int_0^{x^+} |f(y)g(x - y)|\mathrm{d}y = \qquad \text{(Contd)}$$

(Contd)
$$= \int_0^{x^+} f_M(y)g_M(x - y)\, dy$$

$$= \int_{-\infty}^{+\infty} f_M(y)g_M(x - y)\, dy.$$

Since f_M, g_M are integrable, it follows first that $(f*g)(x)$ is defined for almost all $x \leqslant M$, and consequently for almost all the $x \in \mathbb{R}$, and then that

$$(f*g)(x) = \int_0^{x^+} f(y)g(x - y)dy.$$

This shows that $(f*g)(x) = 0$ if $x < 0$. Furthermore, for almost all $x \in [0,M]$

$$(f*g)(x) = (f_M*g_M)(x),$$

which proves that $f*g$ is locally integrable, and hence in L_+.

SOLUTION (b):

$$\left(\int_{-a}^{a} e^{nx}f(a - x)dx\right)^2 = \iint\limits_{\substack{|u|\leqslant a\\|v|\leqslant a}} e^{n(u+v)}f(a - u)f(a - v)dudv$$

$$= \iint\limits_{\substack{u\geqslant -a\\v\geqslant -a\\u+v\leqslant 0}} e^{n(u+v)}f(a - u)f(a - v)dudv$$

$$+ \iint\limits_{\substack{u\leqslant a\\v\leqslant a\\u+v\geqslant 0}} e^{n(u+v)}f(a - u)f(a - v)dudv.$$

Carrying out the changes of variables $u = a - y$, $v = a - x + y$ in the last integral yields

$$\iint_{0\leqslant y\leqslant x\leqslant 2a} e^{n(2a-x)} f(y)f(x - y)\,dx\,dy$$

$$= \int_0^{2a} e^{n(2a-x)}(f*f)(x)\,dx = 0.$$

Since $e^{n(u+v)} \leqslant 1$ if $u + v \leqslant 0$,

$$\left|\int_{-a}^{a} e^{nx} f(a - x)\,dx\right|^2 \leqslant \iint_{\substack{u\geqslant -a\\ v\geqslant -a\\ u+v\leqslant 0}} |f(a - u)f(a - v)|\,du\,dv. \tag{1}$$

Now note that

$$\left|\int_{-a}^{0} e^{nx} f(a - x)\,dx\right| \leqslant \int_{a}^{2a} |f(x)|\,dx. \tag{2}$$

(1) and (2) imply

$$\sup_n \left|\int_0^{a} e^{nx} f(a - x)\,dx\right| \leqslant M < \infty. \tag{3}$$

By exercise 3·72(b) we therefore have $f(a - x) = 0$ for almost all $x \in [0,a]$.

REMARK: For use in the rest of the exercise, note that this shows that $f*f = 0$ implies $f = 0$.

SOLUTION (c): For almost all $x \geqslant 0$

$$(f*g_1)(x) + (f_1*g)(x) = \qquad \text{(Contd)}$$

(Contd)
$$= \int_0^x (x - y)f(y)g(x - y)dy + \int_0^x yf(y)g(x - y)dy$$

$$= x\int_0^x f(y)g(x - y)dy = 0.$$

But then

$$(f*g_1)*(f*g_1) = - (f*g_1)*(f_1*g)$$

$$= - (f*g)*(f_1*g_1) = 0,$$

and by the above remark

$$f*g_1 = 0. \tag{4}$$

SOLUTION (d): Let $g_n(x) = x^n g(x)$, $n \geqslant 0$ an integer. From (4) one deduces by induction

$$f*g_n = 0, \qquad n \geqslant 0.$$

In other words, for almost all x and all $n \geqslant 0$

$$\int_0^x y^n f(x - y)g(y)dy = 0, \tag{5}$$

and consequently for almost all x

$$\int_{-\infty}^{+\infty} |f(x - y)g(y)|dy = \int_0^x |f(x - y)g(y)|dy = 0.$$

By Fubini's Theorem

$$0 = \int_{-\infty}^{+\infty} dx\int_{-\infty}^{+\infty} |f(x - y)g(y)|dy = \int_{-\infty}^{+\infty} |f(x)|dx\int_{-\infty}^{+\infty} |g(y)|dy,$$

and therefore one of the two functions f or g is zero almost everywhere, which proves that L_+ does not admit a divisor of zero.

REMARK: (5) implies that for almost all x there exists a set E_x with measure zero such that

$$f(x - y)g(y) = 0, \qquad y \notin E_x.$$

But since the E_x vary with x, and x runs over a non-denumerable set, the only way of concluding from this that one of the functions is zero almost everywhere is to use Fubini's Theorem.

However, it is interesting to note that this difficulty does not occur if f,g are assumed continuous and that the proof of (1) makes an appeal in this case only to an elementary form of Fubini's Theorem (for functions continuous on rectangles or triangles). Still, the proof of the fact that for a continuous function f,

$$\sup_n \left| \int_0^a e^{nx} f(x) dx \right| < \infty \quad \text{implies } f = 0,$$

is far from being trivial. In Yoshida's *Functional Analysis* there is a proof more elementary than the one we have given in the solution of Exercise 3·72.

When one has proved that $f*g = 0$ implies $f = 0$ or $g = 0$, for continuous functions, one can pass to the general case by regularisation. In fact if (φ_i) is a regularising sequence with the support of φ_i contained in $[0,1/i]$, then

$$(f*\varphi_i)*(g*\varphi_i) = 0;$$

if $g \neq 0$ then $g*\varphi_i \neq 0$ for i large enough, and consequently $f*\varphi_i = 0$. It remains to observe that for all M,

$$\int_0^M |f - f*\varphi_i| \to 0.$$

EXERCISE 8·152: Let $1 < p < \infty$ and $f \in L^p(0,\infty)$. For all $x > 0$ set

$$F(x) = \frac{1}{x}\int_0^x f(t)dt.$$

(a): Show that $F \in L^p(0,\infty)$ and that:

$$\|F\|_p \leq \frac{p}{p-1} \|f\|_p. \qquad \textit{(HARDY'S INEQUALITY)}$$

(b): Prove that equality holds in the above relation only if f is zero almost everywhere.

(c): Show that the constant $p/(p - 1)$ in Hardy's Inequality is the best possible.

(d): Show that if $f \in L^1(0,\infty)$, $f > 0$, then $F \notin L^1(0,\infty)$.

Δ∇Δ = ∇Δ∇ = Δ∇Δ = ∇Δ∇ = ΔV∆

SOLUTION (a): *First Proof:* Let $G(x,t) = f(xt)$. Then

$$F(x) = \int_0^1 G(x,t)dt$$

By exercise 6·110,

$$\|F\|_p \leq \int_0^1 \|G(\cdot,t)\|_p dt$$

$$= \|f\|_p \int_0^1 t^{-1/p}dt = \frac{p}{p-1} \|f\|_p.$$

Second Proof: First assume that f is positive, continuous, and has compact support in $]0,\infty[$. Then F is zero in a neighbourhood of zero, and

$$F(x) = \frac{1}{x}\int_0^\infty f$$

for x sufficiently large. In particular,

$$xF(x)^p \to 0 \quad \text{when } x \to 0 \text{ or } x \to \infty.$$

Differentiating the original expression for $xF(x)$ gives

$$xF'(x) + F(x) = f(x),$$

whence, upon multiplying by $F(x)^{p-1}$ and then integrating,

$$\int_0^\infty xF'(x)F(x)^{p-1}dx + \int_0^\infty F(x)^p dx = \int_0^\infty f(x)F(x)^{p-1}dx.$$

Integrating the first integral by parts, and taking into account that $xF(x)^p$ vanishes at 0 and ∞, one obtain

$$\frac{p-1}{p}\int_0^\infty F(x)^p dx = \int_0^\infty f(x)F(x)^{p-1}dx. \tag{1}$$

Applying Hölder's inequality to the right side yields

$$\|F\|_p^p \leqslant \frac{p}{p-1}\,\|f\|_p\Big(\int_0^\infty F(x)^{(p-1)q}dx\Big)^{1/q}, \qquad \text{where } \frac{1}{p} + \frac{1}{q} = 1.$$

That is

$$\|F\|_p^p \leqslant \frac{p}{p-1}\,\|f\|_p\,\|F\|_p^{p/q}.$$

As $\|F\|_p < \infty$ (because of the value of $F(x)$ in neighbourhoods of

0 and ∞), it follows that

$$\|F\|_p \leqslant \frac{p}{p-1} \|f\|_p. \tag{2}$$

If f is continuous and compactly supported in $]0,\infty[$, then

$$|F(x)| \leqslant \frac{1}{x}\int_0^x |f(t)|\,dt,$$

so

$$\|F\|_p \leqslant \frac{p}{p-1} \||f|\|_p = \frac{p}{p-1} \|f\|_p.$$

Finally, if f is any function in $L^p(0,\infty)$, there exists a sequence (f_i) of continuous functions compactly supported in $]0,\infty[$ such that $\|f - f_i\|_p \to 0$. If

$$F_i(x) = \frac{1}{x}\int_0^x f_i(t)\,dt$$

then

$$F_i(x) \to F(x) \qquad \text{for all } x > 0.$$

By Fatou's Lemma,

$$\|F\|_p \leqslant \limsup_i \|F_i\|_p \leqslant \frac{p}{p-1} \lim_i \|f_i\|_p = \frac{p}{p-1} \|f\|_p.$$

SOLUTION (b): First assume that $f \in L^p(0,\infty)$ and $f \geqslant 0$. Let be a sequence of compactly supported continuous positive functions such that $f_i \to f$ in L^p. Then $F_i \to F$ in L^p, and so by exercises 6·105 $F_i^{p-1} \to F^{p-1}$ in $L^{p/(p-1)} = L^q$. Replacing F by F_i in (1) and passing to the limit, shows that this formula is still valid for f. As (2) results from Hölder's Inequality applied to the right side of (1), equality can hold in (2) only if it holds in this Holder inequality, that is to say, if there exists a constant $A \geqslant 0$ such that

$$F^p = F^{(p-1)q} = Af^p,$$

that is to say $F = Bf$, $B \geqslant 0$ a constant. Let us note that if $f \not\equiv 0$, we necessarily have $F \not\equiv 0$, whence $B > 0$. Since F is to be continuous f would be also, and by differentiation one would obtain

$$Bxf'(x) = (1 - B)f(x),$$

that is to say:

$$f(x) = Cx^{\alpha}, \qquad C > 0, \qquad \alpha = \frac{1 - B}{B}.$$

This is absurd, because whatever α may be such a function does not belong to $L^p(0,\infty)$.

If $f \in L^p(0,\infty)$ without being positive, but if $f \not\equiv 0$ almost everywhere, then setting

$$G(x) = \frac{1}{x}\int_0^x |f(t)|dt$$

yields

$$\|F\|_p \leqslant \|G\|_p < \frac{p}{p-1} \||f|\|_p = \frac{p}{p-1} \|f\|_p.$$

REMARK: The second proof given for part (a) above is more elementary than the first proof, which uses the generalised Minkowski inequality. Further, the proof of part (b) is based upon formula (1), which is the essential point of the second solution. One can avoid using exercise 6·105 to establish the validity of (1) whenever $f \in L^p(0,\infty)$ is only assumed to be positive, by proceeding in the following manner, which furnishes a third proof of Hardy's inequality.

Let

$$\varphi(x) = e^{x/p} f(e^x), \qquad -\infty < x < +\infty ,$$

$$\chi(x) = \begin{cases} e^{-x/q} & \text{if } x \geqslant 0, \\ 0 & \text{if } x < 0, \end{cases} \qquad \frac{1}{p} + \frac{1}{q} = 1.$$

It is not hard to verify that

$$\|\varphi\|_{L^p(-\infty,+\infty)} = \|f\|_{L^p(0,\infty)},$$

$$\chi * \varphi(x) = e^{x/p} F(e^x),$$

$$\|\chi\|_{L^r(-\infty,+\infty)} = \left(\frac{q}{r}\right)^{1/r}, \qquad r > 0,$$

so that

$$\|F\|_{L^p(0,\infty)} = \|\chi * \varphi\|_{L^p(-\infty,+\infty)} \leqslant \|\chi\|_{L^1(-\infty,+\infty)} \|\varphi\|_{L^p(-\infty,+\infty)}$$

$$= q\|f\|_{L^p(0,\infty)},$$

by virtue of the classical properties of the convolution product on $\mathbb{R}$. Moreover, since $\chi \in L^q(-\infty,+\infty)$,

$$\lim_{|x|\to\infty} (\chi * \varphi)(x) = 0,$$

which implies that

$$\lim_{x\to 0} x^{1/p} F(x) = \lim_{x\to\infty} x^{1/p} F(x) = 0. \tag{3}$$

But then, if $f \geqslant 0$, we still have

$$xF'(x) + F(x) = f(x)$$

for almost all x, by the Lebesgue theory of differentiation, and furthermore, by (3), the formula

$$\int_0^\infty xF(x)^{p-1}F'(x)dx = -\frac{1}{p}\int_0^\infty F(x)\, dx$$

is still valid, which ensures the validity of (1). Note that the integrability of $xF(x)^{p-1}F'(x)$ is assured, because F^p and fF^{p-1} are integrable .... since $F^{p-1} \in L^q$. If

$$r \geqslant 1, \quad \frac{1}{p} + \frac{1}{r} \geqslant 1, \quad \text{and} \quad \frac{1}{s} = \frac{1}{p} + \frac{1}{r} - 1,$$

we have (cf. exercise 6·106)

$$\|\chi * \varphi\|_{L^s(-\infty,+\infty)} \leqslant \|\chi\|_{L^r(-\infty,+\infty)} \|\varphi\|_{L^p(-\infty,+\infty)},$$

from which it is easily deduced that

$$\left(\int_0^\infty x^{s/p - 1}|F(x)|^s dx\right)^{1/s} \leqslant \left(\frac{q}{s} + 1\right)^{\frac{1}{s}+\frac{1}{q}} \|f\|_p, \qquad s \geqslant p,$$

a generlization of Hardy's inequality which corresponds to the case $s = p$.

To conclude this remark, let us indicate to the readers who know about the notion of convolution product on the multiplicative group $\mathbb{R}_+^*$ that the preceding is expressed more simply in the following way: If

$$\varphi(x) = x^{1/p}f(x),$$

and

$$\chi(x) = \begin{cases} x^{1/q} & \text{for } x \geqslant 1, \\ 0 & \text{if } 0 < x < 1, \end{cases}$$

then

$$\chi * \varphi(x) = x^{1/p} F(x),$$

where the convolution product this time is defined by

$$\chi * \varphi(x) = \int_0^\infty \chi(t)\varphi(xt^{-1})\,\frac{dt}{t} = \int_0^\infty \chi(t^{-1})\varphi(xt)\,\frac{dt}{t}.$$

SOLUTION (c): If $f(x) = x^{-1/p}$ when $1 \leqslant x \leqslant A$, and 0 otherwise, then simple calculations show that

$$F(x) = \begin{cases} 0 & \text{if } 0 < x \leqslant 1, \\ qx^{-1}(x^{1/q} - 1) & \text{if } 1 \leqslant x \leqslant A, \\ q(A^{1/q} - 1)x^{-1} & \text{if } x \geqslant A, \end{cases}$$

$$\|f\|_p = (\log A)^{1/p},$$

$$\|F\|_p = q\left\{\int_1^A x^{-p}(x^{1/q} - 1)^p dx + \frac{1}{p-1}\,\frac{(A^{1/q} - 1)^p}{A^{p-1}}\right\}^{1/p}.$$

Since

$$x^{-p}(x^{1/q} - 1)^p \sim x^{p/q - p} = x^{-1} \quad \text{as } x \to \infty,$$

we have

$$\int_1^A x^{-p}(x^{1/q} - 1)^p dx \sim \log A,$$

and consequently

$$\frac{\|F\|_p}{\|f\|_p} \to q \qquad \text{as } A \to \infty.$$

SOLUTION (d): In this case

$$\int_0^\infty F(x)dx = \int_0^x \frac{dx}{x}\int_0^x f(t)dt = \int_0^\infty f(t)dt\int_t^\infty \frac{dx}{x} = \infty.$$

EXERCISE 8·153: Let us denote by $L^p(t^{-1}dt)$ the set of measurable functions on $]0,\infty[$ such that

$$\|f\|_{L^p(t^{-1}dt)} = \left(\int_0^\infty |f(t)|^p \frac{dt}{t}\right)^{1/p} < \infty.$$

Also, set

$$L^\infty(t^{-1}dt) = L^\infty(0,\infty).$$

Show that if

$$f \in L^p(t^{-1}dt), \quad g \in L^q(t^{-1}dt), \quad p,q \geqslant 1, \quad \frac{1}{p} + \frac{1}{q} \geqslant 1,$$

then the integral

$$\int_0^\infty f(t)g(xt^{-1})\frac{dt}{t}$$

exists (in the Lebesgue sense) for almost all $x > 0$.

Denoting by $f * g$ the function thus defined almost everywhere on $]0,\infty[$, show next that

$$\|f*g\|_{L^r(t^{-1}dt)} \leqslant \|f\|_{L^p(t^{-1}dt)} \|g\|_{L^q(t^{-1}dt)} \quad \text{if } \frac{1}{r} = \frac{1}{p} + \frac{1}{q} - 1.$$

Lastly, show that if $\frac{1}{p} + \frac{1}{q} = 1$, then $f*g$ is continuous, and

$$\lim_{x\to 0} f*g(x) = \lim_{x\to\infty}(f*g)(x) = 0.$$

Δ∇Δ = ∇Δ∇ = ΔVΔ = ∇Δ∇ = Δ∇Δ

SOLUTION: This is a matter of developing the last part of the remark in the preceding exercise. If we set $F(x) = f(e^x)$, $G(x) = g(e^x)$, it is not hard to verify that $f*g$ is defined almost everywhere if and only if F,G are convolvable in the usual sense, and that then $(F*G)(x) = (f*g)(e^x)$. Then it suffices to use the elementary properties of the convolution product on $\mathbb{R}$, exercise 6·106, and the equality

$$\|F\|_{L^p(\mathbb{R})} = \|f\|_{L^p(t^{-1}dt)}$$

EXERCISE 8·154: Let $p \geqslant 1$ and $1 - \frac{1}{p} < \alpha < 2 - \frac{1}{p}$; for $f \in L^p(0,\infty)$ set

$$F(x) = \int_0^\infty f(t)\, \frac{\sin xt}{t^\alpha}\, dt.$$

Show that for all $r \geqslant p$ there exists a constant $A(r,p,\alpha)$ such that

$$\left(\int_0^\infty x^{r(1-1/p-\alpha)-1} |F(x)|^r dx\right)^{1/r} \leqslant A(r,p,\alpha)\|f\|_p.$$

Show, further, that if $p > 1$

$$F(x) = o(x^{\alpha-1+1/p}) \quad \text{as } x \to 0 \text{ or } x \to +\infty.$$

Deduce from this that

$$\sup_x |F(x+h) - F(x)| = o(|h|^{\alpha-1+1/p}).$$

$$\Delta\nabla\Delta = \nabla\Delta\nabla = \Delta\nabla\Delta = \nabla\Delta\nabla = \Delta\nabla\Delta$$

SOLUTION: Note that

$$F(x) = x^{1/p+\alpha-1} \int_0^\infty t^{1/p} f(t) \frac{\sin xt}{(xt)^{1/p+\alpha-1}} \frac{dt}{t}.$$

Let

$$\varphi(t) = t^{1/p} f(t), \qquad \chi(t) = t^{1-1/p-\alpha} \sin t,$$

so that

$$F(x) = x^{1/p+\alpha-1} \int_0^\infty \varphi(t)\chi(xt) \frac{dt}{t}.$$

$$= x^{1/p+\alpha-1} \int_0^\infty \varphi(t^{-1})\chi(xt^{-1}) \frac{dt}{t}.$$

Setting $\check{\varphi}(t) = \varphi(t^{-1})$ and using the notations of the previous exercise, this becomes

$$F(x) = x^{1/p+\alpha-1} (\check{\varphi} \star \chi)(x).$$

Now observe that

$$\|\check{\varphi}\|_{L^p(t^{-1}dt)} = \|\varphi\|_{L^p(t^{-1}dt)} = \|f\|_p,$$

and that

$$\|\chi\|_{L^r(t^{-1}dt)} < \infty, \qquad 0 < r \leqslant \infty,$$

for $\frac{1}{p} + \alpha - 1 > 0$, which ensures the result for $r = \infty$; and if $r < \infty$

$$t^{-1}|\chi(t)|^r = O\left(\frac{1}{t^{r(1/p+\alpha-1)+1}}\right),$$

$$t^{-1}|\chi(t)|^r \sim \frac{1}{t^{r(1/p+\alpha-2)+1}} \quad \text{as } t \to 0 \quad \text{and} \quad \frac{1}{p} + \alpha - 2 < 0.$$

If

$$s \geqslant 1, \qquad \frac{1}{p} + \frac{1}{s} \geqslant 1, \qquad \frac{1}{r} = \frac{1}{p} + \frac{1}{s} - 1,$$

that is to say, if

$$r \geqslant p \quad \text{and} \quad \frac{1}{s} = \frac{1}{q} + \frac{1}{r}, \quad \text{where } \frac{1}{p} + \frac{1}{q} = 1,$$

then

$$\|\check{\varphi} * \chi\|_{L^r(t^{-1}dt)} \leqslant \|\chi\|_{L^s(t^{-1}dt)} \|f\|_p,$$

which is the desired inequality. When $r = \infty$ then $s = q$, and consequently if $p > 1$ $\check{\varphi} * \chi$ is continuous and tends to zero as $x \to 0$ or $x \to \infty$, which proves that

$$F(x) = o(x^{\alpha-1+1/p}).$$

Finally,

$$F(x+h) - F(x) = 2\int_0^\infty f(t) \frac{\cos(x + \frac{1}{2}h)t \sin\frac{1}{2}ht}{t^\alpha} dt,$$

$$\sup_x |F(x + h) - F(x)| \leqslant 2F(\tfrac{1}{2}|h|) = o(|h|^{\alpha-1+1/p}).$$

EXERCISE 8·155: Let $p > 1$ and $f \in L^p(0,\infty)$. Set

$$F(x) = \int_0^\infty e^{-xt} f(t)\,dt, \qquad x > 0.$$

Show that for $r \geqslant p$

$$\left(\int_0^\infty x^{r/q-1} |F(x)|^r dx\right)^{1/r} < \left(\frac{1}{s}\right)^{1/q} \Gamma\left(\frac{s}{q}\right)^{1/s} \|f\|_p,$$

$$\frac{1}{p} + \frac{1}{q} = 1, \qquad \frac{1}{s} = \frac{1}{r} + \frac{1}{q}.$$

Show also that

$$F(x) = o(x^{-1/q}) \quad \text{as } x \to 0 \text{ or } x \to \infty.$$

Δ∇Δ = ∇Δ∇ = Δ∇Δ = ∇Δ∇ = Δ∇Δ

SOLUTION: Proceed as in the preceding exercise, observing that if

$$\varphi(t) = t^{1/p} f(t), \qquad \chi(t) = t^{1/q} e^{-t},$$

then

$$F(x) = x^{-1/q} (\check{\varphi} * f)(x).$$

Furthermore, if $s > 0$

$$\|\chi\|_{L^s(t^{-1}dt)} = \left(\int_0^\infty t^{s/q-1} e^{-st} dt\right)^{1/s} = \left(\frac{1}{s}\right)^{1/q} \Gamma\left(\frac{s}{q}\right)^{1/s}.$$

EXERCISE 8·156: Show that a subset H of $L^p(\mathbb{R}^n)$, $1 \leqslant p < \infty$, is relatively compact if and only if it satisfies the following conditions:

(i): There exists a number M such that for all $f \in H$:

$$\|f\|_p \leqslant M;$$

(ii): For all $\varepsilon > 0$ there exists a compact set K of $\mathbb{R}^n$ such that for all $f \in H$:

$$\int_{\mathbb{R}^n - K} |f|^p \leqslant \varepsilon;$$

(iii): For any $\varepsilon > 0$ there exists a neighbourhood V of zero in $\mathbb{R}^n$ such that for all $a \in V$ and all $f \in H$:

$$\|f_a - f\|_p \leqslant \varepsilon.$$

(f_a denotes the translation of f by a; to show that the conditions are sufficient prove that for every compact set K of $\mathbb{R}^n$ and every continuous and compactly supported function φ the set of functions $(f1\!\!1_K)*\varphi$, $f \in H$, is equicontinuous).

Δ∇Δ = ∇Δ∇ = ΔVΔ = ∇Δ∇ = Δ∇Δ

SOLUTION : *The Conditions are Necessary:* If H is relatively compact in L^p, H is bounded, which proves (i).

For every $\varepsilon > 0$ there exist functions $f_1, \ldots, f_r \in L^p$ such that

$$\sup_{f \in H}(\min_{1 \leqslant s \leqslant r} \|f - f_s\|_p) \leqslant \varepsilon.$$

There is a compact set K of $\mathbb{R}^n$ such that

$$\int_{\mathbb{R}^n - K} |f_s|^p \leqslant \varepsilon^p, \qquad 1 \leqslant s \leqslant r.$$

Then for all $f \in H$

$$\int_{\mathbb{R}^n - K} |f|^p \leqslant (2\varepsilon)^p, \qquad 1 \leqslant s \leqslant r.$$

which proves (ii).

Similarly one can choose a neighbourhood V of zero in $\mathbb{R}^n$ such that

$$\|(f_s)_a - f_s\|_p \leqslant \varepsilon, \qquad a \in V, \qquad 1 \leqslant s \leqslant r,$$

and then if $f \in H$

$$\|f_a - f\|_p \leqslant 3\varepsilon,$$

since

$$\|f_a - f\|_p \leqslant \|(f - f_s)_a\|_p + \|(f_s)_a - f_s\|_p + \|f_s - f\|_p.$$

The Conditions are Sufficient: For every compact set $K \subset \mathbb{R}^n$ we shall set $f^K = f\mathbb{1}_K$. Let φ be a continuous function with compact support A. For $f \in H$ the functions $f^K * \varphi$ have their support contained in the compact set $K + A$, and if $\frac{1}{p} + \frac{1}{q} = 1$ then

$$\|f^K * \varphi\|_\infty \leqslant \|f^K\|_p \|\varphi\|_q \leqslant M\|\varphi\|_q,$$

$$|(f^K * \varphi)(x_1) - (f^K * \varphi)(x_2)| \leqslant \|f^K\|_p \|\varphi_{-x_1} - \varphi_{-x_2}\|_q$$

$$\leqslant M\|\varphi_{x_2 - x_1} - \varphi\|_q.$$

By (i) the set of functions $f^K * \varphi$, $f \in H$, is therefore bounded and equicontinuous. It follows from Ascoli's Theorem that for all $\alpha > 0$ function $f_1, \ldots, f_r \in H$ can be found such that

$$\sup_{f\in H}\Big(\min_{1\leqslant s\leqslant r}\|f^K*\varphi - f_s^K*\varphi\|_\infty\Big) \leqslant \alpha. \tag{1}$$

Now let $\varepsilon > 0$. Choose K such that (condition (ii)):

$$\|f - f^K\|_p \leqslant \varepsilon, \qquad f\in H, \tag{2}$$

and let V be as in (iii). Assume further that $\varphi \geqslant 0$, $\|\varphi\|_1 = 1$, and that the support of φ is contained in V. Since

$$(f*\varphi)(x) - f(x) = \int_V (f(x - y) - f(x))\varphi(y)dy,$$

the generalised Minkowski inequality (cf., Exercise 6·110) furnishes, for $f \in H$, the bound:

$$\|f*\varphi - f\|_p \leqslant \int_V \|f_y - f\|_p\varphi(y)dy \leqslant \varepsilon, \tag{3}$$

and consequently by (2) and (3):

$$\begin{aligned}\|f^K*\varphi - f^K\|_p &\leqslant \|(f^K - f)*\varphi\|_p + \|f*\varphi - f\|_p + \|f - f^K\|_p\\ &\leqslant 3\varepsilon.\end{aligned} \tag{4}$$

This being so, let us choose $f_1,\dots,f_r \in H$ such that (1) is satisfied with $\alpha = \varepsilon(\text{meas}(K + A))^{-1/p}$. Writing

$$\begin{aligned}\|f - f_s\|_p \leqslant \|f - f^K\|_p &+ \|f^K - f^K*\varphi\|_p + \|f^K*\varphi - f_s^K*\varphi\|_p\\ &+ \|f_s^K*\varphi - f_s^K\|_p + \|f_s^K - f_s\|_p,\end{aligned}$$

it follows from (1),(2), and (4) that

$$\sup_{f\in H}\left(\min_{1\leqslant r\leqslant s}\|f-f_s\|_p\right)\leqslant 9\varepsilon,$$

which proves that H is relatively compact.

EXERCISE 8·157: (a): Let $0 < a_1 < a_2 < \cdots$ and $a_n = O(n^\alpha)$.

Prove that

$$\limsup_{n\to\infty}\frac{a_n}{a_{n+1}} = 1.$$

(b): Let f be a measurable positive function on $\mathbb{R}^p$ such that $0 < \|f\|_\infty < \infty$.

Prove that there exists a sequence (φ_n) of functions such that

(i): $\varphi_n \in \mathcal{D}^\infty(\mathbb{R}^p)$ and $\varphi_n \geqslant 0$;

(ii): $\|f*\varphi_n\|_\infty = 1$;

(iii): $f*\varphi_n \to 1$ uniformly on every compact set of $\mathbb{R}^p$.

(Let $\rho > 0$ be such that

$$\int_{|x|\leqslant 2\rho} f(x)dx > 0.$$

For every integer $n \geqslant 1$ let $\vartheta_n \in \mathcal{D}^\infty(\mathbb{R}^p)$, $0 \leqslant \vartheta_n \leqslant 1$, $\vartheta(x) = 1$ if $|x| \leqslant 2n\rho$ and $\vartheta_n(x) = 0$ if $|x| \geqslant (2n+1)\rho$. Set $a_n = \|f*\vartheta_n\|_\infty$ and consider some $x_n \in \mathbb{R}^p$ such that

$$(f*\vartheta_n)(x_n) \geqslant \left(1-\frac{1}{n}\right)a_n.$$

Show that if

$$h_n(x) = a_{n+1}^{-1}\vartheta_{n+1}(x+x_n)$$

one has $\|f*h_n\|_\infty = 1$ and when $|x| \leqslant \rho$

$$(f*h_n)(x) \geqslant \left(1 - \frac{1}{n}\right)\frac{a_n}{a_{n+1}} .$$

(c): Deduce from this that if $g \in L^1(\mathbb{R}^p)$ is such that $f*g \geqslant 0$, then

$$\int_{\mathbb{R}^p} g(x)dx \geqslant 0.$$

ΔVΔ = VΔV = ΔVΔ = VΔV = ΔVΔ

SOLUTION (a): If

$$\limsup_{n\to\infty} \frac{a_n}{a_{n+1}} < \frac{1}{\lambda} \qquad (\lambda > 1),$$

one would have, starting from some n_0, $a_{n+1} \geqslant \lambda a_n$, and consequently $a_n \geqslant a_{n_0}\lambda^{n-n_0}$ when $n \geqslant n_0$, which contradicts the hypothesis $a_n = O(n^\alpha)$.

SOLUTION (b): With ρ, ϑ_n, a_n being as indicated in the question, note that if $|z| \leqslant \rho$, then for all $x \in \mathbb{R}$

$$\vartheta_n(x) \leqslant \vartheta_{n+1}(x + z). \tag{*}$$

In fact this inequality is clear if $x \geqslant (2n + 1)\rho$, and in the contrary case

$$|x + z| \leqslant (2n + 1)\rho + \rho = 2(n + 1)\rho,$$

and therefore

$$\vartheta_{n+1}(x + z) = 1 \geqslant \vartheta_n(x).$$

In particular $\vartheta_n \leqslant \vartheta_{n+1}$, whence $a_n \leqslant a_{n+1}$. Now note that

$$a_1 \geqslant (f*\varphi_1)(0) \geqslant \int_{|x|\leqslant 2\rho} f(x)\mathrm{d}x > 0,$$

and that, denoting the volume of the ball $|x| \leqslant 1$ by V_p,

$$a_n \leqslant \|\vartheta_n\|_1 \|f\|_\infty \leqslant V_p \rho^p (2n + 1)^p \|f\|_\infty. \tag{**}$$

This being so, if x_n and h_n are as in the statement of the question, then

$$\|f*h_n\|_\infty = a_{n+1}^{-1} \|f*\vartheta_{n+1}\|_\infty = 1,$$

and furthermore, if $|z| \leqslant \rho$ it follows from (*) that

$$(f*h_n)(z) = \frac{1}{a_{n+1}}\int f(y)\vartheta_{n+1}(z + x_n - y)\mathrm{d}y$$

$$\geqslant \frac{1}{a_{n+1}}\int f(y)\vartheta_n(x_n - y)\mathrm{d}y \geqslant \left(1 - \frac{1}{n}\right)\frac{a_n}{a_{n+1}} .$$

By (**) and part (a), for all $\varepsilon > 0$ there exists an integer n_0 such that

$$\left(1 - \frac{1}{n_0}\right)\frac{a_{n_0}}{a_{n_0+1}} > 1 - \varepsilon.$$

Setting $\varphi = h_{n_0}$ one has therefore determined for every ρ large enough and all $\varepsilon > 0$ a function $\varphi \in \mathcal{D}^\infty$, $\varphi \geqslant 0$, such that $\|f*\varphi\|_\infty = 1$ and

$$1 - \varepsilon \leqslant (f*\varphi)(z) \leqslant 1$$

if $|z| \leqslant \rho$. Denoting by φ_n the function corresponding to $\rho = n$ and $\varepsilon = 1/n$, one obtains the desired sequence.

SOLUTION (c): This follows from

$$\int g = \int \check{g} = \lim_n (\check{g}, f*\varphi_n) = \lim_{n\to\infty} (\check{g}*\check{f}, \varphi_n)$$

and the fact that $g*f \geqslant 0$ implies that $(\check{g}*\check{f}, \varphi_n) \geqslant 0$ if $\varphi_n \geqslant 0$.

EXERCISE 8·158: Let $f \in L^1(\mathbb{R}^p)$. Prove that

$$\left|\int f(x)dx\right| = \inf \int |\lambda_1 f(x - a_1) + \cdots + \lambda_n f(x - a_n)| dx,$$

where the infimum is taken over all the systems of elements $a_1, \ldots, a_n \in \mathbb{R}^p$ and positive numbers $\lambda_1, \ldots, \lambda_n$ such that $\lambda_1 + \cdots + \lambda_n = 1$. (Use the result proved in the preceding exercise).

Δ∇Δ = ∇Δ∇ = ΔVΔ = ∇Δ∇ = ΔVΔ

SOLUTION: First let us fix some notations. As usual set $f_a(x) = f(x - a)$; in addition, if $f \in L^1$ we shall set

$$I(f) = \int f(x)dx.$$

Let P be the set of positive functions on $\mathbb{R}^p$ that are zero except at a finite number of points and are such that:

$$\sum_{a \in \mathbb{R}^p} \lambda(a) = 1.$$

If $\lambda \in P$, set:

$$(\lambda*f)(x) = \sum_{a \in \mathbb{R}^p} \lambda(a) f(x - a).$$

There is no difficulty in verifying that if $\lambda_1, \lambda_2 \in P$ then

$$\lambda_1 * (\lambda_2 * f) = (\lambda_1 * \lambda_2) * f,$$

where

$$(\lambda_1 * \lambda_2)(a) = \sum_{b \in \mathbb{R}^p} \lambda_1(b)\lambda_2(a - b)$$

(the reader well versed in measure theory will recognise the notion of the convolution product of f with discrete probability measures). If $\lambda \in P$, $f \in L^1$, it is clear that

$$\|\lambda * f\|_1 \leqslant \|f\|_1.$$

Lastly let us set, for $f \in L^1$,

$$p(f) = \inf \|\lambda * f\|_1, \qquad \lambda \in P.$$

It is then a matter of proving that

$$|I(f)| = p(f). \tag{1}$$

= = = = = = =

If $\alpha \in \mathbb{C}$ and $f, g \in L^1$, then the following properties hold:

$$I(f) = p(f) \quad \text{if } f \geqslant 0,$$

$$p(\alpha f) = |\alpha| p(f),$$

$$|I(f)| \leqslant p(f),$$

$$p(f + g) \leqslant p(f) + p(g).$$

The first two are evident; the third follows from

$$|I(f)| = |I(\lambda * f)| \leqslant \|\lambda * f\|_1,$$

and the last from the fact that for all $\varepsilon > 0$ there exist $\lambda_1, \lambda_2 \in P$ such that

$$\|\lambda_1 * f\|_1 \leqslant p(f) + \varepsilon,$$

$$\|\lambda_2 * g\|_1 \leqslant p(g) + \varepsilon,$$

so that

$$\begin{aligned} p(f + g) &\leqslant \|(\lambda_1 * \lambda_2) * (f + g)\|_1 \\ &\leqslant \|\lambda_2 * (\lambda_1 * f)\|_1 + \|\lambda_1 * (\lambda_2 * g)\|_1 \\ &\leqslant \|\lambda_1 * f\|_1 + \|\lambda_2 * g\|_1 \\ &\leqslant p(f) + p(g) + 2\varepsilon. \end{aligned}$$

Now assume that one had proved that $f \in L^1$, f real, and $I(f) = 0$ implies $p(f) = 0$. Then if $f \in L^1$ is real and if $\varphi \geqslant 0$, $I(\varphi) = 1$, one would have

$$|I(f)| \leqslant p(f) \leqslant p(f - I(f)\varphi) + |I(f)| = |I(f)|,$$

which would prove (1) in this case. To pass to the general case observe that there exists $\alpha \in \mathbb{C}$, $|\alpha| = 1$, such that $|I(f)| = I(\alpha f)$, if $\alpha f = u + iv$, where u, v are real, then

$$p(u) = |I(u)| = I(u),$$

$$p(v) = |I(v)| = 0,$$

and consequently:

$$p(f) \geqslant |I(f)| = I(u) = p(u)$$

$$\geqslant p(u + iv) = p(\alpha f) = p(f).$$

= = = = = = =

Therefore let us assume that $f \in L^1$, f real, and $I(f) = 0$. Note that $p(f)$ is the distance in L^1 from zero to the closed convex envelope of the f_a's, $a \in \mathbb{R}^p$. If one had $p(f) > 0$ the Hahn-Banach Theorem would assure us of the existence of a number $\alpha > 0$ and of a non-zero continuous linear form on L^1, bounded from below by α on this convex envelope, and in particular on the set of f_a's. In other words there would exist $g \in L^\infty$, g non-zero almost everywhere, and such that

$$\int f(x - a)g(x)dx \geqslant \alpha > 0 \tag{2}$$

for all $a \in \mathbb{R}^p$. Since $I(f) = 0$ the left side of (2) is not changed by adding a constant to g, which allows us to assume that $g \geqslant 0$. Let $h(x) = g(-x)$; then $h \geqslant 0$, $h \in L^\infty$, and (2) can be written

$$f*h \geqslant \alpha. \tag{3}$$

Let $M = \|h\|_\infty$. Note that by (3) $M > 0$. Let φ be as above. Then

$$(f - \frac{\alpha}{M}\varphi)*h \geqslant \alpha - \frac{\alpha}{M}M = 0.$$

By the preceding exercise one would have

$$I(f - \frac{\alpha}{M}\varphi) = I(f) - \frac{\alpha}{M} \geqslant 0,$$

which is absurd.

EXERCISE 8·159: If f is integrable on $[a,b]$, and if

$$\int_a^b f(x)e^{i\alpha_n x}\,dx = 0$$

for a sequence (α_n) of complex numbers having at least one finite limit point, then $f = 0$ almost everywhere on $[a,b]$.

ΔVΔ = VΔV = ΔVΔ = VΔV = ΔVΔ

SOLUTION: For all $z \in \mathbb{C}$ let

$$F(z) = \int_a^b f(x)e^{izx}\,dx.$$

This defines an entire function which vanishes at each α_n. If the latter have a finite limit point, then $F(z) \equiv 0$. In particular the Fourier transform of the function equal to f on $[a,b]$ and zero elsewhere is zero. Consequently $f = 0$ almost everywhere on $[a,b]$.

EXERCISE 8·160: (a): Let (α_n) be a sequence of real numbers such that

$$\lim_{n\to\infty} e^{i\alpha_n x}$$

exists for all x's belonging to a measurable set A of $\mathbb{R}$ with meas$(A) > 0$.

Show that the sequence (α_n) is convergent.

(b): Let (c_n) be a sequence of complex numbers and (α_n) a sequence of real nubmers such that

$$\lim_{n\to\infty} c_n e^{i\alpha_n x}$$

exists for all the x's belonging to a measurable set A of $\mathbb{R}$ with $\text{meas}(A) > 0$.

Show that either

$$\lim_{n\to\infty} c_n = 0,$$

or

$$\lim_{n\to\infty} c_n = c \neq 0 \quad \text{and} \quad \lim_{n\to\infty} \alpha_n = \alpha.$$

$$\Delta\nabla\Delta = \nabla\Delta\nabla = \Delta\nabla\Delta = \nabla\Delta\nabla = \Delta\nabla\Delta$$

SOLUTION (a): The set of the $x \in \mathbb{R}$ for which $\lim_{n\to\infty} e^{i\alpha_n x}$ exists is an additive subgroup that is measurable and of strictly positive measure. From this it follows that

$$g(x) = \lim_{n\to\infty} e^{i\alpha_n x} \tag{*}$$

exists for all $x \in \mathbb{R}$ (cf. exercise 8·138). Then for every integrable function f on $\mathbb{R}$,

$$\int_{-\infty}^{+\infty} f(x)g(x)dx = \lim_{n\to\infty} \int_{-\infty}^{+\infty} f(x)e^{i\alpha_n x} dx$$

(Dominated Convergence Theorem). Assume that for a subsequence (β_n) of (α_n) $|\beta_n| \to \infty$. By the Riemann-Lebesgue lemma,

$$\int_{-\infty}^{+\infty} f(x)g(x)dx = 0$$

for every integrable function f, and consequently $g(x) = 0$ for almost all x, which is absurd because $|g(x)| = 1$ for all x. The sequence (α_n) is therefore bounded. Let α, β be two accumulation points of the sequence. By considering subsequences of (α_n) converging towards α and β, one deduces from (*) that

$$e^{i\alpha x} = e^{i\beta x}$$

for all x, whence by differentiation $\alpha = \beta$. This proves that the sequence (α_n) is convergent.

SOLUTION (b): Since

$$|c_n e^{i\alpha_n x}| = |c_n|,$$

we have that

$$\lim_{n\to\infty}|c_n| = \rho$$

exists. Assume that $\rho > 0$. Then from some point on, $c_n \neq 0$. If $x,y \in A$, the sequence $e^{i\alpha_n(x-y)}$ is therefore convergent. Because $\text{meas}(A - A) > 0$ it follows from part (a) above that $\alpha = \lim_{n\to\infty}\alpha_n$ exists. Finally, if $x \in A$ the sequences $c_n e^{i\alpha_n x}$ and $e^{i\alpha_n x}$ are convergent, which proves that $\lim_{n\to\infty} c_n$ exists.

EXERCISE 8·161:(a): Let $f \in L^1(T)$, and

$$f^+(x) = \sup_{h>0} h^{-1}\int_0^h |f(x + u)|\,du,$$

$$f^-(x) = \sup_{h>0} h^{-1}\int_0^h |f(x - u)|\,du,$$

$$f^\Delta(x) = \max(f^+(x), f^-(x)).$$

Show that for all $\lambda > 0$,

$$\text{meas}\{(f^+ > \lambda) \cap [-\pi,\pi]\} \leqslant \frac{4\pi}{\lambda}\|f\|_1, \tag{*}$$

and some analogous relations for f^- and f^Δ. (Use the "Setting Sun Lemma"). What does the application of Markcinkiewicz's Theorem give in this case? (Cf. exercise 6·124).

(b): Let (K_n) be a sequence of functions of $L^1(T)$ and (H_n) a sequence of continuously differentiable functions on $[-\pi,\pi]$ such that

$$\text{(i)}: \quad |K_n(x)| \leqslant H_n(x), \quad -\pi < x \leqslant \pi;$$

$$\text{(ii)}: \quad A = \sup_n \frac{1}{2\pi}\int_{-\pi}^{\pi} H_n(x)dx < \infty;$$

$$\text{(iii)}: \quad B = \sup_n \frac{1}{2\pi}\int_{-\infty}^{+\infty} |xH_n'(x)|dx < \infty$$

If $f \in L^1(T)$ set:

$$K^*f(x) = \sup_n |(K_n * f)(x)|.$$

Show that

$$K^*f(x) \leqslant (A + 2B)f^*(x).$$

What conclusion can be drawn from this?

ΔVΔ = VΔV = ΔVΔ = VΔV = ΔVΔ

SOLUTION (a): Let $\varphi(\lambda)$ be the left side of inequality (*). If $0 < \lambda \leqslant 2\|f\|_1$, then

$$\varphi(\lambda) \leqslant 2\pi \leqslant \frac{4\pi}{\lambda}\|f\|_1.$$

Furthermore, if $|x| \leqslant \pi$ and $h > 0$ are such that

$$2\|f\|_1 < \lambda < h^{-1}\int_0^h |f(x+u)|\,du,$$

and if N is the integer such that $2\pi(N-1) < h \leqslant 2\pi N$, then because f has period 2π,

$$\int_0^h |f(x+u)|\,du \leqslant N\int_0^{2\pi} |f(u)|\,du = 2\pi N\|f\|_1 < (h+2\pi)\|f\|_1,$$

and consequently $h < 2\pi$. From this it follows that for $\lambda > 2\|f\|_1$,

$$\varphi(\lambda) = \operatorname{meas}\left(x : |x| < \pi, \frac{1}{z-x}\int_x^z |f| > \lambda \text{ for a } z \text{ such that } x < z < x + 2\pi\right\}$$

$$\leqslant \operatorname{meas}\left(x : -\pi < x < 3\pi, \frac{1}{z-x}\int_x^z |f| > \lambda \text{ for a } z \text{ such that } x < z \leqslant 3\pi\right\}.$$

By introducing the continuous function

$$F(x) = \int_0^x |f(u)|\,du - \lambda x$$

the last inequality may be written

$$\varphi(\lambda) \leqslant \operatorname{meas}(V),$$

where

$$V = \{x : -\pi < x < 3\pi, F(z) > F(x) \text{ for a } z \text{ such that } x < z \leqslant 3\pi\}.$$

By the "Setting Sun Lemma" V is the disjoint union of a sequence of intervals (a_i, b_i) such that $F(a_i) \leqslant F(b_i)$, and consequently

$$\lambda\varphi(\lambda) \leqslant \sum_i \lambda(b_i - a_i) \leqslant \sum \int_{a_i}^{b_i} |f| \leqslant \int_{-\pi}^{3\pi} f = 4\pi\|f\|_1,$$

which accomplishes the proof.

Proceeding similarly for f^- and observing that

$$(f^\Delta > \lambda) = (f^+ > \lambda) \cup (f^- > \lambda),$$

one obtains

$$\text{meas}\{(f^\Delta > \lambda) \cap [-\pi, \pi]\} < \frac{8\pi}{\lambda} \|f\|_1\}.$$

Let us note that:

$$f^\Delta(x) = \sup_{h \neq 0} \left| h^{-1} \int_0^h |f(x+u)| du \right|,$$

so

$$(f_1 + f_2)^\Delta \leqslant f_1^\Delta + f_2^\Delta.$$

Furthermore, if $f \in L^\infty(T)$ then

$$\|f^\Delta\|_\infty \leqslant \|f\|_\infty.$$

By Marcinkiewicz's Theorem (cf. exercise 6·124) for all p, $1 < p \leqslant \infty$, there exists a constant A_p such that

$$\|f^\Delta\|_p \leqslant A_p \|f\|_p, \qquad f \in L^\infty(T).$$

The relation above is still valid if $f \in L^p(T)$. In fact, if g_i is an increasing sequence of simple functions that tends to $|f|$ at every point, then for all x and all $h > 0$

$$\sup_i h^{-1} \int_0^h g_i(x+u) du = h^{-1} \int_0^h |f(x+u)| du,$$

so

$$\sup_i g_i^+(x) = f^+(x).$$

From this it is deduced that

$$\sup_i g_i^\Delta(x) = f^\Delta(x),$$

and therefore

$$\|f^\Delta\|_p = \sup_i \|g_i^\Delta\|_p \leq A_p \sup_i \|g_i\|_p = A_p \|f\|_p.$$

(Actually the argument above is valid for every measurable function with period 2π). In particular, $f^\Delta \in L^p$ if $f \in L^p$, $1 < p \leq \infty$.

Using the last part of Exercise 6·124, one can show similarly that there exist constants A, B, A_p $(0 < p < 1)$ such that

$$\|f^\Delta\|_1 \leq A + B\int_{-\pi}^{\pi} |f| \log^+ |f|,$$

$$\|f^\Delta\|_p \leq A_p \|f\|_1, \qquad 0 < p < 1.$$

SOLUTION (b): As each function H_n is bounded, so is each K_n, and the convolution product $K_n * f$ is therefore defined everywhere, moreover

$$|(K_n * f)(x)| < \frac{1}{2\pi}\int_{-\pi}^{\pi} H_n(y)|f(x - y)|dy. \qquad (1)$$

Writing

$$H_n(y) = H_n(\pi) - \int_y^{\pi} H_n'(u)du$$

and changing the orders of integration, this yields

$$\int_0^\pi H_n(y)|f(x - y)|dy = H_n(\pi)\int_0^\pi |f(x - y)|dy$$

$$- \int_0^\pi H_n'(u)du\int_0^u |f(x - y)|dy.$$

For $u \geq 0$

$$\int_0^u |f(x - y)|dy \leq uf^-(x),$$

so

$$\int_0^\pi H_n(y)|f(x - y)|dy \leq (\pi H_n(\pi) + \int_0^\pi |uH_n'(u)|du)f^-(x).$$

Similarly,

$$\int_{-\pi}^0 H_n(y)|f(x - y)|dy \leq (\pi H_n(-\pi) + \int_{-\pi}^0 |uH_n'(u)|du)f^+(x),$$

so that by (iii) and (1)

$$|(K_n * f)(x)| \leq \tfrac{1}{2}(H_n(\pi) + H_n(-\pi) + 2B)f^\Delta(x). \tag{2}$$

Furthermore, the relation:

$$yH_n(y) = \int_0^y H_n(u)du + \int_0^y uH_n'(u)du$$

show that

$$\pi H_m(\pi) \leq \int_0^\pi H_n(u)du + \int_0^\pi |uH_m'(u)|du,$$

$$\pi H_m(-\pi) \leqslant \int_{-\pi}^{0} H_m(u)du + \int_{-\pi}^{0} |uH_m'(u)|du,$$

and consequently, by (ii) and (iii):

$$\tfrac{1}{2}(H_n(\pi) + H_n(-\pi)) \leqslant A + B.$$

Returning to (2) one obtains

$$|(K_n * f)(x)| \leqslant (A + 2B)f^{\Delta}(x).$$

One can then deduce for the operator $f \mapsto K*f$ the same properties as for $f \mapsto f^{\Delta}$.

EXERCISE 8·162: In what follows set $z = x + iy$, where $x \in \mathbb{R}$ and $y > 0$, and

$$P(x,y) = \frac{1}{\pi}\frac{y}{x^2 + y^2}, \qquad Q(x,y) = \frac{1}{\pi}\frac{x}{x^2 + y^2},$$

$$K(x,y) = P(x,y) + iQ(x,y) = \frac{i}{\pi z}.$$

(a): Show that the functions $P(\cdot,y)$ form, as $y \to 0$, an approximate identity in $L^1(\mathbb{R})$.

(b): If $f \in L^p(\mathbb{R})$, $1 \leqslant p \leqslant \infty$, set:

$$Pf(z) = \int_{-\infty}^{+\infty} P(x - t,y)f(t)dt.$$

Determine $\lim_{y\to 0} Pf(x + iy)$ when f is the characteristic function of an interval $[a,b]$.

(c): Drawing inspiration from the preceding exercise, show

that if $f \in L^1(\mathbb{R})$ and $\lambda > 0$, then

$$\text{meas}\{x : \sup_{y>0} |Pf(x + iy)| > \lambda\} \leqslant 2\lambda^{-1} \|f\|_1.$$

From this deduce that

$$\lim_{y \to 0} Pf(x + iy) = f(x) \text{ for almost all } x.$$

(Use part (b) above). Generalise this result to the case where $f \in L^p(\mathbb{R})$, $1 \leqslant p \leqslant \infty$.

(d): Let g be a function bounded and holomorphic in the half-plane $\Pi = \{z : \text{Im}(z) > 0\}$.

Show that there exists $\tilde{g} \in L^\infty(\mathbb{R})$ such that $g = P\tilde{g}$. (Use the fact that from every bounded sequence of L^∞ one can extract a subsequence which converges weakly in this space).

(e): When $f \in L^p(\mathbb{R})$, $1 \leqslant p < \infty$, set:

$$Kf(z) = \int_{-\infty}^{+\infty} K(x - t, y) f(t) \mathrm{d}t,$$

$$Qf(z) = \int_{-\infty}^{+\infty} Q(x - t, y) f(t) \mathrm{d}t.$$

Show that Kf is holomorphic in the half-plane Π. Next show that $\lim_{y \to 0} Kf(x + iy)$ exists for almost all x. (Reduce to the case where $f \leqslant 0$, set $g(z) = \exp Kf(z)$, and use parts (c) and (d) above. Deduce from this that

$$Hf(x) = \lim_{y \to 0} Qf(x + iy).$$

exists for almost all x. (Hf is called the *HILBERT TRANSFORM* of f).

(f): Calculate the Fourier transform $\hat{Q}(\cdot,y)$ of $Q(\cdot,y)$. From this deduce that $\|Hf\|_2 = \|f\|_2$ when $f \in L^2(\mathbb{R})$.

(g): Let $f \in L^1(\mathbb{R})$, $f \geq 0$. Show that for all $\varepsilon > 0$ and all $\mu > 0$

$$\frac{y}{\pi}\int_{-\infty}^{+\infty} \frac{\log|1 + \mu Kf(t + i\varepsilon)|}{(x - t)^2 + y^2}\, dt = \log|1 + \mu Kf(z + i\varepsilon)|.$$

Deduce from this that

$$\int_{-\infty}^{+\infty} \log^+|\mu Qf(t + i\varepsilon)|dt \leq \mu\|f\|_1.$$

Next show that for every $f \in L^1(\mathbb{R})$

$$\text{meas}\{t:|Qf(t + i\varepsilon)| > \lambda\} \leq \frac{4e}{\lambda}\|f\|_1.$$

(h): Using parts (f) and (g) and Marcinkiewicz's Theorem (cf. exercise 6•124) show that if $1 < p \leq 2$ there exists a constant A_p such that

$$\|Hf\|_p \leq A_p\|f\|_p \quad \text{for all } f \in L^p(\mathbb{R}).$$

Extend this result to the case where $2 < p < \infty$.

ΔVΔ = VΔV = ΔVΔ = VΔV = ΔVΔ

SOLUTION (a): This follows from $P(x,y) > 0$ and

$$\int_a^b P(x,y)dx = \frac{1}{\pi}\left(\tan^{-1}\frac{b}{y} - \tan^{-1}\frac{a}{x}\right). \tag{1}$$

SOLUTION (b): Note that $P(\cdot,y) \in L^q(\mathbb{R})$ for $1 \leq q \leq \infty$, which assures the existence of $Pf(z)$. Moreover, (1) shows that if f is the characteristic function of $[a,b]$ then

$$\lim_{y\to 0} Pf(x,y) = \begin{cases} 1 & \text{if } a < x < b, \\ \frac{1}{2} & \text{if } x = a \text{ or } x = b, \\ 0 & \text{otherwise.} \end{cases}$$

SOLUTION (c): First note that $-P'_x(u,y) \geqslant 0$ if $u \geqslant 0$, and that

$$\int_0^\infty -uP'_x(u,y)\,du = \frac{2}{\pi}\int_0^\infty \frac{u^2\,du}{(u^2+1)^2} = \frac{1}{2}\,.$$

Using the notations of the preceding exercise, this yields

$$\int_0^\infty P(t,y)|f(x \pm t)|\,dt = \int_0^\infty |f(x \pm t)|\,dt\int_t^\infty -P'_x(u,y)\,dy$$

$$= \int_0^\infty -uP'_x(u,y)\left(\frac{1}{u}\int_0^u |f(x \pm t)|\,dt\right)du$$

$$\leqslant \tfrac{1}{2}f^{\pm}(x).$$

Since

$$Pf(z) = \int_0^\infty P(t,y)\{f(x+t) + f(x-t)\}dt,$$

it follows that

$$|Pf(z)| \leqslant f^{\Delta}(x).$$

Proceeding as in the preceding exercise, one shows that for all $\lambda > 0$

$$\text{meas}(f^{\pm} > \lambda) \leqslant \lambda^{-1}\|f\|_1,$$

from which the desired inequality follows. This being so, let (φ_k) be a sequence of step functions such that $\|\varphi_k\|_1 = O(k^{-4})$ and $f = \sum \varphi_k$ in $L^1(\mathbb{R})$. There exists a set N_1 of measure zero such that

$$f(x) = \sum_k \varphi_k(x), \qquad x \notin N_1. \tag{2}$$

Furthermore, for all $z \in \Pi$,

$$Pf(z) = \sum_k P\varphi_k(z). \tag{3}$$

Finally, by part (b) there exists a set N_2 of measure zero such that

$$\lim_{y\to 0} P\varphi_k(x + iy) = \varphi_k(x) \quad \text{for all } k \text{ and all } x \notin N_2. \tag{4}$$

Since

$$\sum_k \operatorname{meas}\{x : \sup_{y>0} |P\varphi_k(x + iy)| > k^{-2}\} = \sum_k O(k^{-2}) < \infty,$$

there exists a set N_3 of measure zero such that if $x \notin N_3$:

$$\sup_{y>0} |P\varphi_k(x + iy)| \leqslant k^{-2}$$

starting from some k_0 (which may depend on x). In other words, when $x \notin N_3$ the series (3) is uniformly convergent with respect to y. It then follows from (2),(3) that if $x \notin N_1 \cup N_2 \cup N_3$,

$$\lim_{y\to 0} Pf(x + iy) = \sum_k \lim_{y\to 0} P\varphi_k(x + iy) = \sum_k \varphi_k(x) = f(x).$$

When $f \in L^p(\mathbb{R})$, $1 < p \leqslant \infty$, for every integer $N \geqslant 1$ let us set

$f_N(x) = f(x)$ or 0 according as $|x| < N$ or not, and $f^N = f - f_N$. Since $f_N \in L^1(\mathbb{R})$,

$$\lim_{y \to 0} Pf_N(x + iy) = f(x)$$

for almost all x such that $|x| < N$. Furthermore,

$$|Pf^N(x + iy)| \leqslant \frac{y}{\pi}\int_{|t| \geqslant N} \frac{|f(t)|}{(x - t)^2}\, dt.$$

From this one concludes that

$$\lim_{y \to 0} Pf(x + iy) = f(x)$$

for almost all $x \in]-N,N[$, and consequently for almost all $x \in \mathbb{R}$.

SOLUTION (d): For $\varepsilon > 0$ and $R > 0$ let $\Gamma_{\varepsilon,R}$ be the loop formed by the segment $[-R + i\varepsilon, R + i\varepsilon]$ and the semi-circle $\xi \mapsto i\varepsilon + Re^{i\vartheta}$, $0 \leqslant \vartheta \leqslant \pi$. For all $z \in \Pi$, whenever $1/\varepsilon$ and R are large enough,

$$g(z) = \frac{1}{2\pi i}\int_{\Gamma_{\varepsilon,R}} \frac{g(\xi)}{\xi - z}\, d\xi, \qquad 0 = \frac{1}{2\pi i}\int_{\Gamma_{\varepsilon,R}} \frac{g(\xi)}{\xi - \bar{z}}\, d\xi,$$

so by subtraction it is deduced that

$$g(z) = \frac{y}{\pi}\int_{\Gamma_{\varepsilon,R}} \frac{g(\xi)}{(\xi - z)(\xi - \bar{z})}\, d\xi.$$

As the function g is bounded, the integral along the semi-circle is $O(1/R)$, and consequently

$$g(z) = \frac{y}{\pi}\int_{-\infty}^{+\infty} \frac{g(t + i\varepsilon)}{(t + i\varepsilon - z)(t + i\varepsilon - \bar{z})}\, dt. \tag{5}$$

Now, let (ε_k) be a sequence that decreases to zero. Extracting a subsequence, one can assume that the functions $t \mapsto g(t + i\varepsilon_k)$ converge weakly in $L^\infty(\mathbb{R})$ to a function $\tilde{g}$. One can also assume that $\varepsilon_0 < y$. Since

$$|t + i\varepsilon_k - z| \geqslant |t + i\varepsilon_0 - z|, \qquad |t + i\varepsilon_k - \bar{z}| > |t - \bar{z}|,$$

Lebesgue's Theorem shows that the functions

$$t \mapsto (t + i\varepsilon_k - z)^{-1}(t + i\varepsilon_k - \bar{z})^{-1}$$

tend in *norm* in $L^1(\mathbb{R})$ to the function

$$t \mapsto ((x - t)^2 + y^2)^{-1},$$

and consequently one deduces from (5) that

$$g(z) = P\tilde{g}(z).$$

Note that by the preceding part of the exercise, for the sequence under consideration,

$$\lim_{y \to 0} g(x + iy) = \tilde{g}(x)$$

for almost all x.

SOLUTION (e): The functions $K(\cdot, y)$ belong to $L^q(\mathbb{R})$ for $1 < q \leqslant \infty$, which ensures the existence of $Kf(z)$ when $f \in L^p(\mathbb{R})$, $1 \leqslant p < \infty$. Furthermore, if C is a compact set contained in Π there exists $0 < R < \alpha$ such that the disc with centre ia and radius R contains C. But then

$$\left|\frac{f(t)}{(z - t)}\right| \leqslant \frac{1}{((a^2 + t^2)^{\frac{1}{2}} - R)} |f(t)| \qquad \text{when } z \in C.$$

The function of t which appears in the right side of this inequality is integrable, for the first factor belongs to $L^q(\mathbb{R})$ for $1 < q \leqslant \infty$. From this it follows that Kf is holomorphic in Π. Further, if $f \leqslant 0$ then

$$\mathrm{Re}Kf = Pf \leqslant 0,$$

and the function $g = \exp Kf$ is holomorphic and bounded in Π. By what we have just seen, $\lim_{y \to 0} g(x + iy)$ exists for almost all x, as well as $\lim_{y \to 0} Pf(x + iy)$, moreover. Since

$$\exp iQf = g\exp(-Pf),$$

it follows from this that

$$\lim_{y \to 0} Qf(x + iy) = Hf$$

exists for almost all x. When f is not negative it is written as a linear combination of four negative functions.

SOLUTION (f): The function $Q(\cdot,y)$ belongs to $L^2(\mathbb{R})$. Its Fourier transform is, by definition, the limit in $L^2(\mathbb{R})$ of the functions:

$$\xi \to \int_{-M}^{M} e^{-2\pi i\xi} Q(x,y)dx = -\frac{i}{\pi}\int_{-M}^{M} \frac{x\sin 2\pi\xi x}{x^2 + y^2}\, dx \quad \text{as } M \to \infty.$$

As the integral on the right has a limit as $M \to \infty$, one deduces from this that

$$\hat{Q}(\xi,y) = -\frac{i}{\pi}\int_{-\infty}^{+\infty} \frac{x\sin 2\pi\xi x}{x^2 + y^2}\, dx = -\frac{i}{\pi}\int_{-\infty}^{+\infty} \frac{x\sin 2\pi y\xi x}{x^2 + 1}\, dx.$$

It is a classic result that for $u \geqslant 0$,

$$\frac{1}{\pi}\int_{-\infty}^{+\infty} \frac{\cos ux}{x^2 + 1}\, dx = e^{-u}.$$

One easily justifies the differentiation with respect to u under the integral sign, which, taking into account that

$$\hat{Q}(-\xi,y) = -\hat{Q}(\xi,y),$$

yields

$$\hat{Q}(\xi,y) = -i\,\mathrm{sign}(\xi)e^{-2\pi y|\xi|}.$$

Let

$$Y(\xi) = -i\,\mathrm{sign}(\xi).$$

When $f \in L^2(\mathbb{R})$ we have

$$Qf(\cdot,y) = Q(\cdot,y)*f,$$

and therefore

$$\widehat{Qf(\cdot,y)} = \hat{Q}(\cdot,y)\hat{f},$$

functions which tend to $Y\hat{f}$ in $L^2(\mathbb{R})$ as $y \to 0$. As the Fourier transformation is an isometry of $L^2(\mathbb{R})$, it follows that $Qf(\cdot,y)$ tends in $L^2(\mathbb{R})$ to the inverse Fourier transform of $Y\hat{f}$. Since

$$Qf(\cdot,y) \to Hf \quad \text{almost everywhere,}$$

it follows that $Hf \in L^2(\mathbb{R})$, $\widehat{Hf} = Y\hat{f}$, and therefore that

$$\|Hf\|_2 = \|\widehat{Hf}\|_2 = \|Y\hat{f}\|_2 = \|\hat{f}\|_2 = \|f\|_2.$$

SOLUTION (g): Since

$$|K(x,y)| \leq \frac{1}{\pi y},$$

it is clear that

$$\pi y|Kf(z)| \leqslant \|f\|_1.$$

If, furthermore, $f \geqslant 0$, then

$$\mathrm{Re}Kf = Pf \geqslant 0,$$

and the function $1 + \mu Kf(z)$ takes its values in the half-plane $\mathrm{Re}(z) \geqslant 1$. By considering the principal value of the logarithm in this half-plane, the function $\log(1 + \mu Kf(z))$, which is holomorphic in Π, is defined unambiguously; moreover, its modulus is bounded by

$$\frac{\pi}{2} + \log|1 + \mu Kf(z)| \leqslant \frac{\pi}{2} + \mu|Kf(z)| \leqslant \frac{\pi}{2} + \frac{\mu}{\pi y}\|f\|_1$$

(since $\log|1 + \mu Kf(z)| \geqslant 0$). For all $\varepsilon > 0$ the function

$$g(z) = \log(1 + \mu Kf(z + i\varepsilon))$$

is therefore bounded and holomorphic in Π. Since

$$\lim_{y \to 0} g(x + iy) = \log(1 + \mu Kf(x + i\varepsilon))$$

we have (cf., part (d))

$$\frac{y}{\pi}\int_{-\infty}^{+\infty} \frac{\log(1 + \mu Kf(t + i\varepsilon))}{(x - t)^2 + y^2}\,dt = \log(1 + \mu Kf(z + i\varepsilon)),$$

and on taking real parts,

$$\frac{y}{\pi}\int_{-\infty}^{+\infty} \frac{\log|1 + \mu Kf(t + i\varepsilon)|}{(x - t)^2 + y^2}\,dt = \log|1 + \mu Kf(z + i\varepsilon)|.$$

We have observed that

$$\pi\log|1 + \mu Kf(z + i\varepsilon)| \leqslant \frac{\mu}{(y + \varepsilon)}\|f\|_1,$$

furthermore, since $Kf = Pf + iQf$, as Pf, Qf are real and $Pf \geqslant 0$, we have

$$\log^+|\mu Qf| \leqslant \log|1 + \mu Kf|,$$

so that

$$\int_{-\infty}^{+\infty} \frac{y^2}{(x - t)^2 + y^2} \log^+|\mu Qf(t + i\varepsilon)|dt \leqslant \mu \frac{y}{y + \varepsilon} \|f\|_1.$$

When $y \to \infty$Fatou's Lemma gives

$$\int_{-\infty}^{+\infty} \log^+|\mu Qf(t + i\varepsilon)|dt \leqslant \mu\|f\|_1.$$

But then for all $\lambda > 0$

$$\text{meas}\{t : |Qf(t + i\varepsilon)| > \lambda\}.\log^+(\mu\lambda) \leqslant \mu\|f\|_1,$$

and on making $\mu = e/\lambda$,

$$\text{meas}\{t : |Qf(t + i\varepsilon)| > \lambda\} \leqslant \frac{e}{\lambda} \|f\|_1.$$

If $f \in L^1(\mathbb{R})$ is complex, one deduces that

$$\text{meas}\{t : |Qf(t + i\varepsilon)| > \lambda\} \leqslant \frac{4e}{\lambda} \|f\|_1. \qquad (6)$$

SOLUTION (h): Note that if $f \in L^2(\mathbb{R})$

$$\|Qf(\cdot,\varepsilon)\|_2 = \|\hat{Q}(\cdot,\varepsilon)\hat{f}\|_2 \leqslant \|f\|_2,$$

and consequently

$$\text{meas}(|Qf(\cdot,\varepsilon)| > \lambda) \leqslant \frac{1}{\lambda^2} \|f\|_2^2.$$

Coupled with (6) this allows us to use Marcinkiewicz's Theorem (cf., exercise 6·134). For $1 < p \leqslant 2$ there exists a constant A_p (*independent of* ε) such that

$$\|Qf(\cdot,\varepsilon)\|_p \leqslant A_p \|f\|_p$$

if $f \in E$, the space of step functions on $\mathbb{R}$. To extend this result to the case where $2 < p < \infty$, let us observe that if $f,g \in L^2(\mathbb{R})$, then

$$(Qf(\cdot,\varepsilon)|g) = (\hat{Q}(\cdot,\varepsilon)\hat{f}|\hat{g}) = -(\hat{f}|\hat{Q}(\cdot,\varepsilon)\hat{g}) = -(f|Qg(\cdot,\varepsilon)).$$

If $\frac{1}{p} + \frac{1}{q} = 1$, then $1 < q < 2$, and if $f \in E$,

$$\|Qf(\cdot,\varepsilon)\|_p = \sup\{|(Qf(\cdot,\varepsilon)|g)| : \in E, \|g\|_q \leqslant 1\}$$

$$= \sup\{|(f|Qg(\cdot,\xi))| : g \in E, \|g\|_q \leqslant 1\}$$

$$\leqslant A_q \|f\|_p .$$

Now assume that $f \in L^p(\mathbb{R})$, $1 < p < \infty$. There exists a sequence (f_k) of step functions such that $f_k \to f$ in $L^p(\mathbb{R})$. For all $t \in \mathbb{R}$

$$|Qf(t,\varepsilon) - Qf_k(t,\varepsilon)| \leqslant \|Q(\cdot,\varepsilon)\|_q \|f - f_k\|_p,$$

and Fatou's Lemma allows us to conclude that again in this case

$$\|Qf(\cdot,\varepsilon)\|_p \leqslant A_p \|f\|_p$$

(setting $A_p = A_q$ if $2 < p < \infty$). By part (e), for almost all t

$$\lim_{\varepsilon \to 0} Qf(t,\varepsilon) = Hf(t),$$

and another application of Fatou's Lemma yields

$$\|Hf\|_p \leqslant A_p\|f\|_p.$$

CHAPTER 9

Functions with Bounded Variation: Absolutely Continuous Functions: Differentiation and Integration

EXERCISE 9·163: Show that if f is absolutely continuous on $[a,b]$ then so are the functions $|f|^p$ for $p \geqslant 1$.

$$\Delta\nabla\Delta = \nabla\Delta\nabla = \Delta\nabla\Delta = \nabla\Delta\nabla = \Delta\nabla\Delta$$

SOLUTION: This follows immediately from the fact that for $p \geqslant 1$ and $|z_1| \leqslant M$, $|z_2| \leqslant M$,

$$||z_1|^p - |z_2|^p| \leqslant pM^{p-1}|z_1 - z_2|.$$

EXERCISE 9·164: Let E be a subset of $[0,1]$ with measure zero. Construct an increasing and absolutely continuous function on $[0,1]$ such that $f'(x) = +\infty$ at every point $x \in E$.

$$\Delta\nabla\Delta = \nabla\Delta\nabla = \Delta\nabla\Delta = \nabla\Delta\nabla = \Delta\nabla\Delta$$

SOLUTION: Let (V_n) be a sequence of open sets such that

$$E \subset V_n, \qquad \text{meas}(V_n) < 2^{-n}.$$

The function

$$g = \sum_n \mathbb{1}_{V_n}$$

is integrable. Let

$$f(x) = \int_0^x g(t)dt.$$

The function f is increasing and absolutely continuous. Let $x \in E$. For every integer N there exists $\alpha > 0$ such that $|y - x| < \alpha$ implies $y \in V_n$ if $1 \leqslant n \leqslant N$. Then

$$\frac{f(y) - f(x)}{y - x} \geqslant N,$$

which proves that $f'(x) = +\infty$.

EXERCISE 9·165: Let (c_n) be a sequence of strictly positive numbers such that

$$\sum_{n=1}^{\infty} c_n = \infty.$$

(a): Show that one can construct a sequence of intervals $I_n =]x_n - c_n, x_n + c_n[$ such that $x_n \neq x_m$ if $n \neq m$, the sequence x_n is dense in $\mathbb{R}$ and every point of $\mathbb{R}$ belongs to an infinite number of I_n's.

(b): With the I_n's the same as in part (a), consider the function f such that $f(x_n) = c_n$ and $f(x) = 0$ if x is distinct from all the x_n's.

Show that if $\lim c_n = 0$ f is differentiable at no point of $\mathbb{R}$.

ΔVΔ = VΔV = ΔVΔ = VΔV = ΔVΔ

SOLUTION (a): One of the two series $\sum c_{2n}$, $\sum c_{2n+1}$ is divergent. Assume, for example, that $\sum c_{2n} = \infty$. Let (z_n) be a sequence of points everywhere dense in $\mathbb{R}$ and such that $z_n \neq z_m$ if $n \neq m$. Let

$$I_{2n+1} =]z_n - c_{2n+1}, z_n + c_{2n+1}[.$$

Next, consider integers $n_1 < n_2 < \cdots$ such that

$$\sum_{n_p \leqslant n < n_{p+1}} c_{2n} > p.$$

By induction, for all p one can determine some distinct points y_n ($n_p \leqslant n < n_{p+1}$) distinct from all the z_n and from the $y_1, y_2, \ldots, y_{n_p - 1}$ such that the intervals $I_{2n} =]y_n - c_{2n}, y + c_{2n}[$ cover $[-p,p]$. The sequence of I_n's so obtained satisfies the required conditions.

SOLUTION (b): It is clear that

$$\liminf_{x \to x_n} f(x) = 0 < f(x_n) = c_n.$$

The function f is therefore not continuous at x_n, and in particular it is not differentiable. If x is distinct from all the x_n's then

$$\liminf_{y \to x} \left| \frac{f(y) - f(x)}{y - x} \right| = 0.$$

Furthermore, $|x - x_n| < c_n$ for an infinite number of integers n, and consequently

$$\limsup_{y \to x} \left| \frac{f(y) - f(x)}{y - x} \right| \geqslant \limsup_{n \to \infty} \frac{c_n}{|x - x_n|} \geqslant 1.$$

EXERCISE 9·166: Let E be a measurable set of $[0,1]$ such that for some $\alpha > 0$

$$\operatorname{meas}(E \cap [a,b]) \geqslant \alpha(b - a)$$

for any $0 \leqslant a \leqslant b \leqslant 1$.

Show that $\text{meas}(E) = 1$.

ΔVΔ = VΔV = ΔVΔ = VΔV = ΔVΔ

FIRST SOLUTION: On setting $f = \mathbb{1}_E$ the condition of the problem becomes

$$\frac{1}{b-a}\int_a^b f(x)\,dx \geqslant \alpha.$$

By using Lebesgue's Differentiation Theorem, one deduces from this that for almost all $x \in [0,1]$ $f(x) \geqslant \alpha$, and consequently $f(x) = 1$. This shows that $\text{meas}(E) = 1$.

SECOND SOLUTION: Let

$$\varphi_n = n\mathbb{1}_{[0,1/n]}, \qquad n = 1,2,\ldots .$$

The functions φ_n form an approximate identity in $L^1(\mathbb{R})$. There exists therefore, a subsequence (n_s) such that $f*\varphi_{n_s} \to f$ almost everywhere. If $0 < x \leqslant 1$, and if n is large enough, then

$$f*\varphi_n(x) = n\int_{x-1/n}^{x} f(y)\,dy \geqslant \alpha.$$

From this one concludes, as above, that $f(x) = 1$ for almost all the $x \in [0,1]$.

EXERCISE 9·167: Let (f_n) be a sequence of increasing functions on $[a,b]$. Assume that the series

$$f(x) = \sum_{n=1}^{\infty} f_n(x)$$

converges for all x $(a \leqslant x \leqslant b)$.

Prove that for almost all $x \in [a,b]$

$$f'(x) = \sum_{n=1}^{\infty} f_n'(x). \qquad \textit{(FUBINI'S THEOREM)}$$

$\Delta\nabla\Delta = \nabla\Delta\nabla = \Delta\nabla\Delta = \nabla\Delta\nabla = \Delta\nabla\Delta$

SOLUTION: By considering the functions $f_n - f_n(a)$ one can assume that $f_n \geqslant 0$ for every integer n. Let

$$s_n = f_1 + \cdots + f_n.$$

By a theorem by Lebesgue which asserts that an increasing function is differentiable almost everywhere, one is assured that at almost all the points of $[a,b]$ all the numbers $s_n'(x)$ and $f'(x)$ exist. When $a \leqslant x < y \leqslant b$

$$s_n(y) - s_n(x) \leqslant s_{n+1}(y) - s_{n+1}(x) \leqslant f(y) - f(x),$$

and consequently

$$s_n'(x) \leqslant s_{n+1}'(x) \leqslant f'(x)$$

almost everywhere. This proves that almost everywhere:

$$\sum_{n=1}^{\infty} f_n'(x) < \infty.$$

Now consider a sequence $n_1 < n_2 < \cdots$ such that

$$f(b) - s_{n_p}(b) < 2^{-p}.$$

Then for all $x \in [a,b]$

$$\sum_p (f(x) - s_{n_p}(x)) \leqslant \sum_p (f(b) - s_{n_p}(b)) < \infty.$$

As the functions $f - s_{n_p}$ are increasing and positive, it follows from what we have proved above that:

$$\sum_p (f'(x) - s'_{n_p}(x)) < \infty$$

for almost all x. In particular,

$$f'(x) - s'_{n_p}(x) \to 0$$

for almost all x. As the sequence $s'_n(x)$ is increasing almost everywhere,

$$s'_n(x) \to f'(x)$$

which accomplishes the proof of Fubini's Theorem.

EXERCISE 9·168: Let E be a set of $\mathbb{R}$ that is not necessarily measurable.

Show that at almost all the points $x \in E$

$$\lim_{h\to 0} \frac{m^*(E \cap [x - h, x + h]}{2h} = 1,$$

where $m^*(A)$ denotes the outer measure of A.

Δ∇Δ = ∇Δ∇ = ΔVΔ = ∇Δ∇ = ΔVΔ

SOLUTION: Recall that the outer measure of a set A is defined by

$$m^*(A) = \inf\{\text{meas}(V) : V \text{ an open set, and } A \subset V\},$$

and that

$$m^*(A\cup B) \leqslant m^*(A) + m^*(B).$$

First let us show that if $A \subset [a,c]$ and $B \subset [c,b]$, then

$$m^*(A\cup B) = m^*(A) + m^*(B).$$

In fact, for all $\varepsilon > 0$ there exists an open set V such that $A\cup B \subset V$ and

$$\text{meas}(V) < m^*(A\cup B) + \varepsilon.$$

Let

$$V^- = V\cap]-\infty, c + \varepsilon[,$$

$$V^+ = V\cap]c - \varepsilon, +\infty[.$$

Then

$$V = V^-\cup V^+, \qquad \text{meas}(V^-\cap V^+) \leqslant 2\varepsilon, \qquad A\subset V^-, \qquad B\subset V^+,$$

and consequently

$$m^*(A) + m^*(B) \leqslant \text{meas}(V^-) + \text{meas}(V^+) = \text{meas}(V) + \text{meas}(V^-\cap V^+)$$

$$\leqslant m^*(A\cup B) + 3\varepsilon,$$

which proves the result.

Second, note that generality is not lost by assuming that E is contained in a compact interval $[a,b]$. Let V_n be an open set containing E and such that

$$\text{meas}(V_n) < m^*(E) + 2^{-n}.$$

For all $a \leqslant x \leqslant b$ let

$$f(x) = m^*(E \cap [a,x]), \qquad f_n(x) = \text{meas}(V_n \cap [a,x]).$$

Everything comes down to proving that $f'(x) = 1$ at almost all the points of E. If $a \leqslant x \leqslant y \leqslant b$, then by the preceding

$$(f_n(y) - f(y)) - (f_n(x) - f(x))$$

$$= \text{meas}(V_n \cap [x,y]) - m^*(E \cap [x,y])) \geqslant 0.$$

From this it follows that the functions $f_n - f$ are positive and increasing, and that $f_n - f \leqslant 2^{-n}$. Fubini's Theorem (cf., the preceding exercise) applied to the series

$$\sum_n (f_n - f)$$

implies that for almost all x

$$f_n'(x) - f'(x) \to 0$$

It remains to observe that if $x \in E$ then $f_n'(x) = 1$ for every integer n.

EXERCISE 9·169: Let f be a continuous function on $[a,b]$ and let V be its variation on this interval $(0 \leqslant V \leqslant \infty)$.

Show that for every $\lambda < V$ there exists $\delta > 0$ such that

$$\sum_{i=0}^{n-1} |f(x_{i+1} - f(x_i)| > \lambda \tag{*}$$

if

$$a = x_0 < x_1 < \cdots < x_n = b \quad \text{and} \quad \max(x_{i+1} - x_i) < \delta.$$

Δ∇Δ = ∇Δ∇ = ΔΔΔ = ∇Δ∇ = Δ∇Δ

SOLUTION: Let $a = y_0 < \cdots < y_p = b$ be such that

$$A = \sum_{i=0}^{p-1} |f(y_{i+1}) - f(y_i)| > \lambda.$$

There exists $\delta > 0$ such that $|x - y| < \delta$ implies

$$|f(x) - f(y)| < \frac{(A - \lambda)}{2p}.$$

If $a = x_0 < \cdots < x_n = b$ and $\max(x_{i+1} - x_i) < \delta$ the contribution to the right side of (*) of the terms corresponding to i's such that $y_j \leqslant x_i < x_{i+1} \leqslant y_{j+1}$ is greater than

$$|f(y_{j+1}) - f(y_j)| - \frac{(A - \lambda)}{p}.$$

From this it follows that

$$\sum_{i=0}^{n-1} |f(x_{i+1}) - f(x_i)| > A - (A - \lambda) = \lambda.$$

EXERCISE 9·170: Let f be a real continuous function on $[a,b]$. Show that

$$V(f;a,b) = \int_{-\infty}^{+\infty} n(t)dt,$$

where $n(t)$ denotes the (finite or infinite) number of solutions of the equation $f(x) = t$ (*BANACH'S THEOREM*).

$\Delta\nabla\Delta = \nabla\Delta\nabla = \Delta\nabla\Delta = \nabla\Delta\nabla = \Delta\nabla\Delta$

SOLUTION: If $\Delta = (a = x_0 < x_1 < \cdots < x_p = b)$ is a subdivision of $[a,b]$, set as usual

$$\Delta(f) = \sum_{i=0}^{p-1} |f(x_{i+1}) - f(x_i)|,$$

$$|\Delta| = \max(x_{i+1} - x_i : 0 \leqslant i \leqslant p - 1).$$

Let f_Δ be the function that coincides with f on Δ and which is affine linear on each of the intervals $[x_{i+1}, x_i]$. Let $n_\Delta(t)$ be the number of solutions of the equation $f_\Delta(x) = t$.

First show that

$$\int_{-\infty}^{+\infty} n_\Delta(t)\mathrm{d}t = \Delta(f), \tag{1}$$

$$n_\Delta(t) \leqslant n(t) \quad \text{if } t \notin f(\Delta). \tag{2}$$

Let us denote the numbers of solutions of the equations $f(x) = t$ and $f_\Delta(t) = t$ in $]x_{i+1}, x_i[$ by $\nu_i(t)$ and $\rho_i(t)$. When $t \notin f(\Delta)$ we have

$$n(t) = \sum_{i=0}^{p-1} \nu_i(t), \qquad n_\Delta(t) = \sum_{i=0}^{p-1} \rho_i(t),$$

and furthermore, $\rho_i(t) = 1$ or 0 according as t does or does not belong to the open interval with end points $f(x_i)$ and $f(x_{i+1})$. Therefore

$$\int \rho_i(t)\mathrm{d}t = |f(x_{i+1}) - f(x_i)| \quad \text{and} \quad \nu_i(t) \geqslant \rho_i(t)$$

by the Intermediate Value theorem. This proves (1) and (2). Now note that if $\Delta \subset \Delta'$ then

$$f_\Delta = (f_{\Delta'})_\Delta \quad \text{and} \quad f(\Delta) = f_{\Delta'}(\Delta)$$

so that (2) gives

$$n_\Delta(t) \leqslant n_{\Delta'}(t) \quad \text{if } t \notin f(\Delta). \tag{3}$$

Consider a sequence (Δ_s) of subdivisions of $[a,b]$ such that

$$\Delta_s \subset \Delta_{s+1}, \qquad \lim_s |\Delta_s| = 0. \tag{4}$$

Let D be the union of the $f(\Delta\ )$'s and of the maxima and minima of f on the subintervals of an arbitrary Δ_s. In other words, if $t \notin D$, and if α,β are two consecutive points of a subdivision Δ_s, then t is distinct from $f(\alpha)$ and $f(\beta)$ and from the maximum and minimum of f on $[\alpha,\beta]$. The set D is denumerable, and we are going to show that

$$\lim_s n_{\Delta_s}(t) = n(t) \quad \text{if } t \notin D. \tag{5}$$

Since, by the preceding exercise

$$\lim_s \Delta_s(f) = V(f;a,b), \tag{6}$$

the equation to be proved will follows from (1),(3),(5),(6) and Lebesgue's monotone convergence theorem. To prove (5), consider $t \notin D$ and let k be an integer such that $k \leqslant n(t)$. Let $\xi_1 < \xi_2 < \cdots < \xi_k$ be solutions of $f(x) = t$. By choosing s_0 large enough, Δ_{s_0} can be assume to separate the ξ , that is to say in each interval corresponding to Δ_{s_0} there is nor more than one ξ_i. If $\alpha < \xi_1 < \beta$, where α and β are two consecutive points of Δ_{s_0}, two cases can arise: either $f(\alpha) - t$ and $f(\beta) - t$ (which are not zero) have oppositc signs, in which case the equation $f_{\Delta_{s_0}}(x) = t$ has a solution in $]\alpha,\beta[$; or these two numbers have the same sign. If they are positive, for example, then since t is not the minimum of f on $[\alpha,\beta]$ there exist γ,δ such that $\alpha < \gamma < \delta < \beta$ and $f(x) - t < 0$ for $\gamma \leqslant x \leqslant \delta$. One can choose $s_1 > s_0$ so that Δ_{s_1} has a point in $[\gamma,\delta]$, and then the equation $f_{\Delta_{s_1}}(x) = t$ has at least one solution in $]\alpha,\beta[$. Observing that this property is not lost by increasing s_1, one can thus operate step by step and obtain an integer $s \geq s_0$ such that $n_{\Delta_s}(t) \geq k$. Taking (2) into account, this certainly implies (5).

EXERCISE 9·171: Let F be a function defined on $[a,b]$ and $1 < p < \infty$. Show that in order for there to exist $f \in L^p(a,b)$ such that:

$$F(x) = F(a) + \int_a^x f(t)dt, \qquad a \leq x \leq b, \tag{*}$$

it is necessary and sufficient that:

$$\sup_{\Delta} \sum_{i=0}^{n-1} \frac{|F(x_{i+1}) - F(x_i)|^p}{(x_{i+1} - x_i)^{p-1}} < \infty, \tag{**}$$

where the supremum is taken over all partitions $\Delta = (a = x_0 < x_1 < \cdots < x_n = b)$ of the interval $[a,b]$.

Show also that under these conditions the left side of (**) equals $\|f\|_p^p$.

$\Delta\nabla\Delta = \nabla\Delta\nabla = \Delta\nabla\Delta = \nabla\Delta\nabla = \Delta\nabla\Delta$

SOLUTION: In what follows, with Δ being given, set

$$\Delta_i = x_{i+1} - x_i, \qquad F_i = F(x_{i+1}) - F(x_i).$$

Now assume that (*) holds, with $f \in L^p(a,b)$. If $\frac{1}{p} + \frac{1}{q} = 1$ then

$$|\Delta F_i| = \left|\int_{x_i}^{x_{i+1}} f\right| \leq \Delta_i^{1/q}\left(\int_{x_i}^{x_{i+1}} |f|^p\right)^{1/p},$$

and since $p/q = p - 1$ this yields

$$\sum_{i=0}^{n-1} |\Delta F_i|^p \Delta_i^{-(p-1)} \leqslant \sum_{i=0}^{n-1} \int_{x_i}^{x_{i+1}} |f|^p = \int_a^b |f|^p. \tag{1}$$

Next assume that (**) is satisfied and let M be the value of the left side of this relation. Let us first show that F is absolutely continuous. For that, consider a finite sequence (a_i, b_i) of intervals contained in $[a,b]$ and mutually disjoint. Using Hölder's inequality for finite sequences this yields

$$\sum_i |F(b_i) - F(a_i)| \leqslant \left(\sum_i \frac{|F(b_i) - F(a_i)|^p}{(b_i - a_i)^{p-1}}\right)^{1/p} \left(\sum_i (b_i - a_i)\right)^{1/q}$$

$$\leqslant M^{1/p} \left(\sum (b_i - a_i)\right)^{1/q},$$

so that

$$\lim_{\Sigma(b_i - a_i) \to 0} \sum_i |F(b_i) - F(a_i)| - 0.$$

Having thus proved that F is absolutely continuous, it can be asserted that (*) holds for some function $f \in L^1(a,b)$. It remains to show that $f \in L^p(a,b)$. If Δ is a decomposition of $[a,b]$, let f_Δ be the function equal to $\Delta F_i / \Delta_i$ on the $(i + 1)$-th subinterval. As $F'(t) = f(t)$ for almost all t, it is clear that if Δ^s is a sequence of partitions such that $\max_i \Delta_i^s \to 0$ then $f_{\Delta_s} \to f$ almost everywhere. Furthermore

$$\int_a^b |f_\Delta|^p = \sum_{i=0}^{n-1} |\Delta F_i|^p \Delta_i^{-(p-1)} \leqslant M.$$

Therefore by Fatou's Lemma

$$\int_a^b |f|^p \leqslant M,$$

which, taking (1) into account, completes the proof.

EXERCISE 9·172: Let $1 < p < \infty$ and $\frac{1}{p} + \frac{1}{q} = 1$.

Show, using the preceding exercise, that for every continuous linear function u on $L^p(\mathbb{R})$ there exists $g \in L^q(\mathbb{R})$ such that

$$u(f) = \int fg, \qquad f \in L^p(\mathbb{R}).$$

Δ∇Δ = ∇Δ∇ = ΔVΔ = ∇Δ∇ = Δ∇Δ

SOLUTION: Let

$$F(x) = \begin{cases} u(\mathbb{1}_{[0,x]}) & \text{if } x \geqslant 0, \\ -u(\mathbb{1}_{[x,0]}) & \text{if } x \leqslant 0, \end{cases}$$

so that

$$F(\beta) - F(\alpha) = u(\mathbb{1}_{[\alpha,\beta]}), \qquad \alpha \leqslant \beta. \tag{1}$$

Let $A > 0$ and $\Delta = (-A = x_0 < x_1 < \cdots < x_n = A)$. Observe that if

$$g = \sum_{i=0}^{n-1} c_i \mathbb{1}_{[x_i,x_{i+1}]},$$

then using the notations of the previous exercise,

$$\left|\sum_{i=0}^{n-1} c_i \Delta F_i\right| = |u(g)| \leqslant \|u\| \left(\sum_{i=0}^{n-1} |c_i|^p \Delta_i\right)^{1/p}.$$

In this inequality substitute

$$c_i = \begin{cases} \dfrac{\overline{\Delta F_i}\cdot|\Delta F_i|^{q-2}}{\Delta_i^{q-1}} & \text{if } F_i \neq 0, \\ 0 & \text{otherwise} \end{cases}$$

This yields

$$\sum_{i=0}^{n-1} \frac{|\Delta F_i|^q}{\Delta_i^{q-1}} \leq \|u\| \left(\sum_{i=0}^{n-1} \frac{|\Delta F_i|^{p(q-1)}}{\Delta_i^{p(q-1)-1}} \right)^{1/p}$$

$$= \|u\| \left(\sum_{i=0}^{n-1} \frac{|\Delta F_i|^q}{\Delta_i^{q-1}} \right)^{1/p},$$

so

$$\sum_{i=0}^{n-1} \frac{|\Delta F_i|^q}{\Delta_i^{q-1}} \leq \|u\|^q.$$

By the preceding exercise $g(t) = F'(t)$ exists for almost all $t \in [-A,A]$, and

$$F(x) = \int_0^x g(t)dt, \qquad |x| \leq A,$$

$$\int_{-A}^{+A} |g(t)|^q dt \leq \|u\|^q.$$

In view of the arbitrariness of A, $g(t) = F'(t)$ exists for almost all $t \in \mathbb{R}$, $\|g\|_q \leq \|u\|$, and

$$F(x) = \int_0^x g(t)dt, \qquad x \in \mathbb{R}. \tag{2}$$

Let

$$u_g(f) = \int_{-\infty}^{+\infty} f(t)g(t)\mathrm{d}t, \qquad f \in L^p(\mathbb{R}).$$

This defines a continuous linear function on $L^p(\mathbb{R})$, and taking (2) and (3) into account (1) can be written

$$u(\mathbb{1}_{[\alpha,\beta]}) = u_g(\mathbb{1}_{[\alpha,\beta]}), \qquad \alpha \leqslant \beta.$$

As the functions $\mathbb{1}_{[\alpha,\beta]}$ form a total set in $L^p(\mathbb{R})$ we therefore have $u = u_g$.

EXERCISE 9·173: Two functions f,g that are positive and measurable on $]0,1[$ are called *EQUIMEASURABLE FUNCTIONS* if

$$\text{meas}(f > y) = \text{meas}(g > y) \quad \text{for all } y > 0.$$

(a): Show that if f,g are equimeasurable, then

$$\int_0^1 f(x)\mathrm{d}x = \int_0^1 g(x)\mathrm{d}x.$$

(b): Let s be a positive Borel function on $[0,\infty[$. Show that if f,g are equimeasurable, then so are $s\circ f$ and $s\circ g$.

(c): With f positive and measurable on $]0,1[$ show that there exists an unique function f^* that is positive, decreasing, right-continuous on $]0,1[$, and equimeasurable with f. (Show that $\{y : y > 0 \text{ and meas}(f > y) > x\} =]0,f^*(x)[$).

(d): Show that for every measurable set $E \subset]0,1[$

$$\int_E f(x)\mathrm{d}x \leqslant \int_0^{\text{meas}(E)} f^*(x)\mathrm{d}x,$$

and that for every positive and decreasing function φ on $]0,1[$

$$\int_0^1 f(x)\varphi(x)dx \leqslant \int_0^1 f^*(x)\varphi(x)dx.$$

(e): If f is positive and integrable on $]0,1[$, then for $0 < x < 1$ set

$$\vartheta f(x) = \sup\left\{\frac{1}{x-u}\int_u^x f(t)dt,\ 0 < u < x\right\}.$$

Show that:

$$(\vartheta f)^*(x) \leqslant \frac{1}{x}\int_0^x f^*(t)dt.$$

(f): Let s be a positive and increasing function on $[0,\infty[$. Show that

$$\int_0^1 s(\vartheta f(x))dx \leqslant \int_0^1 s\left(\frac{1}{x}\int_0^x f^*(t)dt\right)dx.$$

(g): Show from part (f) that if $p > 1$ then

$$\|\vartheta f\|_p \leqslant \frac{p}{p-1}\|f\|_p.$$

(Use Hardy's Inequality, cf. exercise 8·152).

(h): If $0 < p < 1$, show that

$$\|\vartheta f\|_p < \left(\frac{1}{1-p}\right)^{1/p}\|f\|_1.$$

(i): Show that for all $\varepsilon > 0$ there exists a number A depending only upon ε such that

$$\int_0^1 \vartheta f(x)\mathrm{d}x \leqslant 2\int_0^1 f(x)\log^+ f(x)\mathrm{d}x + A\int_0^1 f(x)\mathrm{d}x + \varepsilon.$$

ΔVΔ = VΔV = ΔVΔ = VΔV = ΔVΔ

SOLUTION (a): This property follows from the equation (cf. exercise 5·98)

$$\int_0^1 f(x)\mathrm{d}x = \int_0^\infty \mathrm{meas}(f > y)\mathrm{d}y.$$

SOLUTION (b): It is easily verified that if f,g are positive on $]0,1[$ then the set of sets $A \subset [0,\infty[$ such that $f^{-1}(A)$ and $g^{-1}(A)$ are measurable, and

$$\mathrm{meas}(f^{-1}(A)) = \mathrm{meas}(g^{-1}(A))$$

is a σ-algebra. When f,g are equimeasurable this σ-algebra contains all the intervals $]y,\infty[$, $y > 0$; therefore it contains all the Borel sets of $[0,\infty[$. If s is positive and Borel on $[0,\infty[$, then for all $y > 0$

$$\mathrm{meas}(f^{-1}(s^{-1}(]y,\infty[))) = \mathrm{meas}(g^{-1}(s^{-1}(]y,\infty[))),$$

which proves that $s \circ f$ and $s \circ g$ are equimeasurable.

SOLUTION (c): Let us begin with some considerations about decreasing functions. Let F be the set of positive functions on $]0,\infty[$ that are decreasing, right-continuous, and tend to zero at infinity. If $\varphi \in F$ and $x > 0$, the set $(\varphi > x)$ is a bounded interval which, if it is not empty, has 0 as its left end point, and is open on the right. In other words there exists $\varphi^*(x) \geqslant 0$ such that

$$(\varphi > x) =]0,\varphi^*(x)[. \tag{1}$$

We have $\varphi^* \in F$; in fact φ^* is decreasing and right continuous, for $x_n > x_{n+1}$, $x_n \to x$ implies

$$(\varphi > x) = \bigcup_n (\varphi > x_n),$$

i.e.

$$]0,\varphi^*(x)[= \bigcup_n]0,\varphi^*(x_n)[,$$

which means that $\varphi^*(x_n) \to \varphi^*(x)$. Finally, if one had $\varphi^* > \varepsilon > 0$ then, for $0 < y < \varepsilon$, one would have $\varphi(y) \geqslant x$ for all $x > 0$, which is absurd. Note that (1) can be expressed in the following manner: for $x > 0$ and $y > 0$, the conditions $\varphi(y) > x$ and $\varphi^*(x) > y$ are equivalent. From this it follows that

$$(\varphi^*)^* = \varphi. \tag{2}$$

Now return to our problem. If we set $\varphi(y) = \text{meas}(f > y)$ for $y > 0$, it is easily verified that $\varphi \in F$. Let g be a positive, decreasing, right-continuous functions on $]0,1[$, and let g_1 be the function obtained by extending g by 0 on $[1,\infty[$. Then $g_1 \in F$, and in order that g be equimeasurable with f it is necessary and sufficient that for all $y > 0$

$$\varphi(y) = \text{meas}(f > y) = \text{meas}(g > y) = \text{meas}(g_1 > y) = g_1^*(y),$$

that is to say, by (2), that $g_1 = \varphi^*$.

SOLUTION (d):

$$\int_E f = \int_0^1 \mathbb{I}_E \cdot f = \int_0^1 (\mathbb{I}_E \cdot f)^*.$$

For all $y > 0$

$$\mathrm{meas}(\mathbb{1}_E f > y) \leqslant \mathrm{meas}(E),$$

and consequently

$$(\mathbb{1}_E f)^*(x) = 0 \quad \text{if} \quad x \geqslant \mathrm{meas}(E).$$

Also,

$$\mathrm{meas}(\mathbb{1}_E f > y) \leqslant \mathrm{meas}(f > y),$$

so that

$$(\mathbb{1}_E f)^* \leqslant f^*.$$

Therefore

$$\int_E f = \int_0^{\mathrm{meas}(E)} (\mathbb{1}_E f)^* \leqslant \int_0^{\mathrm{meas}(E)} f^*.$$

In particular, if $0 < c < 1$, then

$$\int_0^c f \leqslant \int_0^c f^*.$$

From this it is immediately deduced that

$$\int_0^1 f\varphi < \int_0^1 f^*\varphi \tag{3}$$

for every positive decreasing step function φ. If φ is positive and decreasing on $]0,1[$ it is the limit on this interval of an increasing sequence of positive decreasing step functions, and (3) is therefore still valid in this case.

SOLUTION (e): Let

$$F(x) = \int_0^x f(t)dt, \qquad 0 < x < 1.$$

The set $(\vartheta f > y)$ is formed of the x's for which there exists u such that $0 < u < x$ and $F(u) - yu < F(x) - yx$. By the "Setting Sun lemma", this set is the disjoint union of a sequence of intervals $]a_n, b_n[$ such that

$$F(a_n) - ya_n \leqslant F(b_n) - yb_n.$$

Thus

$$\text{meas}(\vartheta f > y) = \sum_n (b_n - a_n)$$

$$\leqslant \frac{1}{y} \sum \int_{a_n}^{b_n} f = \frac{1}{y}\int_{(\vartheta f > y)} f$$

$$\leqslant \frac{1}{y}\int_0^{\text{meas}(\vartheta f > y)} f^*.$$

Therefore when $\text{meas}(\vartheta f > y) > 0$

$$y \leqslant \frac{1}{\text{meas}(\vartheta f > y)}\int_0^{\text{meas}(\vartheta f > y)} f^*. \tag{4}$$

Note that since f^* is decreasing, $a^{-1}\int_0^a f^*$ decreases as a increases. Consider, then, an x such that $(\vartheta f)^*(x) > 0$ (otherwise there would be nothing to prove). Since

$$\{y : \text{meas}(\vartheta f > y) > x\} =]0, (\vartheta f)^*(x)[,$$

it follows from (4) and from the remark above that

$$y < (\vartheta f)^*(x) \quad \text{implies} \quad y \leqslant \frac{1}{x}\int_0^x f^*,$$

which shows that

$$(\vartheta f)^*(x) \leqslant \frac{1}{x}\int_0^x f^*.$$

SOLUTION (f): As the functions $s\circ\vartheta f$ and $s\circ(\vartheta f)^*$ are equimeasurable,

$$\int_0^1 s\circ\vartheta f = \int_0^1 s\circ(\vartheta f)^* \leqslant \int_0^1 s\left(\frac{1}{x}\int_0^x f^*\right)dx.$$

SOLUTION (g): In particular, if $s(x) = x^p$, $p > 0$, then

$$\|\vartheta f\|_p^p \leqslant \int_0^1 \left(\frac{1}{x}\int_0^x f^*\right)^p dx. \tag{5}$$

After extending f^* by zero on $[1,\infty[$ and setting

$$G(x) = x^{-1}\int_0^x f^*,$$

(5) and Hardy's inequality gives

$$\|\vartheta f\|_p \leqslant \|G\|_p \leqslant \frac{p}{p-1}\|f^*\|_p = \frac{p}{p-1}\|f\|_p,$$

since f^p and $(f^*)^p$ are equimeasurable.

SOLUTION (h): If $0 < p < 1$ then Hölder's inequality for the pair $\left(\frac{1}{p}, \frac{1}{1-p}\right)$ applied to (5) yields

$$\|\vartheta f\|_p^p \leqslant \int_0^1 \left(\frac{1}{x^p}\int_0^x f^*\right)^p \frac{dx}{x^{p(1-p)}}$$

$$\leqslant \left(\int_0^1 \frac{dx}{x^p}\right)^{1-p} \left(\int_0^1 \frac{dx}{x^p}\int_0^x f^*\right)^p$$

$$\leqslant \|f\|_1^p \int_0^1 \frac{dx}{x^p},$$

so

$$\|\vartheta f\|_p \leqslant \left(\frac{1}{1-p}\right)^{1/p} \|f\|_1.$$

SOLUTION (i): Finally, if $p = 1$, then starting from (5), one obtains

$$\int_0^1 \vartheta f \leqslant \int_0^1 \frac{dx}{x}\int_0^x f^* = \int_0^1 f^*(t)dt \int_t^1 \frac{dx}{x} = \int_0^1 f^*(t)\log\frac{1}{t}dt.$$

Let $\varepsilon > 0$ and choose α such that $0 < \alpha < 1$ and

$$\int_0^\alpha \log\frac{1}{t}\cdot\frac{dt}{\sqrt{t}} < \varepsilon.$$

Then

$$\int_0^1 \vartheta f \leqslant \int_0^\alpha f^*(t)\log\frac{1}{t}dt + \log\frac{1}{\alpha}\int_0^1 f(t)dt.$$

Let E be the set of t's such that $0 < t \leqslant \alpha$ and $f^*(t) \leqslant t^{-\frac{1}{2}}$. Then

$$\int_0^\alpha f^*(t)\log\frac{1}{t}dt \leqslant \int_E \log\frac{1}{t}\,\frac{dt}{\sqrt{t}} + 2\int_0^1 f^*(t)\log^+ f^*(t)dt \leqslant \text{(Contd)}$$

(Contd) $$\leqslant \varepsilon + 2\int_0^1 f(t)\log^+ f(t)dt,$$

whence, at last,

$$\int_0^1 \vartheta f \leqslant 2\int_0^1 f\log^+ f + \log\frac{1}{\alpha}\int_0^1 f + \varepsilon.$$

CHAPTER 10

Summation Processes: Trigonometric Polynomials

EXERCISE 10·174: (a): If the series

$$\sum_{n=0}^{\infty} u_n \tag{1}$$

converges, and if

$$\sum_{n=0}^{\infty} |\lambda_n - \lambda_{n+1}| < \infty, \tag{2}$$

show that the series

$$\sum_{n=0}^{\infty} \lambda_n u_n \tag{3}$$

converges.

(b): Conversely, prove that if the series (3) converges whenever the series (1) converges, then (2) is satisfied.

Δ∇Δ = ∇Δ∇ = ΔVΔ = ∇Δ∇ = ΔVΔ

SOLUTION (a): (2) implies that $\lambda_n = \lambda + \varepsilon_n$, with $\varepsilon_n \to 0$. As

$$\sum |\varepsilon_n - \varepsilon_{n+1}| < \infty \quad \text{and} \quad \sup|u_0 + \cdots + u_n| < \infty,$$

the Abel-Dirichlet Theorem shows that the series $\sum \varepsilon_n u_n$ is convergent. The same then holds for the series $\sum \lambda_n u_n$.

SOLUTION (b): *First Proof:* Let E be the Banach space formed by convergent sequences $s = (s_n)$, with the norm

$$\|s\| = \sup_n |s_n|.$$

For every integer $N \geqslant 1$ define a linear function f_N on E by

$$f_N(s) = \sum_{n=0}^{N-1} (\lambda_n - \lambda_{n+1})s_n + \lambda_N s_N.$$

This linear functional is continuous, and it is a classical result that its norm is

$$\|f_N\| = \sum_{n=0}^{N-1} |\lambda_n - \lambda_{n+1}| + |\lambda_N|.$$

Since

$$f_N(s) = \lambda_0 s_0 + \sum_{n=1}^{N} \lambda_n(s_n - s_{n-1}),$$

and as the series $\sum (s_n - s_{n-1})$ is convergent, $\lim_N f_N(s)$ exists for every $s \in E$. By the Banach-Steinhaus Theorem one therefore has

$$\sup_N \|f_N\| < \infty,$$

which implies that:

$$\sum_{n=0}^{\infty} |\lambda_n - \lambda_{n+1}| < \infty.$$

Second Proof: Let

$$\lambda_n - \lambda_{n+1} = r_n \xi_n, \qquad r_n \geqslant 0, \qquad |\xi_n| = 1,$$

$$s_n = \frac{\bar{\xi}_n}{1 + r_0 + r_1 + \cdots + r_n},$$

$$u_0 = s_0, \qquad u_n = s_n - s_{n-1} \quad \text{if } n \geqslant 1.$$

If it is assumed that

$$\sum_{n=0}^{\infty} |\lambda_n - \lambda_{n+1}| = \sum_{n=0}^{\infty} r_n = \infty,$$

then $s_n \to 0$, and the series $\sum u_n$ is therefore convergent. Furthermore,

$$\sum_{n=0}^{N} \lambda_n u_n = \lambda_N s_N + \sum_{n=0}^{N-1} (\lambda_n - \lambda_{n+1}) s_n.$$

The first term of the right side is bounded by

$$\frac{|\lambda_0| + r_0 + \cdots + r_{N-1}}{1 + r_0 + \cdots + r_N},$$

a quantity that remains bounded as $N \to \infty$; the second term is equal to

$$\sum_{n=0}^{N-1} \frac{r_n}{1 + r_0 + \cdots + r_n},$$

and tends to infinity with N by a classical property of positive divergent series. It follows that the series $\sum \lambda_n u_n$ is divergent.

EXERCISE 10·175: Let $(u_n)_{n\geq 0}$ be a decreasing sequence of positive real numbers.

Show that if

$$\sum_{n=0}^{\infty} u_n < \infty$$

then

$$u_n = o\left(\frac{1}{n}\right).$$

Δ∇Δ = ∇Δ∇ = Δ∇Δ = ∇Δ∇ = Δ∇Δ

SOLUTION: Let

$$\varepsilon_N = \sum_{n=N+1}^{\infty} u_n.$$

If $p > N$ then

$$(p - N)u_p \leqslant u_{N+1} + \cdots + u_p \leqslant \varepsilon_N,$$

and consequently

$$\limsup_{p\to\infty} pu_p \leqslant \varepsilon_N,$$

which implies the desired result upon making $N \to \infty$.

EXERCISE 10·176: Let $(a_n)_{n\geq 0}$ be a sequence of real numbers. Set:

$$\Delta a_n = a_n - a_{n+1}, \qquad \Delta^2 a_n = \Delta a_n - \Delta a_{n+1}.$$

The sequence is called a *CONVEX SEQUENCE* if

$$\Delta^2 a_n \geqslant 0 \quad \text{for all } n \geqslant 0.$$

(a): Let φ be a convex function on $[0,\infty[$. Show that the sequence $a_n = \varphi(n)$ is convex.

(b): Show that if the sequence $(a_n)_{n\geqslant 0}$ is convex and bounded, then

(i): It is decreasing;

(ii): $\Delta a_n = o(1/n)$;

(iii): $\sum_{n=0}^{\infty} (n + 1)\Delta^2 a_n = a_0 - \lim_n a_n$.

(c): Show that for every sequence $(c_n)_{n\geqslant 0}$ which tends to zero there exists a convex sequence $(a_n)_{n\geqslant 0}$ which tends to zero, and which is such that $|c_n| \leqslant a_n$ for all $n \geqslant 0$.

Δ∇Δ = ∇Δ∇ = Δ∇Δ = ∇Δ∇ = Δ∇Δ

SOLUTION (a): This follows from

$$\Delta^2 a_n = a_n - 2a_{n+1} + a_{n+2}$$

$$= 2[\tfrac{1}{2}\varphi(n) + \tfrac{1}{2}\varphi(n + 2) - \varphi(\tfrac{1}{2}n + \tfrac{1}{2}(n + 2))].$$

SOLUTION (b): The condition $\Delta^2 a_n \geqslant 0$ for all n means that the sequence (Δa_n) is decreasing. If, for some integer r $\Delta a_r < 0$, then for every integer $n > r$

$$a_n = a_r - \sum_{s=r}^{n-1} \Delta a_s \geqslant a_r + (n - r)(-\Delta a_r),$$

and so the sequence (a_n) would tend to $+\infty$, which is absurd. Therefore $\Delta a_n \geqslant 0$ for all n, which proves (i).

But then the sequence (a_n) is decreasing and bounded, and thus has a limit. Since:

$$\sum_{n=0}^{N-1} \Delta a_n = a_0 - a_N,$$

the series $\sum \Delta a_n$ is therefore convergent, and since the Δa_n's are positive and decreasing, (ii) follows by the preceding exercise.

Finally,

$$\sum_{n=0}^{N-1} (n+1)\Delta^2 a_n = \sum_{n=0}^{N-1} \Delta a_n - N\Delta a_n = a_0 - a_N - N\Delta a_N,$$

which implies (iii), since by (ii) the last term tends to zero.

SOLUTION (c): On replacing c_n by $\sup(|c_s| : s \geqslant n)$ it can be assumed that the sequence (c_n) is positive and decreases to zero. By part (a) it suffices to construct a convex function φ on $[0,\infty[$ that tends to zero at infinity, and that is such that $\varphi(n) \geqslant c_n$ for all n.

To do that, consider a sequence (φ_k) such that

$$\varphi_k > \varphi_{k+1} \quad \text{if } k \geqslant 0, \qquad \varphi_1 = c_0, \qquad \varphi_k \to 0. \tag{1}$$

Let $n_0 = 0$, and let n_1 be such that

$$n_1 > n_0, \qquad c_{n_1} < \varphi_2, \tag{2}$$

then construct by induction some integers n_k such that for $k \geqslant 2$

$$n_k > n_{k-1}, \tag{3}$$

$$c_{n_k} \leqslant \varphi_{k+1}, \tag{4}$$

$$\frac{\varphi_{k-1} - \varphi_k}{n_k - n_{k-1}} \leqslant \frac{\varphi_{k-2} - \varphi_{k-1}}{n_{k-1} - n_{k-2}} . \tag{5}$$

This is possible because the sequence (c_n) tends to zero. Note that (4) is valid for all $k > 0$. This being so, let φ be such that $\varphi(n_k) = \varphi_k$ and that is linear on each of the intervals $[n_k, n_{k+1}]$. This function is decre sing and tends to zero at infinity by (1), and it is convex by (5). If $n_K < n < n_{K+1}$, then by virtue of Relation (4)

$$c_n \leqslant c_{n_k} \leqslant \varphi(n_{k+1}) \leqslant \varphi(n).$$

EXERCISE 10·177: Let $(u_{n,p})_{n\geqslant 0, p\geqslant 0}$ be a double sequence of complex numbers, such that

(i): $\sum_{n=0}^{\infty} |u_{n,p}| \leqslant M$ for all $p \geqslant 0$;

(ii): $\lim_{p\to\infty} \sum_{n=0}^{\infty} u_{n,p} = 1$;

(iii): $\lim_{p\to\infty} u_{n,p} = 0$ for all $n \geqslant 0$.

Show that if $s_n \to s$, then

$$\lim_{p\to\infty} \sum_{n=0}^{\infty} u_{n,p} s_n = s.$$

ΔVΔ = VΔV = ΔVΔ = VΔV = ΔVΔ

SOLUTION: If $s_n \to s$, then on setting $s_n = s + c_n$ it is seen that by (ii) one is reduced to the case where $s = 0$. Let c_0 be the Banach space formed by the sequences $c = (c_n)$ that tend to zero, with the norm

$$\|c\| = \sup_n |c_n|,$$

and let us consider on c_0 the linear functionals

$$f_p(c) = \sum_{n=0}^{\infty} u_{n,p} c_n.$$

It is clear that these latter are continuous and that

$$\|f_p\| \leqslant \sum_{n=0}^{\infty} |u_{n,p}| \leqslant M. \tag{1}$$

It is a question of proving that $f_p(c) \to 0$ for all $c \in c_0$. Let e^n be the sequence such that $e_m^n = \delta_{nm}$. From the relation

$$\|c - \sum_{n=0}^{N} c_n e^n\| = \sup_{n>N} |c_n|$$

one deduces that the e_n's form a total set in c_0. By (1) it therefore suffices to prove that for all n

$$\lim_{p\to\infty} f_p(e^n) = 0.$$

But this is nothing other than (iii), because

$$f_p(e^n) = u_{n,p}.$$

In the following six exercises the following notations are used:

If $(c_n)_{n\in\mathbb{Z}}$ is a sequence of complex numbers, set:

$$s_N = \sum_{|n|\leqslant N} c_n, \qquad \sigma_N = \frac{s_0 + s_1 + \cdots + s_N}{N + 1},$$

$$t_N = \sum_{|n|\leqslant N} |n| c_n.$$

It will be recalled that the series $\sum_{-\infty}^{+\infty} c_n$ is called a *CONVERGENT SERIES* if $\lim_N s_N$ exists. It is called a *CESARO CONVERGENT SERIES* if $\lim_N \sigma_N$ exists. By the preceding exercise convergence implies Cesaro convergence. (Let $u_{N,p} = \dfrac{1}{p + 1}$ for $N \leqslant p$ and 0 otherwise.)

EXERCISE 10·178: Prove that if $s_0 \leqslant s_1 \leqslant \cdots \leqslant s_n < \cdots$ and $\sigma_N \to s$, then $s_N \to s$.

Δ∇Δ = ∇Δ∇ = ΔVΔ = ∇Δ∇ = Δ∇Δ

SOLUTION: Let $s' = \lim s_N$. If $s' < \infty$ then $s = s'$, by what was recalled in the note above. If $s' = \infty$, then for $p < N$

$$\sigma_N \geqslant \frac{s_0 + \cdots + s_p}{N + 1} + \frac{N - p}{N + 1} s_p,$$

which implies that $s \geqslant s_p$, and consequently $s = s' = \infty$.

EXERCISE 10·178: (a): Show that $\lim_N \sigma_N$ exists if and only if the series

$$\sum_{N=1}^{\infty} N^{-2} t_N$$

converges. (Use exercise 10·174).

(b): Assume that $\sigma_N \to s$. Show that $s_N \to s$ if and only if $t_N = o(N)$.

(c): Again assume that $\sigma_N \to s$. Show that if for some real number $p \geqslant 1$

$$\sum_{-\infty}^{+\infty} |n|^{p-1} |c_n|^p < \infty,$$

then $s_N \to s$.

(d): Prove the following result (*HARDY'S LEMMA*): If $c_n = O(1/n)$ then $\sigma_N \to s$ implies $s_N \to s$. (Assume that for a sequence $N_1 < N_2 < \cdots\ t_{N_s} \geqslant BN_s$, $B > 0$, for example, and show that

$$t_{N_s+\nu} \geqslant \frac{1}{2} BN_s \quad \text{if} \quad 0 \leqslant \nu \leqslant \frac{1}{2}\frac{B}{A} N_s,$$

where A is a constant such that $|c_n + c_{-n}| \leqslant A/n$ for $n \geqslant 1$).

$$\Delta\nabla\Delta = \nabla\Delta\nabla = \Delta\nabla\Delta = \nabla\Delta\nabla = \Delta\nabla\Delta$$

SOLUTIONS (a);(b):

$$t_N = \sum_{|n| \leqslant N} n\, c_n = \sum_{n=1}^{N} n(c_n + c_{-n}) = \sum_{n=1}^{N} n(s_n - s_{n-1}),$$

from which it is easily deduced that

$$t_N = N(s_N - \sigma_{N-1}), \tag{1}$$

which proves (b). Now

$$s_N = (N + 1)\sigma_N - N\sigma_{N-1},$$

and substituting this expression in (1) yields

$$t_N = N(N + 1)(\sigma_N - \sigma_{N-1}). \tag{2}$$

Therefore, σ_N converges if and only if the series

$$\sum_{N=1}^{\infty} (\sigma_N - \sigma_{N-1}) = \sum_{N=1}^{\infty} \frac{t_N}{N(N + 1)}$$

converges. Since the sequences

$$\lambda_N^1 = N(N + 1)^{-1} \quad \text{and} \quad \lambda_N^2 = N^{-1}(N + 1)$$

satisfy

$$\sum_{N=1}^{\infty} |\lambda_N^i - \lambda_{N+1}^i| < \infty, \quad i = 1,2,$$

the convergence of σ_N is equivalent to that of the series

$$\sum_{N=1}^{\infty} N^{-2} t_N$$

(cf. exercise 10·174).

SOLUTION (c): If $p = 1$ the condition becomes

$$\sum_{-\infty}^{+\infty} |c_n| < \infty,$$

in which case s_N converges to a value that can only be s. Therefore assume that $p > 1$, and let q be such that $\frac{1}{p} + \frac{1}{q} = 1$; it is now a matter of proving that $N^{-1}t_N \to 0$ (cf., the part (b)). Now, if $r < N$ then

$$\left|\frac{t_N}{N}\right| \leqslant \frac{1}{N}\left|\sum_{|n|\leqslant r} |n|c_n\right| + \frac{1}{N}\left|\sum_{r<|n|\leqslant N} |n|^{1/p}|n|^{1-1/p}c_n\right|$$

$$\leqslant \frac{r}{N}\sum_{|n|\leqslant r} |c_n| + \frac{1}{N}(2\sum_{n=1}^{N} n^{q/p})^{1/q}(\sum_{|n|>r} |n|^{p-1}|c_n|^p)^{1/p}.$$

Since

$$\sum_{n=1}^{N} n^{q/p} \leqslant \frac{1}{q}(N+1)^q,$$

it follows that

$$\left|\frac{t_N}{N}\right| \leqslant \frac{r}{N}\sum_{|n|\leqslant r} |c_n| + \left(\frac{2}{q}\right)^{1/q}\frac{N+1}{N}(\sum_{|n|>r} |n|^{p-1}|c_n|^p)^{1/p},$$

and consequently

$$\limsup_{N\to\infty}\left|\frac{t_N}{N}\right| < \left(\frac{2}{q}\right)^{1/q}(\sum_{|n|>r} |n|^{p-1}|c_n|^p)^{1/p}.$$

By making $r \to \infty$ the desired result is obtained.

SOLUTION (d): Assume that $\sigma_N \to s$ and that

$$|c_n + c_{-n}| \leqslant \frac{A}{n}, \qquad n \geqslant 1.$$

Let us show that $N^{-1}t_N \to 0$ under these conditions, which by part (b) will prove that $s_N \to s$. Otherwise there would exist a $B > 0$ and an increasing sequence of integers N_s such that

$$t_{N_s} > BN_s.$$

If ν_s is the integral part of $BN_s/2A$, then for $0 \leqslant \nu \leqslant \nu_s$

$$t_{N_s+\nu} = t_{N_s} + \sum_{N_s<n\leqslant N_s+\nu} n(c_n + c_{-n})$$

$$\geqslant BN_s - \nu A$$

$$\geqslant BN_s - \frac{BN_s}{2} = \frac{BN_s}{2} .$$

Furthermore, by part (a), the series $\sum N^{-2}t_N$ must be convergent, and in particular

$$\sum_{N_s\leqslant N<N_s+\nu_s} N^{-2}t_N$$

must tend to zero as $s \to \infty$. But this quantity is bounded from below by

$$\frac{BN_s}{2}\int_{N_s}^{N_s+\nu_s} x^{-2}dx = \frac{B}{2}\frac{\nu_s}{N_s + \nu_s} \to \frac{B^2}{2B + 4A} > 0,$$

a contradiction.

REMARK: We have implicitly assumed that the c_n's were real. If this were not the case, decompose them into their real and imaginary parts. Moreover, in assuming that $N^{-1}t_N$ did not tend to zero we supposed that $\limsup N^{-1}t_N > 0$. If this supremum were zero then necessarily $\liminf N^{-1}t_N < 0$, and it would suffice to change the signs of all the c_n's.

EXERCISE 10·180: Set

$$\sigma_{p,N} = \frac{s_p + s_{p+1} + \cdots + s_{p+N}}{N + 1}.$$

(a): Express $\sigma_{p,N}$ as a function of the σ_N's.

Deduce from this that if $\sigma_N \to s$ then $\sigma_{p,N_p} \to s$ for every sequence of integers (N_p) such that $N_p \geqslant ap$, $a > 0$.

(b): Prove that if $|c_n| \leqslant A|n|$ for $n \geqslant 1$, then

$$|\sigma_{p,N} - s_p| \leqslant A \frac{N}{p}.$$

(c): Deduce from the preceding a new proof of Hardy's Lemma (cf. the preceding exercise).

$$\Delta\nabla\Delta = \nabla\Delta\nabla = \Delta\nabla\Delta = \nabla\Delta\nabla = \Delta\nabla\Delta$$

SOLUTION (a):

$$\sigma_{p,N} = \frac{(p + N + 1)\sigma_{p+N} - p\sigma_{p-1}}{N + 1}$$

$$= \sigma_{p+N} + \frac{p}{N + 1}(\sigma_{p+N} - \sigma_{p-1}). \qquad (1)$$

If the sequence (N_p) is such that $N_p \geqslant ap$, $a > 0$, then

$$\frac{p}{N_p + 1} \leqslant \frac{p}{ap + 1} \leqslant \frac{1}{a},$$

so that if $\sigma_N \to s$ then $\sigma_{p,N_p} \to s$.

SOLUTION (b):

$$(N + 1)\sigma_{p,N} = (N + 1)s_p + \sum_{n=p+1}^{p+N} (N + p + 1 - n)(c_n + c_{-n}),$$

so that

$$|\sigma_{p,N} - s_p| = \frac{1}{N + 1}\left|\sum_{n=p+1}^{p+N} (N + p + 1 - n)(c_n + c_{-n})\right|$$

$$\leqslant \frac{2A}{p(N + 1)} \sum_{r=1}^{N} r = \frac{AN}{p} .$$

SOLUTION (c): If $|c_n| \leqslant A|n|^{-1}$ for $n \geqslant 1$ and $\sigma_N \to s$, let

$$\varepsilon_p = \sup_{N \geqslant p-1} |\sigma_N - \sigma_{p-1}|,$$

so that $\varepsilon_p \to 0$. Let N be the integer such that

$$N_p \leqslant p\sqrt{\varepsilon_p} < N_p + 1.$$

Then by equation (1) proved in part (a)

$$|\sigma_{p,N_p} - \sigma_{p+N_p}| \leqslant \sqrt{\varepsilon_p}.$$

On the other hand, by the bound obtained in part (b),

$$|\sigma_{p,N_p} - s_p| \leqslant A\sqrt{\varepsilon_p}.$$

This proves that $s_p \to s$.

EXERCISE 10·181: Assume that $c_n \to 0$ and that $c_n = 0$ except when $n = 0, \pm n_1, \pm n_2, \ldots$ with

$$\frac{n_{p+1}}{n_p} \geqslant q > 1.$$

Show that if $\sigma_N \to s$ then $s_N \to s$. (Let $a_p = \max(|c_{n_p}|, |c_{-n_p}|)$ and show that for $n_p \leqslant N < n_{p+1}$

$$N^{-1} t_N \leqslant 2 \sum_{r=1}^{p} a_r q^{r-p};$$

then use exercises 10·177 and 10·179).

Δ∇Δ = ∇Δ∇ = ΔVΔ = ∇Δ∇ = Δ∇Δ

SOLUTION:

$$\frac{n_p}{n_r} \geqslant q^{p-r} \quad \text{if } p \geqslant r,$$

so that if $n_p \leqslant N < n_{p+1}$, then

$$|N^{-1} t_N| = |N^{-1} \sum_{r=1}^{p} n_r (c_{n_r} + c_{-n_r})|$$

$$\leqslant 2 \sum_{r=1}^{p} a_r \frac{n_r}{n_p} \leqslant 2 \sum_{r=1}^{p} a_r q^{r-p}.$$

Now note that $a_r \to 0$, and that

$$\lim_{p\to\infty} \sum_{r=1}^{p} q^{r-p} = \frac{q}{q-1},$$

$$\lim_{p\to\infty} q^{r-p} = 0.$$

By exercise 10·177 it follows that

$$\lim_{p\to\infty} \sum_{r=1}^{p} a_r q^{r-p} = 0,$$

and therefore that $N^{-1}t_N \to 0$. Then by exercise 10·179 $s_N \to s$ if $\sigma_N \to s$.

EXERCISE 10·182: Let $(a_n)_{n\geqslant 0}$ be a sequence of strictly positive real numbers that is convex and tends to zero.

Show that if

$$|\sigma_N| \leqslant A, \qquad s_N = o\left(\frac{1}{a_N}\right),$$

then the series

$$\sum_{-\infty}^{+\infty} a_{|n|} c_n$$

is convergent.

Δ∇Δ = ∇Δ∇ = Δ∇Δ = ∇Δ∇ = Δ∇Δ

SOLUTION: Using the notations $\Delta a_n, \Delta^2 a_n$ of exercise 10·176, we have

$$\sum_{n=-N}^{N} a_{|n|} c_n = a_0 c_0 + \sum_{n=1}^{N} a_n (c_n + c_{-n})$$

$$= a_0 s_0 + \sum_{n=1}^{N} a_n (s_n - s_{n-1}) = \qquad \text{(Contd)}$$

(Contd)
$$= a_N s_N + \sum_{n=0}^{N-1} \Delta a_n . s_n$$

$$= a_N s_N + \sum_{n=0}^{N-1} \Delta a_n ((n + 1)\sigma_n - n\sigma_{n-1})$$

$$= a_N s_N + N\Delta a_{N-1}\sigma_{N-1} + \sum_{n=0}^{N-2} (n + 1)\Delta^2 a_n . \sigma_n .$$

By hypothesis $a_N s_N \to 0$. Moreover, as the sequence (a_n) is convex and bounded $N\Delta a_{N-1} \to 0$, and further (cf. exercise 10·176)

$$\sum_{n=0}^{\infty} (n + 1)|\Delta^2 a_n . \sigma_n| \leqslant A \sum_{n=0}^{\infty} (n + 1)\Delta^2 a_n = a_0 A.$$

From this it follows that

$$\lim_{N\to\infty} \sum_{n=-N}^{N} a_{|n|} c_n = \sum_{n=0}^{\infty} (n + 1)\Delta^2 a_n . \sigma_n .$$

EXERCISE 10·183: Denote by E_n the set of trigonometric polynomials of the form

$$f(x) = c_0 + c_1 \cos x + \cdots + c_n \cos nx,$$

where

$$c_0 \geqslant c_1 \geqslant \cdots \geqslant c_n > 0.$$

Set

$$m(f) = \sup\{|f(x)| : \tfrac{1}{2}\pi \leqslant x \leqslant \pi\},$$

$$M(f) = \sup\{|f(x)| : 0 \leqslant x \leqslant 2\pi\}.$$

Prove that

$$\left(1 - \frac{2}{\pi}\right)\frac{1}{n + 1} \leqslant \inf\left\{\frac{m(f)}{M(f)} : f \in E_n\right\} \leqslant \left(\frac{1}{2} + \frac{1}{\sqrt{2}}\right)\frac{1}{n + 1} .$$

Δ∇Δ = ∇Δ∇ = Δ∇Δ = ∇Δ∇ = ΔΔ

SOLUTION: We can restrict ourselves to the case where

$$c_0 + c_1 + \cdots + c_n = 1.$$

We then have

$$M(f) = 1 \quad \text{and} \quad c_0 \geqslant (n + 1)^{-1}.$$

$$\int_{\pi/2}^{\pi} f(x)\mathrm{d}x = \frac{\pi}{2} c_0 - c_1 + \frac{c_3}{3} - \frac{c_5}{5} + \cdots \leqslant \frac{\pi}{2} m(f).$$

By noticing that

$$\frac{\pi}{2} c_0 - c_1 \geqslant \left(\frac{\pi}{2} - 1\right)c_0 \geqslant \left(\frac{\pi}{2} - 1\right)\frac{1}{n + 1}$$

and that:

$$\frac{c_3}{3} - \frac{c_5}{5} + \cdots \geqslant 0,$$

we obtain:

$$m(f) \geqslant \left(1 - \frac{2}{\pi}\right)\frac{1}{n + 1} .$$

On the other hand, if:

$$f(x) = \frac{1 + \cos x + \cdots + \cos nx}{n + 1}$$

an elementary calculation shows that:

$$f(x) = \frac{1}{n+1}\left(\frac{1}{2} + \frac{\sin(n+\frac{1}{2})x}{2\sin\frac{1}{2}x}\right).$$

It follows from this that if $\pi/2 \leqslant x \leqslant \pi$:

$$|f(x)| \leqslant \frac{1}{2(n+1)}\left(1 + \frac{1}{\sin\frac{1}{4}\pi}\right) = \left(\frac{1}{2} + \frac{1}{\sqrt{2}}\right)\frac{1}{n+1}.$$

EXERCISE 10·184: Let f be a trigonometric polynomial of degree at most N.

(a): Show that if $f \not\equiv 0$ then f has at most $2N$ zeros in the interval $[0,2\pi[$, counting multiplicity.

(b): Assume that f is real and that

$$A = \|f\|_\infty = f(x_0).$$

Show that if $|x| < \pi/N$ then

$$f(x_0 + x) \geqslant A\cos Nx.$$

(c): f is no longer to be assumed to be real. Show that

$$\|f'\|_\infty \leqslant N\|f\|_\infty. \qquad \text{(BERNSTEIN'S INEQUALITY)}$$

Δ∇Δ = ∇Δ∇ = ΔVΔ = ∇Δ∇ = Δ∇Δ

SOLUTION (a): If

$$f(x) = \sum_{|n|\leqslant N} c_n e^{inx},$$

consider the polynomial

$$P(z) = z^N \sum_{|n|\leqslant N} c_n z^n.$$

Then $f(x) = 0$ if and only if $P(e^{ix}) = 0$. The result follows from the fact that $P \not\equiv 0$ and $\deg(P) \leqslant 2N$.

SOLUTION (b): By translation we reduce to the case where $x_0 = 0$. If the property were not true, then possibly after replacing $f(x)$ by $f(-x)$, there would exist a y such that

$$0 < y < \frac{\pi}{N}, \qquad f(y) - A\cos Ny < 0.$$

For $\varepsilon \geqslant 0$ let

$$g_\varepsilon(x) = f(x) - (A + \varepsilon)\cos Nx.$$

If ε is small enough then $g_\varepsilon(y) < 0$. Furthermore, if $x_r = r\pi/N$ then $g_\varepsilon(x_r) \geqslant \varepsilon$ if r is odd and $g_\varepsilon(x_r) \leqslant -\varepsilon$ if r is even. From this it follows that for $\varepsilon > 0$ g_ε has at least one zero in each of the intervals $]y,x_1[,]x_1,x_2[,\ldots,]x_{2N-2},x_{2N-1}[$. By making $\varepsilon \to 0$ one concludes from this that g_0 has at least $2N - 1$ zeros in the interval $[y,2\pi[$. Moreover, $g_0(0) = g_0'(0) = 0$; g_0 would therefore be a trigonometric polynomial of degree less than or equal to $2N$ which would have at least $2N + 1$ zeros in the interval $[0,2\pi[$, which is absurd (because $g_0 \not\equiv 0$, since $g_0(y) < 0$).

SOLUTION (c): First assume that f is real. Replacing f by $-f$, if necessary, it may be assumed that for some x

$$m = \|f'\|_\infty = f'(x_0).$$

By part (b)

(Contd)

$$2\|f\|_\infty \geqslant f\left(x_0 + \frac{\pi}{2N}\right) - f\left(x_0 - \frac{\pi}{2N}\right) = \int_{-\pi/2N}^{\pi/2N} f'(x_0 + x)dx \geqslant$$

(Contd) $$m\int_{-\pi/2N}^{\pi/2N} \cos Nx\,dx = \frac{2m}{N},$$

which proves the result in this case.

In the general case let x be a real number. There exists an α such that $|\alpha| = 1$ and $|f'(x)| = \alpha f'(x)$. Let us set $\alpha f = u + iv$, where u and v are real. Then

$$|f'(x)| = u'(x) \leqslant N\|u\|_\infty$$

$$\leqslant N\|\alpha f\|_\infty = N\|f\|_\infty.$$

CHAPTER 11
Trigonometric Series

EXERCISE 11·185: Let $z \in \mathbb{C}$, $z \notin \mathbb{Z}$. Prove the following formulas

$$e^{zx} = \frac{e^{2\pi z} - 1}{\pi}\left(\frac{1}{2z} + \sum_{n=1}^{\infty} \frac{z\cos nx - n\sin nx}{z^2 + n^2}\right), \qquad 0 < x < 2\pi;$$

$$e^{zx} = \frac{e^{\pi z} - 1}{\pi z} + \frac{2}{\pi}\sum_{n=1}^{\infty} [(-1)^n e^{\pi z} - 1]\,\frac{z\cos nx}{z^2 + n^2}, \qquad 0 < x < \pi;$$

$$e^{zx} = \frac{2}{\pi}\sum_{n=1}^{\infty} [1 - (-1)^n e^{\pi z}]\,\frac{n\sin nx}{z^2 + n^2}, \qquad 0 < x < \pi.$$

What are the values of these series when $x = 0$?

ΔVΔ = VΔV = ΔVΔ = VΔV = ΔVΔ

SOLUTION: Let f be a function of period 2π such that $f(x) = e^{zx}$ if $0 < x < 2\pi$. Then $[f'] = zf$, and consequently

$$in\hat{f}(n) = z\hat{f}(n) - \frac{1}{2\pi}(e^{2\pi z} - 1)$$

(because the jump in f at 0 is $1 - e^{2\pi z}$). Therefore

$$\hat{f}(n) = \frac{e^{2\pi z} - 1}{2\pi} \frac{1}{z - in}, \qquad n \neq 0.$$

Since

$$\hat{f}(0) = \frac{e^{2\pi z} - 1}{2\pi z},$$

the first formula is proved (Jordan-Dirichlet Theorem). For $x = 0$ the sum of the series is

$$\tfrac{1}{2}(f(x + 0) + f(x - 0) = \tfrac{1}{2}(e^{2\pi z} + 1)$$

from which it is easily deduced that

$$\pi \coth \pi z = \frac{1}{z} + \sum_{n=1}^{\infty} \frac{2z}{z^2 + n^2}.$$

Now consider the functions g and h, with period 2π, respectively even and odd, and which coincide with f on $]0,\pi[$. Then

$$[g'] = zh, \qquad [h'] = zg,$$

so

$$in\hat{g}(n) = z\hat{h}(n),$$

$$in\hat{h}(n) = z\hat{g}(n) + \frac{1}{2\pi}(2 - 2e^{\pi z}e^{-\pi n}).$$

Therefore if $n \neq 0$ then

$$\hat{g}(n) = \frac{(-1)^n e^{\pi z} - 1}{\pi} \frac{z}{z^2 + n^2},$$

$$\hat{h}(n) = \frac{1 - (-1)^n e^{\pi z}}{i\pi} \frac{n}{z^2 + n^2}.$$

Moreover

$$\hat{g}(0) = \frac{e^{\pi z} - 1}{\pi z}, \qquad \hat{h}(0) = 0,$$

which proves the two other formulas. In fact the second is valid for $x = 0$. From this it is easily deduced that

$$\frac{\pi}{\sinh \pi z} = \frac{1}{z} + \sum_{n=1}^{\infty} (-1)^n \frac{2z}{z^2 + n^2}.$$

EXERCISE 11·186: Find the sums of the following series:

$$\sum_{n=1}^{\infty} \frac{a \cos nx}{n^2 + a^2}, \qquad \sum_{n=1}^{\infty} \frac{n \sin nx}{n^2 + a^2}, \qquad (a \text{ real}, a \neq 0).$$

Δ∇Δ = ∇Δ∇ = Δ∇Δ = ∇Δ∇ = Δ∇Δ

SOLUTION: Let

$$g(x) = \sum_{n=1}^{\infty} \frac{a \cos nx}{n^2 + a^2}, \qquad -\infty < x < \infty, \tag{1}$$

$$G(z) = \sum_{n=1}^{\infty} \frac{a e^{inz}}{n^2 + a^2}, \qquad z = x + iy, \quad y \geqslant 0. \tag{2}$$

The series in (2) converges absolutely for $y \geqslant 0$ and defines on the half-plane $y > 0$ a holomorphic function that satisfies the differential equation

$$G'' - a^2 G = \frac{a e^{iz}}{e^{iz} - 1}, \qquad \mathrm{Re}(z) > 0. \tag{3}$$

But the G can be analytically continued to $\mathbb{C}$ minus the two real half-lines $]-\infty,0]$ and $]2\pi,+\infty[$. From this it follows that g is infinitely differentiable (and even analytic) on $]0,2\pi[$ and that on this interval satisfies the differential equation:

$$g'' - a^2 g = a\mathrm{Re}\left(\frac{e^{ix}}{e^{ix} - 1}\right) = \frac{a}{2}.$$

Hence

$$g(x) = \alpha\cosh a(x - \pi) + \beta\sinh a(x - \pi) - \frac{1}{2a}$$

for $0 < x < 2\pi$, and hence also, by continuity, for $0 \leqslant x \leqslant 2\pi$. Since $g(0) = g(2\pi)$, $\beta = 0$.

Now,

$$0 = \int_0^{2\pi} g(x)dx = \frac{2\alpha\sinh a\pi}{a} - \frac{\pi}{a},$$

so that

$$\sum_{n=1}^{\infty} \frac{a\cos nx}{a^2 + n^2} = \frac{\pi\cosh a(\pi - x)}{2\sinh \pi a} - \frac{1}{2a}, \qquad 0 \leqslant x \leqslant 2\pi.$$

The derivative of the right side is continuous on $[0,2\pi]$, hence one can differentiate term by term, giving

$$\sum_{n=1}^{\infty} \frac{n\sin nx}{a^2 + n^2} = \frac{\pi\sinh a(\pi - x)}{2\sinh \pi a}, \qquad 0 < x < 2\pi.$$

(The formula is false at $x = 0$, for at this point the function with period 2π that coincides with the right side for $0 < x < 2\pi$ possesses right and left limits equal, respectively, to $\pi/2$ and $-\pi/2$).

EXERCISE 11·187: Find an expansion of x^2 as a trigonometric series that is valid for $0 < x < 2\pi$. Also, determine expansions of x^2 as a series in cosines, then in sines, valid fo $0 < x < \pi$.

Δ∇Δ — ∇Δ∇ — Δ∇Δ — ∇Δ∇ — Δ∇Δ

SOLUTION: If f has period 2π and $f(x) = x^2$ for $0 < x < 2\pi$, at zero this functions has a step equal to $-4\pi^2$. We have $[f'](x) = 2x$, $0 < x < 2\pi$, so therefore

$$\widehat{[f']}(n) = -\frac{2}{in}, \qquad n \neq 0,$$

and

$$in\hat{f}(n) = -\frac{2}{in} - 2\pi,$$

so

$$\hat{f}(n) = \frac{2}{n^2} - \frac{2\pi}{in}.$$

Moreover

$$\hat{f}(0) = \frac{4\pi^2}{3},$$

so that

$$x^2 = \frac{4\pi^2}{3} + 4\sum_{n=1}^{\infty} \frac{\cos nx - \pi n \sin nx}{n^2}, \qquad 0 < x < 2\pi.$$

If $x = 0$ the right side is equal to

$$\tfrac{1}{2}(f(x + 0) + f(x - 0)) = 2\pi^2,$$

which recovers the well known formula

$$\sum_{n=1}^{\infty} \frac{1}{n^2} = \frac{\pi^2}{6} .$$

═ ═ ═ ═ ═ ═

If g is even and $g(x) = x^2$ for $0 \leqslant x \leqslant \pi$, then

$$[g'](x) = 2x, \qquad -\pi < x < \pi,$$

hence if $n \neq 0$

$$\widehat{[g']}(n) = (-1)^{n+1} \frac{2}{in} ,$$

and consequently

$$\hat{g}(n) = (-1)^n \frac{2}{n^2} .$$

Furthermore,

$$g(0) = \frac{\pi^2}{3} .$$

Thus

$$x^2 = \frac{\pi^2}{3} + 4 \sum_{n=1}^{\infty} (-1)^{3n} \frac{\cos nx}{n^2}, \qquad 0 \leqslant x \leqslant \pi.$$

For $x = 0$ we recover the formula

$$\sum_{n=1}^{\infty} (-1)^{n+1} n^{-2} = \frac{\pi^2}{12} .$$

═ ═ ═ ═ ═ ═

Lastly, if h is odd and $h(x) = x^2$ for $0 \leqslant x < \pi$, then

$$[h'](x) = 2|x|,$$

and h has a step equal to $-2\pi^2$ at the point π. Since

$$\widehat{[h']}(n) = -\frac{2}{\pi}\frac{1-(-1)^n}{n^2}, \qquad n \neq 0,$$

and

$$in\hat{h}(n) = \widehat{[h']}(n) - \pi e^{-in\pi},$$

we have

$$\hat{h}(n) = -\frac{2}{\pi}\frac{1-(-1)^n}{in^3} + (-1)^{n+1}\frac{\pi}{in},$$

and consequently

$$x^2 = 2\pi\sum_{n=1}^{\infty}(-1)^{n+1}\frac{\sin nx}{n} - \frac{8}{\pi}\sum_{n=1}^{\infty}\frac{\sin(2n+1)x}{(2n+1)^3}, \qquad 0 \leqslant x < \pi.$$

REMARK: It is known that the first series has $\frac{1}{2}x$ as its sum; therefore

$$\sum_{n=1}^{\infty}\frac{\sin(2n+1)x}{(2n+1)^3} = \frac{\pi}{8}(\pi x - x^2), \qquad 0 \leqslant x < \pi.$$

EXERCISE 11·188: Prove the following formulas:

$$\sum_{n=1}^{\infty}\sin nx \overset{(c)}{=\!=} \tfrac{1}{2}\cot\tfrac{1}{2}x, \qquad 0 < x < 2\pi,$$

$$\sum_{n=-\infty}^{+\infty} e^{inx} \overset{(c)}{=\!=} 0, \qquad 0 < x < 2\pi.$$

(The (c) over the equals sign indicates that the equation holds when the sum is taken in the sense of Cesaro).

$\Delta\nabla\Delta = \nabla\Delta\nabla = \Delta\nabla\Delta = \nabla\Delta\nabla = \Delta\nabla\Delta$

SOLUTION: If $0 < x < 2\pi$,

$$s_N = \sum_{n=1}^{N} \sin nx = \frac{\cos\frac{1}{2}x - \cos(N + \frac{1}{2})x}{2\sin\frac{1}{2}x},$$

and consequently

$$\sigma_N = \tfrac{1}{2}\cot\tfrac{1}{2}x - \frac{1}{2(N + 1)\sin\frac{1}{2}x} \sum_{n=0}^{N} \cos(n + \tfrac{1}{2})x$$

$$= \tfrac{1}{2}\cot\tfrac{1}{2}x - \frac{\sin(N + 1)x}{4(N + 1)\sin^2\frac{1}{2}x},$$

which proves that

$$\sigma_N \to \tfrac{1}{2}\cot\tfrac{1}{2}x.$$

Similarly, for the second sum

$$s_N = \sum_{n=-N}^{N} e^{inx} = D_N(x) = \frac{\sin(N + \frac{1}{2})x}{\sin\frac{1}{2}x},$$

and

$$\sigma_N = F_N(x) = \frac{1}{N + 1}\left(\frac{\sin(N + 1)\frac{1}{2}x}{\sin\frac{1}{2}x}\right)^2,$$

so $\sigma_N \to 0$. Not that if $x = 0$, then

$$s_N = 2N + 1,$$

so

$\sigma_N = N + 1 \to \infty.$

EXERCISE 11·189: Let f be a continuous function.

Show that the following two conditions are equivalent:

(i): There exists a function f continuous on the closed disc $|z| \leqslant 1$ and holomorphic in the open disc $|z| < 1$, such that $f(x) = F(e^{ix})$, $x \in \mathbb{R}$;

(ii): f is periodic with period 2π, and $\hat{f}(n) = 0$ if $n < 0$.

ΔVΔ = VΔV = ΔVΔ = VΔV = ΔVΔ

SOLUTION: (i) => (ii): For all $0 < r < 1$ and all $n > 0$,

$$0 = \int_{|z|=r} z^n F(z)dz = ir^{n+1}\int_0^{2\pi} e^{i(n+1)x}F(re^{ix})dx.$$

As the function F is continuous on $|z| \leqslant 1$ the limit $r \to 1$ may be taken, yielding

$$\hat{f}(-n-1) = 0, \qquad n \geqslant 0.$$

(ii) => (i): If:

$$P_N(z) = \sum_{n=0}^{N}\left(1 - \frac{n}{N+1}\right)\hat{f}(n)z^n,$$

then by Féjer's Theorem

$$P_N(e^{ix}) \to f(x) \text{ uniformly on } \mathbb{R}.$$

By the Maximum Principle, it follows that

$$\lim_{N,M\to\infty}\left(\sup_{|z|\leqslant 1}|P_M(z) - P_N(z)|\right) = \lim_{N,M\to\infty}\left(\sup_{|z|=1}|P_M(z) - P_N(z)|\right) = 0.$$

Therefore

$$P_N(z) \to F(z) \text{ uniformly on } z \leqslant 1,$$

is continuous on $|z| \leqslant 1$, holomorphic on $|z| < 1$, and

$$f(x) = F(e^{ix}) \text{ if } x \in \mathbb{R}.$$

EXERCISE 11·190: Let $f \in L^1(T)$. Show that the following conditions are equivalent:

(i): $|\hat{f}(n)| = O(e^{-\varepsilon|n|})$, $\varepsilon > 0$;

(ii): f is the restriction to $\mathbb{R}$ of a function holomorphic on the strip $|\mathrm{Im}(z)| < \delta$, $\delta > 0$;

(iii): The series

$$\sum_{n=-\infty}^{+\infty} \hat{f}(n)e^{inz} \qquad (*)$$

converges on a strip $|\mathrm{Im}(z)| < \delta$, $\delta > 0$;

(iv): The series (*) converges at two points, $z_1 = x_1 + i\delta_1$ and $z_2 = x_2 - i\delta_2$, with $\delta_1, \delta_2 > 0$.

Δ∇Δ = ∇Δ∇ = ΔVΔ = ∇Δ∇ = Δ∇Δ

SOLUTION: Let

$$a_n(z) = \hat{f}(n)e^{inz} + \hat{f}(-n)e^{-inz}.$$

Then

$$(e^{2inz_1} - e^{2inz_2})\hat{f}(n) = a_n(z_1)e^{inz_1} - a_n(z_2)e^{inz_2}.$$

If (iv) is satisfied there exists a constant M such that

$$|a_n(z_1)| \leqslant M, \qquad |a_n(z_2)| \leqslant M, \qquad n \geqslant 0,$$

so that if $n \neq 0$

$$|\hat{f}(n)| \leqslant M \frac{e^{n\delta_2} + e^{-n\delta_1}}{|e^{2n\delta_2} - e^{-2n\delta_1}|} = O(e^{-\varepsilon|n|}),$$

if $\varepsilon = \min(\delta_1, \delta_2)$. This proves that (iv) => (i).

If (i) is satisfied, the series (*) converges in norm on every strip $|\mathrm{Im}(z)| \leqslant \delta$, $\delta < \varepsilon$ Therefore (i) => (ii).

Since it is trivial that (iii) => (iv), it remains to prove that (ii) => (iii).

Let F be a function holomorphic on the strip $\mathrm{Im}(z) \leqslant \delta$, $\delta > 0$, and such that $f(x) = F(x)$ if x is real. As the relation

$$F(x + 2\pi) = F(z)$$

is true on $\mathbb{R}$, it holds on the entire strip $|\mathrm{Im}(z)| < \delta$ (principle of analytic continuation). Hence there exists a function G defined on the annulus

$$e^{-\delta} < |\xi| < e^{\delta},$$

such that

$$F(z) = G(\xi) \quad \text{if } \xi = e^{iz}.$$

As the function e^{iz} is locally invertible, it is clear that G is holomorphic on its domain of definition. Let

$$G(\xi) = \sum_{-\infty}^{+\infty} a_n \xi^n$$

be its Laurent expansion. Since the circle $|z| = 1$ is contained in the annulus of definition of G,

$$\sum_{-\infty}^{+\infty} |a_n| < \infty,$$

$$f(x) = \sum_{-\infty}^{+\infty} a_n e^{inx}, \qquad x \in \mathbb{R}.$$

and consequently $a_n = \hat{f}(n)$. It follows that the series (*) converges (and even converges absolutely) on the strip $|\mathrm{Im}(z)| < \delta$.

EXERCISE 11·191: (a): Let $f \in C^\infty(T)$. Show that if

$$\|f^{(s)}\|_\infty = O(R^s \Gamma(\alpha s + 1)), \qquad R > 0, \quad \alpha > 0,$$

$$\hat{f}(n) = O(|n|^{1/2\alpha} e^{-(|n|/R)^{1/\alpha}}). \tag{*}$$

(b): Now assume that (*) is satisfied. Show that

$$\|f^{(s)}\|_\infty = O(s^\beta R^s \Gamma(\alpha s + 1)), \qquad \beta = (\alpha - \tfrac{1}{2})_+. \tag{**}$$

(c): Deduce from this that f is analytic if and only if

$$|\hat{f}(n)| = O(e^{-\varepsilon|n|}) \quad \text{for some } \varepsilon > 0.$$

ΔVΔ = VΔV = ΔVΔ = VΔV = ΔVΔ

SOLUTION: (a): If A is such that

$$\|f^{(s)}\|_\infty \leqslant AR^s \Gamma(\alpha s + 1), \qquad s \geqslant 0,$$

then for all $n \neq 0$

$$|\hat{f}(n)| = |(in)^{-s}\widehat{f^{(s)}}(n)| \leqslant A\left(\frac{R}{|n|}\right)^{s}\Gamma(\alpha s + 1).$$

Let us take for s the integral part of $(1/\alpha)(|n|/R)^{1/\alpha}$. Then

$$\left(\frac{R}{|n|}\right)^{s}\Gamma(\alpha s + 1) \leqslant (\alpha s)^{-\alpha s}\Gamma(\alpha s + 1) \sim e^{-\alpha s}\sqrt{2\pi\alpha s}.$$

Since

$$\left(\frac{|n|}{R}\right)^{1/\alpha} - \alpha < \alpha s \leqslant \left(\frac{|n|}{R}\right)^{1/\alpha},$$

(*) certainly holds.

SOLUTION: (b):

$$f^{(s)}(x) = \sum_{-\infty}^{+\infty} (in)^{s}\hat{f}(n)e^{inx}, \qquad s \geqslant 0,$$

and consequently

$$\|f^{(s)}\|_{\infty} = O\Big(\sum_{n=1}^{\infty} n^{s+1/2\alpha}e^{-(n/R)^{1/\alpha}}\Big).$$

Let $\sigma = s + 1/2\alpha$ and

$$\varphi_{\sigma}(x) = x^{\sigma}e^{-(x/R)^{1/\alpha}}, \qquad x \geqslant 0.$$

φ_{σ} attains its maximum at

$$x_{\sigma} = R(\alpha\sigma)^{\alpha}.$$

Let N_{σ} be the integral part of x_{σ}. Then

$$\sum_{n=1}^{\infty} \varphi_\sigma(n) \leqslant \int_1^{N_\sigma} \varphi_\sigma(x)dx + \varphi_\sigma(N_\sigma) + \varphi_\sigma(N_\sigma + 1) + \int_{N_\sigma+1}^{\infty} \varphi_\sigma(x)dx$$

$$\leqslant \int_0^{\infty} \varphi_\sigma(x)dx + 2\varphi_\sigma(x_\sigma).$$

Now,

$$\int_0^{\infty} \varphi_\sigma(x)dx = \alpha R^{\sigma+1}\Gamma(\alpha\sigma + \alpha),$$

$$\varphi_\sigma(x_\sigma) = R^\sigma(\alpha\sigma)^{\alpha\sigma}e^{-\alpha\sigma}.$$

Since $\Gamma(x + h) \sim x^h\Gamma(x)$ as $x \to \infty$,

$$\alpha R^{\sigma+1}\Gamma(\alpha\sigma + \alpha) = O(\sigma^\alpha R^\sigma\Gamma(\alpha\sigma)).$$

Furthermore,

$$R^\sigma(\alpha\sigma)^{\alpha\sigma}e^{-\alpha\sigma} = O(\sigma^{\frac{1}{2}}R^\sigma\Gamma(\alpha\sigma))$$

Therefore

$$\sum_{n=1}^{\infty} \varphi_\sigma(n) = O(\sigma^\gamma R^\sigma\Gamma(\alpha\sigma)), \qquad \gamma = \max(\alpha,\tfrac{1}{2}),$$

so

$$\|f^{(s)}\|_\infty = O(s^\gamma R^s\Gamma(\alpha s + \tfrac{1}{2})) = O(s^{\gamma-\frac{1}{2}}R^s\Gamma(\alpha s + 1),$$

which is precisely (**).

SOLUTION (c): It is known that f is analytic if and only if

$$\|f^{(s)}\|_\infty = O(R^s s!), \qquad R > 0. \tag{1}$$

By part (a) above, (1) implies that

$$|\hat{f}(n)| = O(|n|^{\frac{1}{2}} e^{-|n|/R}) = O(e^{-\varepsilon|n|}) \quad \text{if } 0 < \varepsilon < \frac{1}{R}.$$

On the other hand, if

$$|\hat{f}(n)| = O(e^{-\varepsilon|n|}),$$

then by part (b),

$$\|f^{(s)}\|_\infty = O(s^{\frac{1}{2}} \varepsilon^{-s} s!) = O(R^s s!) \quad \text{when } R > \frac{1}{\varepsilon}.$$

EXERCISE 11·192: Show that if $f \in L^1(T)$ and:

$$g(x) = \sum_{n=-\infty}^{\infty} \hat{f}(n) e^{inx} \quad \text{almost everywhere,}$$

then $f = g$ almost everywhere.

ΔVΔ = VΔV = ΔVΔ = VΔV = ΔVΔ

SOLUTION: By the Lebesgue-Féjer theorem

$$\sigma_N(x) \to f(x) \quad \text{almost everywhere,}$$

But $s_N(x) \to g(x)$ implies that $\sigma_N(x) \to g(x)$, so $f = g$ almost everywhere.

EXERCISE 11·193: Let $f \in L^1(T)$. Show that if

$$\int_0^\pi |f(x+t) + f(x-t) - 2s| \frac{dt}{t} < \infty,$$

then

$$\sum_{n=-\infty}^{+\infty} \hat{f}(n)e^{inx} = s. \qquad \textit{(DINI'S TEST)}$$

ΔVΔ = VΔV = ΔVΔ = VΔV = ΔVΔ

SOLUTION: In fact

$$s_N(f;x) - s = \frac{1}{\pi}\int_0^{\pi} (f(x + t) + f(x - t) - 2s) \frac{\sin(N + \frac{1}{2})t}{t} dt$$

$$+ \varepsilon_N(x),$$

with $\varepsilon_N(x) \to 0$. Since the function

$$t \mapsto t^{-1}(f(x + t) + f(x - t) - 2s)$$

is integrable, the integal tends to zero by the Riemann-Lebesgue Theorem.

EXERCISE 11·194: Let f be a complex function with period 2π. It is said that $f \in \text{Lip}(\alpha)$ if there exists a constant M such that

$$|f(x) - f(y)| \leq M|x - y|^{\alpha}$$

for all x and all y. (Evidently, $0 < \alpha \leq 1$. Why?).

Show that then

$$\hat{f}(n) = O(n^{-\alpha}).$$

ΔVΔ = VΔV = ΔVΔ = VΔV = ΔVΔ

SOLUTION: In fact

$$\hat{f}(n) = \frac{1}{4\pi}\int_0^{2\pi}\left(f(x) - f\left(x + \frac{\pi}{n}\right)\right)e^{-inx}dx,$$

so

$$|\hat{f}(n)| \leqslant \frac{M}{2}\left(\frac{\pi}{n}\right)^{\alpha}.$$

EXERCISE 11·195: Let $f \in L^1(T)$.

(a): If at a point x where $f(x + 0)$ and $f(x - 0)$ exist, one has for $0 < t < \delta$,

$$|f(x + t) + f(x - t) - f(x + 0) - f(x - 0)| \leqslant Mt^{\alpha}, \qquad \alpha > 0,$$

then the Fourier series of f converges at this point to

$$\tfrac{1}{2}(f(x + 0) + f(x - 0)).$$

(b): If $f \in \text{Lip}(\alpha)$ the Fourier series of f converges uniformly to f.

Δ∇Δ = ∇Δ∇ = Δ∇Δ = ∇Δ∇ = Δ∇Δ

SOLUTION: (a): Let

$$\varphi_x(t) = f(x + t) + f(x - t) - f(x + 0) - f(x - 0).$$

It is known that

$$s_N(f;x) - \tfrac{1}{2}(f(x + 0) + f(x - 0)) = \frac{1}{\pi}\int_0^{\delta}\varphi_x(t)\,\frac{\sin(N + \frac{1}{2})t}{t}\,dt$$

$$+ \varepsilon_N(x),$$

where

$$\lim_{N} \varepsilon_N(x) = 0.$$

By the hypothesis made, $t^{-1}\varphi_x(t)$ is integrable, so the result follows from the Riemann-Lebesgue Theorem.

SOLUTION: (b): If $f \in \mathrm{Lip}(\alpha)$, f is continuous and it is known that in this case $\varepsilon_N(x) \to 0$ uniformly in x. Moreover,

$$|\varphi_x(t)| \leqslant Mt^{\alpha},$$

which proves the result.

EXERCISE 11·196: If

$$f(x) = |x|^{-\alpha}\phi(x), \qquad 0 < \alpha < 1,$$

where ϕ has bounded variation on $[-\pi,+\pi]$, show that

$$|\hat{f}(n)| = O(n^{\alpha-1}).$$

ΔVΔ = VΔV = ΔVΔ = VΔV = ΔVΔ

SOLUTION: One can assume that ϕ is increasing and positive on $[-\pi,+\pi]$. Then

$$\left|\int_{-\pi}^{\pi} \phi(x) \frac{e^{-inx}}{|x|^{\alpha}}\, dx\right| \leqslant \phi(\pi) \sup_{|\xi|<\pi} \left|\int_{\xi}^{\pi} \frac{e^{-inx}}{|x|^{\alpha}}\, dx\right|.$$

Now,

$$\int_{\xi}^{\pi} \frac{e^{-inx}}{|x|^{\alpha}}\, dx = n^{\alpha-1}(G(\pi n) - G(\xi n)),$$

where

$$G(x) = \int_0^x \frac{e^{-it}}{|t|^{\alpha}}\,dt.$$

It is known that $G(x)$ tends to finite limits as $x \to \pm\infty$; $|G|$ is therefore bounded by a constant M, and consequently

$$|\hat{f}(n)| \leq \frac{M\phi(\pi)}{\pi}\, n^{\alpha-1}.$$

EXERCISE 11·197: Let $f \in L^p(T)$, $1 \leq p \leq \infty$. Set

$$f_1 = f, \qquad f_k = f*f_{k-1} \text{ for } k \geq 2.$$

(a): Show that:

$$\rho(f) = \lim_{k\to\infty} \|f_k\|_p^{1/k}$$

exists.

(b): Show that if $\rho(f) = 0$ then $f = 0$ almost everywhere.

Δ∇Δ = ∇Δ∇ = Δ∇Δ = ∇Δ∇ = Δ∇Δ

SOLUTION: Note that if $f,g \in L^p(T)$ then

$$\|f*g\|_p \leq \|f\|_1\|g\|_p \leq \|f\|_p\|g\|_p.$$

SOLUTION: (a): It follows from what was noted above, that, setting

$$a_k = \|f_k\|_p,$$

one has

$$a_{k_1+k_2} \leqslant a_{k_1}a_{k_2}, \qquad a_k \leqslant \|f\|_p^k.$$

It is then classical that

$$\rho(f) = \lim_k a_k^{1/k}$$

exists.

SOLUTION: (b): For $n \in \mathbb{Z}$ and $k = 1$

$$|\hat{f}(n)|^k = |\hat{f}_k(n)| \leqslant \|f_k\|_1 \leqslant \|f_k\|_p,$$

which implies that

$$|\hat{f}(n)| \leqslant \rho(f), \qquad n \in \mathbb{Z}.$$

Therefore, when $\rho(f) = 0$ one has $\hat{f} = 0$, and consequently $f = 0$ almost everywhere.

EXERCISE 11·198: Let $f \in L^1(T)$. Assume that for some $\alpha > 0$

$$\int_0^{2\pi} |f(x + h) - f(x)|dx = O(|h|^\alpha).$$

(a): Show that if

$$f_1 = f \quad \text{and} \quad f_k = f*f_{k-1}, \quad k \geqslant 2,$$

then

$$f_k \in L^2(T) \quad \text{when } k > \frac{1}{2\alpha}.$$

(b): Show that f_k coincides almost everywhere with a continuous function when $k > 1/\alpha$.

Δ∇Δ = ∇Δ∇ = ΔVΔ = ∇Δ∇ = ΔVΔ

SOLUTION: Note that

$$\hat{f}(n) = \frac{1}{4\pi}\int_0^{2\pi}\{f(x) - f\left(x + \frac{\pi}{n}\right)\}e^{-inx}dx,$$

so

$$|\hat{f}(n)| = O(|n|^{-\alpha}). \tag{1}$$

SOLUTION: (a): It follows from (1) that

$$|\hat{f}_k(n)| = O(|n|^{-\alpha k}),$$

and consequently that

$$\sum_n |\hat{f}_k(n)|^2 < \infty \quad \text{if } k > \frac{1}{2\alpha}.$$

But then $f_k \in L^2(T)$.

SOLUTION: (b): If $k > 1/\alpha$, then

$$\sum_n |\hat{f}_k(n)| < \infty,$$

in which case f_k coincides almost everywhere with the sum of its Fourier series, which is absolutely convergent, and hence continuous.

EXERCISE 11·199: Let $f \in L^1(T)$ be such that $\hat{f}(n) = O(|n|^{-\alpha})$, $\alpha > 0$.

Show that if p is an integer such that $\alpha - p > \frac{1}{2}$, then $f^{(p)}$ is defined almost everywhere and belongs to $L^2(T)$.

ΔVΔ = VΔV = ΔVΔ = VΔV = ΔVΔ

SOLUTION: It suffices to carry out the proof for $p = 1$. Indeed, the result is evident if $p = 0$, and if $p > 2$ then

$$\sum |n|^{p-1}|\hat{f}(n)| < \infty,$$

which ensures that f is $(p - 1)$ times continuously differentiable, and that

$$|\widehat{f^{(p-1)}}(n)| = O(|n|^{-(\alpha-p+1)}) \quad \text{with } (\alpha - p + 1) - 1 > \tfrac{1}{2}.$$

Assume, therefore, that $\alpha > 3/2$, and let us prove that f' exists almost everywhere and belongs to $L^2(T)$. We have

$$\sum |n\hat{f}(n)|^2 < \infty.$$

Hence there exists $g \in L^2(T)$ such that

$$\hat{g}(n) = in\hat{f}(n), \qquad n \in \mathbb{Z}.$$

Since it is possible to integrate a Fourier series term by term,

$$\int_0^x g(t)dt = \sum (\hat{f}(n)e^{inx} - \hat{f}(n)).$$

Since

$$\sum |\hat{f}(n)| < \infty,$$

the right side is equal to $f(x) - f(0)$, which proves that $f'(x) = g(x)$ almost everywhere.

EXERCISE 11·200: If f is absolutely continuous and $f' \in L^2$, then

$$\sum_n |\hat{f}(n)| \leqslant \|f\|_1 + \frac{\pi}{\sqrt{3}} \|f'\|_2.$$

Δ∇Δ = ∇Δ∇ = ΔΔ∇ = ∇Δ∇ = ΔΔ∇

SOLUTION: In fact, $|\hat{f}(0)| \leqslant \|f\|_1$, and

$$\sum_{n \neq 0} |\hat{f}(n)| \leqslant \Big(\sum_{n \neq 0} n^{-2}\Big)^{\frac{1}{2}} \Big(\sum_{n \neq 0} |n\hat{f}(n)|^2\Big)^{\frac{1}{2}}$$

$$= \frac{\pi}{\sqrt{3}} \|f'\|_2.$$

EXERCISE 11·201: For every finite set $A \subset \mathbb{Z}$ and all $\varepsilon > 0$ show that there exists $f \in L^1$ such that

(i): $0 \leqslant \hat{f}(n) \leqslant 1$ for all $n \in \mathbb{Z}$;

(ii): $\hat{f}(n) = 1$ if $n \in A$;

(iii): $\|f\|_1 < 1 + \varepsilon$.

ΔΔ∇ = ∇Δ∇ = ΔΔ∇ = ∇Δ∇ = ΔΔ∇

SOLUTION: If $A \subset [-r, r]$ let

$$f(x) = (2N + 1)^{-1} \Big(\sum_{|n| \leqslant N} e^{inx}\Big)\Big(\sum_{|n| \leqslant N+r} e^{inx}\Big),$$

where N is an integer to be determined later. It is easy to see that

$$\hat{f}(n) = \begin{cases} 1 & \text{if } |n| \leqslant r, \\ 1 - \dfrac{|n| - r}{2N + 1} & \text{if } r < |n| \leqslant r + 2N + 1, \\ 0 & \text{if } |n| \geqslant r + 2N + 1. \end{cases}$$

Using the Cauchy-Schwarz inequality and Bessel's equation, it follows that

$$\frac{1}{2\pi}\int_0^{2\pi} |f(x)|\,dx \leqslant (2N + 1)^{-1}(2N + 1)^{\frac{1}{2}}(2N + 2r + 1)^{\frac{1}{2}}$$

$$= \left(\frac{2N + 2r + 1}{2N + 1}\right)^{\frac{1}{2}}.$$

If N is chosen sufficiently large, then

$$\|f\|_1 < 1 + \varepsilon.$$

EXERCISE 11·202: Let (φ_i) be an orthonormal basis of $L^2(T)$. Prove that

$$\sum_i \|\varphi_i\|_1 = \infty.$$

$\Delta\nabla\Delta = \nabla\Delta\nabla = \Delta\nabla\Delta = \nabla\Delta\nabla = \Delta\nabla\Delta$

SOLUTION: If $f \in L^2(T)$ then

$$\|f\|_2^2 = \sum_i |(f|\varphi_i)|^2.$$

In particular, letting $f(x) = e^{inx}$,

$$\sum_i |\widehat{\varphi_i}(n)|^2 = 1, \qquad n \in \mathbb{Z}.$$

Now consider a sequence $(c_n)_{n\in\mathbb{Z}}$ such that

$$c_n > 0, \qquad \sum_n c_n = \infty, \qquad \sum_n c_n^2 < \infty,$$

and let $g \in L^2(T)$ be such that $\hat{g}(n) = c_n$. If $\sum \|\varphi_i\|_1 < \infty$ were to hold, then we could write

$$\infty = \sum_n c_n = \sum_n \sum_i c_n |\hat{\varphi}_i(n)|^2$$

$$= \sum_i \sum_n c_n \hat{\varphi}_i(n)\overline{\hat{\varphi}_i(n)}$$

$$= \sum_i (g*\varphi_i|\varphi_i)$$

$$= \sum_i (g|\varphi_i*\tilde{\varphi}_i)$$

$$\leqslant \|g\|_2 \sum_i \|\varphi_i\|_1,$$

which is absurd.

EXERCISE 11·203: Let $1 < n_1 < n_2 < \cdots < n_p < \cdots$ be a strictly increasing sequence of integers. Set

$$f_N(x) = \frac{1}{N} \sum_{p=1}^{N} e^{in_p x}.$$

Show that for almost all x

$$\lim_N f_N(x) = 0.$$

Δ∇Δ = ∇Δ∇ = Δ∇Δ = ∇Δ∇ = Δ∇Δ

SOLUTION: By Plancherel's Formula

$$\frac{1}{2\pi}\sum_{m=1}^{\infty}\int_0^{2\pi}|f_{m^2}(x)|^2 dx = \sum_{m=1}^{\infty}\frac{1}{m^2} < \infty,$$

and consequently for almost all x:

$$\lim_m f_{m^2}(x) = 0. \tag{1}$$

If $m^2 \leqslant N < (m+1)^2$, then

$$|f_N(x) - \frac{m^2}{N} f_{m^2}(x)| = \frac{1}{N}\Big|\sum_{p=m^2+1}^{N} e^{in_p x}\Big|$$

$$\leqslant \frac{(m+1)^2 - 1 - m^2}{N} \leqslant \frac{2}{\sqrt{N}}.$$

Since $m^2/N \to 1$ as $N \to \infty$, it follows that

$$\lim_N f_N(x) = 0$$

for every x for which (1) holds.

EXERCISE 11·204: Let E be a subset of $]0,2\pi[$ with measure zero.

Show that there exists a function $f \in L^p(T)$ for all $p < \infty$, and such that for all $x \in E$

$$\lim_N \sigma_N f(x) = \infty.$$

ΔVΔ = VΔV = ΔVΔ = VΔV = ΔVΔ

SOLUTION: For every integer $k > 1$ let V_k be an open set such that

$$E \subset V_k \subset]0,2\pi[, \qquad \text{meas}(V_k) \leqslant 2^{-k}.$$

Let φ_k be the characteristic function of V_k and f the function, with period 2π, equal to $\sum k\varphi_k$ on $[0,2\pi]$. If $0 < p < \infty$ then

$$\sum_k \|k\varphi_k\|_p \leqslant \left(\frac{1}{2\pi}\right)^{1/p} \sum_k k2^{-k/p} < \infty,$$

which shows that $f \in L^p(T)$ for $0 < p < \infty$. As Féjer's kernel is positive, the inequality $f \geqslant k\varphi_k$ implies that

$$\sigma_N f \geqslant k\sigma_N f_k.$$

By Féjer's Theorem, if $x \in E \subset V_k$, then

$$\liminf_N \sigma_N f(x) \geqslant k,$$

so

$$\lim_N \sigma_N f(x) = \infty.$$

EXERCISE 11·205: Let (φ_i) be a sequence of positive elements of $L^1(T)$ such that

$$\lim_i \hat{\varphi}_i(n) = 1 \quad \text{for all } n \in \mathbb{Z}.$$

(a): Show that if $f \in C(T)$ then $\varphi_i * f \to f$ uniformly.

(b): Deduce from this that (φ_i) is an approximate identity for $L^1(T)$.

ΔVΔ = VΔV = ΔVΔ = VΔV = ΔVΔ

SOLUTION: (a): If $f \in C(T)$ let $T_i(f) = \varphi_i * f$. The T_i's are continuous linear operators on $C(T)$, and

$$M = \sup_i \|T_i\| = \sup_i \|\varphi_i\|_1 = \sup_i \hat{\varphi}_i(0) < \infty.$$

If $e_n(x) = \exp(inx)$, then

$$T_i(e_n) = \hat{\varphi}_i(n) e_n \to e_n$$

in the sense of convergence in $C(T)$. As the e_n's form a total set in $C(T)$, it follows that $T_i(f) \to f$ in $C(T)$ for all $f \in C(T)$.

SOLUTION: (b): Let $0 < a < \pi$, and let f be the function in $C(T)$ that is zero on $[-\frac{1}{2}a, \frac{1}{2}a]$, equal to unity on $[-\pi, -a]$ and $[a, \pi]$, and linear on $[-a, -\frac{1}{2}a]$ and $[\frac{1}{2}a, a]$. Then

$$\int_{a \leqslant |x| \leqslant \pi} \varphi_i(t) dt \leqslant \int_{-\pi}^{\pi} \varphi_i(t) f(t) dt = 2\pi(\varphi_i * \check{f})(0)$$

$$\to 2\pi \check{f}(0) = 0.$$

This shows that (φ_i) is an approximate identity, because, furthermore, $\varphi_i \geqslant 0$ and

$$\|\varphi_i\|_1 = \hat{\varphi}_i(0) \to 1.$$

EXERCISE 11·206: Let $(\lambda_n)_{n>0}$ be such that $\lambda_n > 0$ for all n, and $\lambda_n \to \infty$.

Show that there exists a continuous function f such that

$$\limsup_{|n| \to \infty} \lambda_{|n|} |\hat{f}(n)| = \infty, \qquad \sum_{-\infty}^{+\infty} \lambda_{|n|} |\hat{f}(n)|^2 = \infty.$$

(Consider a series $\sum \lambda_{n_s}^{-\frac{1}{2}} e^{in_s x}$, where the integers n_s increase quickly enough).

$$\Delta\nabla\Delta = \nabla\Delta\nabla = \Delta\nabla\Delta = \nabla\Delta\nabla = \Delta\nabla\Delta$$

SOLUTION: Let $n_1 < n_2 < \cdots$ be a sequence of integers such that

$$\lambda_{n_s} > s^4.$$

The series

$$f(x) = \sum \lambda_{n_s}^{-\frac{1}{2}} e^{in_s x}$$

is then absolutely convergent, and thus f is continuous. If $n \neq n_s$ for all s then $\hat{f}(n) = 0$, while

$$\hat{f}(n_s) = \lambda_{n_s}^{-\frac{1}{2}}.$$

Therefore

$$\limsup_{|n|\to\infty} \lambda_{|n|} |\hat{f}(n)| = \lim_{s\to\infty} \lambda_{n_s}^{\frac{1}{2}} = \infty,$$

$$\sum_n \lambda_{|n|} |\hat{f}(n)|^2 = \sum_s \lambda_{n_s} |\hat{f}(n_s)|^2 = \sum_s 1 = \infty.$$

EXERCISE 11·207: Let $(\lambda_n)_{n\in\mathbb{Z}}$ be a sequence of complex numbers such that

$$\sum_n |\lambda_n \hat{f}(n)| < \infty \quad \text{for all } f \in L^1(T).$$

Show that

$$\sum_n |\lambda_n| < \infty.$$

Δ∇Δ = ∇Δ∇ = Δ∇Δ = ∇Δ∇ = Δ∇Δ

SOLUTION: Consider the linear mapping of $L^1(T)$ into $\ell^1(\mathbb{Z})$ which associates to $f \in L^1(T)$ the sequence

$$\tilde{f} = (\lambda_n \hat{f}(n))_{n\in\mathbb{Z}}.$$

If $f_i \to 0$ in $L^1(T)$ and $\tilde{f}_i \to \gamma$ in $\ell^1(\mathbb{Z})$, then

$$\gamma_n = \lim_i \tilde{f}_i(n) = \lim_i \lambda_n \hat{f}_i(n) = 0.$$

By the Closed Graph Theorem the mapping $f \to \tilde{f}$ is continuous. Hence there exists a constant $A > 0$ such that:

$$\sum_n |\lambda_n \hat{f}(n)| \leqslant A\|f\|_1, \qquad f \in L^1(T).$$

If in the preceding inequality we take for f the Féjer kernel F_N, one obtains

$$\sum_{|n|\leqslant N} \left(1 - \frac{n}{N+1}\right)|\lambda_n| \leqslant A,$$

and by making $N \to \infty$ this yields

$$\sum_n |\lambda_n| \leqslant A.$$

EXERCISE 11·208: Let $(\lambda_n)_{n\in\mathbb{Z}}$ be a sequence of real numbers such that

$$\lambda_n \geqslant 0, \qquad \sum_n \lambda_n = \infty, \qquad \sum_n \lambda_n^2 < \infty.$$

Show that there exists a function $f \in L^2(T)$ such that

(i): $|\hat{f}(n)| = o(\lambda_n)$;

(ii): For any $a < b$, $\operatorname{ess\,sup}_{a \leqslant x \leqslant b} |f(x)| = \infty$.

Δ∇Δ = ∇Δ∇ = Δ∇Δ = ∇Δ∇ = Δ∇Δ

SOLUTION: Let c_0 be the vector space of sequences $u = (u_n)_{n\in\mathbb{Z}}$ of complex numbers such that

$$\lim_{|n|\to\infty} |u_n| = 0,$$

provided with the norm

$$\|u\| = \sup_n |u_n|.$$

c_0 is a Banach space. If $u \in c_0$, there exists $\tilde{u} \in L^2(T)$ such that

$$\tilde{u}(x) \sim \sum_n \lambda_n u_n e^{inx},$$

and

$$\|\tilde{u}\|_2^2 = \sum_n \lambda_n^2 |u_n|^2 \leqslant \|u\|^2 \sum_n \lambda_n^2.$$

The mapping $u \to \tilde{u}$ is therefore continuous from c_0 into $L^2(T)$. If I is a compact interval of length greater than zero, let

$$p_I(u) = \operatorname{ess\,sup}_{x\in I} |\tilde{u}(x)|.$$

It is clear that p_I is a semi-norm on c_0; this semi-norm, furthermore, is lower semi-continuous. In fact, for every function g, continuous on I, the mapping

$$u \mapsto \int_I \tilde{u}g$$

is continuous on c_0 and

$$p_I(u) = \sup\{|\int_I \tilde{u}g| : g \text{ continuous}, \int_I |g| \leqslant 1\}.$$

For every integer $s \geqslant 1$ let us consider the closed set

$$A_s = \{u : p_I(u) \leqslant s\}.$$

We are going to show that $\mathring{A}_s = \emptyset$, which will prove that $p_I(u) = \infty$ for all the u belonging to an everywhere dense G_δ set (Baire's Theorem). If one had $\mathring{A}_s \neq \emptyset$ one would deduce, taking the convexity of A_s into account, that for some $\rho > 0$,

$$\|u\| \leqslant \rho \Rightarrow p_I(u) \leqslant s. \tag{1}$$

Let (a_r), $1 \leqslant r \leqslant k$, be a sequence of real numbers such that the intervals $I + a_r$ cover $[0,2\pi]$. If

$$\|u\| \leqslant 1, \quad \text{and} \quad u^{(r)} = (u_n e^{ia_r n})_{n \in \mathbb{Z}},$$

then also

$$\|u^{(r)}\| \leqslant 1,$$

and by (1)

$$\operatorname*{ess\,sup}_{I+a_r} |\tilde{u}(x)| = \operatorname*{ess\,sup}_{I} |\tilde{u}(x + a_r)| = p_I(u^{(r)}) \leqslant \frac{s}{\rho}$$

(since $\tilde{u}(x+a_r) = \widetilde{u^{(r)}}(x)$). Hence $\|\tilde{u}\| \leqslant s/\rho$ would hold if $\|u\| \leqslant 1$. For every function $f \in L^1(T)$ the mapping

$$u \mapsto \frac{1}{2\pi}\int_0^{2\pi} \tilde{u}(x)f(-x)\mathrm{d}x$$

would be a continuous linear functional on c_0. Thus there would exist a sequence $(a_n)_{n\in\mathbb{Z}}$ such that

$$\sum_n |a_n| < \infty \quad \text{and} \quad \sum_n u_n a_n = \frac{1}{2\pi}\int_0^{2\pi} \tilde{u}(x)f(-x)\mathrm{d}x, \qquad u \in c_0.$$

By taking for u the sequence whose n-th term equals 1 and all of whose other terms are zero, one obtains

$$a_n = \lambda_n \hat{f}(n),$$

and consequently

$$\sum_n \lambda_n |\hat{f}(n)| < \infty,$$

and thus would hold for all $f \in L^1(T)$. By the preceding exercise this would imply

$$\sum_n \lambda_n < \infty,$$

contrary to the hypothesis.

Let (I_k) be a sequence of compact intervals with lengths greater than zero such that every interval I of length greater than zero contains an I_k. For every k there exists a G_δ everywhere dense in c_0, say E_k, such that $p_{I_k}(u) = \infty$ if $u \in E_k$. The set

$$E = \bigcap_k E_k$$

is then a G_δ everywhere dense in c_0, and if $u \in E$ then

$$\tilde{u} \in L^2(T),$$

$$\hat{\tilde{u}}(n) = \lambda_n u_n = o(\lambda_n),$$

$$p_I(u) = \operatorname{ess\,sup}_I |\tilde{u}(x)| \geqslant \operatorname{ess\,sup}_{I_k} |\tilde{u}(x)| = \infty.$$

EXERCISE 11·209: Assume that

$$c_n \geqslant c_{n+1} \geqslant 0, \qquad c_n \leqslant \frac{A}{n}, \qquad n \geqslant 1.$$

(a): Show that

$$\left|\sum_{n=1}^{N} c_n \sin nx\right| < 2A\sqrt{\pi}.$$

(b): Deduce from this that the functions

$$f(x) = \sum_{n=1}^{\infty} c_n \sin nx$$

belongs to $L^1(T)$, and that the right side is its Fourier series.

Δ∇Δ = ∇Δ∇ = Δ∇Δ = ∇Δ∇ = Δ∇Δ

SOLUTION: (a): One can assume that $0 < x < \pi$. Let M be an integer between 1 and $N - 1$. Noting that

$$|\sin nx| \leqslant nx \quad \text{and} \quad |\sin px + \cdots + \sin qx| \leqslant \frac{1}{\sin\frac{1}{2}x} \leqslant \frac{\pi}{x},$$

one obtains

$$\left|\sum_{n=1}^{N} c_n \sin nx\right| \leqslant \left|\sum_{n=1}^{M}\right| + \left|\sum_{n=M+1}^{N}\right| \leqslant AMx + \frac{A\pi}{(M+1)x} .$$

If $x > \sqrt{\pi}$ let $M = 0$, so that the first sum disappears and a bound $A\sqrt{\pi}$ is obtained. If $x \leqslant \sqrt{\pi}/N$ let $M = N$, so the second sum disappears, which again gives the bound $A\sqrt{\pi}$. Finally, if $1 \leqslant \sqrt{\pi}/x < N$ make $M = [\sqrt{\pi}/x]$, which leads to the bound $2A\sqrt{\pi}$.

SOLUTION: (b): It is known that the series giving $f(x)$ converges for all x (Abel-Dirichlet Theorem). By Lebesgue's Dominated Convergence theorem, $f \in L^1(T)$ and the partial sums of the series converge to f in $L^1(T)$, so $\hat{f}(n) = c_n$.

REMARK 1: f is continuous on $]0,\pi]$, since

$$\left|\sum_{n=N+1} c_n \sin nx\right| \leqslant \frac{\pi}{x} c_{N+1}.$$

REMARK 2: Plancherel's Theorem shows directly that there exists a function $g \in L^2(T) \subset L^1(T)$ such that:

$$g(x) = \sum_{n=1}^{\infty} c_n \sin nx,$$

in the sense of convergence in $L^2(T)$. By Remark 1 $f = g$ almost everywhere, whence the result.

EXERCISE 11·210: Consider the trigonometric series

$$\sum_{n=1}^{\infty} a_n \sin nx, \qquad (*)$$

where (a_n) is a sequence of positive numbers *which decreases to zero.*

(a): Show that the series (*) converges for all x.

(b): Show that the following conditions are equivalent:

(i): The series (*) is uniformly convergent;

(ii): The series (*) is the Fourier series of a continuous function;

(iii): $a_n = o(1/n)$.

(c): Show that the following conditions are equivalent:

(i): There exists a constant M such that:

$$\left|\sum_{n=1}^{N} a_n \sin nx\right| \leqslant M, \qquad x \in \mathbb{R}, \qquad N \geqslant 1;$$

(ii): The series (*) is the Fourier series of a function of $L^\infty(T)$;

(iii): $a_n = O(1/n)$.

(d): Show that the following conditions are equivalent:

(i): The series (*) is the Fourier series of a function of $L^1(T)$;

(ii): The sum of the series (*) is a function in $L^1(T)$;

(iii): $\sum n^{-1} a_n < \infty$.

Show that in this case the partial sums of the series (*) converge in $L^1(T)$.

ΔVΔ = VΔV = ΔVΔ = VΔV = ΔVΔ

SOLUTION: First recall that the formula:

$$2\sin nx . \sin\tfrac{1}{2}x = \cos(n - \tfrac{1}{2})x - \cos(n + \tfrac{1}{2})x$$

shows that for $N \leqslant M$ and $0 < |x| \leqslant \pi$

$$\tilde{D}_{N,M}(x) = \sum_{n=N}^{N} \sin nx = \frac{\cos(N - \frac{1}{2})x - \cos(M + \frac{1}{2})x}{2\sin\frac{1}{2}x}, \tag{1}$$

and consequently that

$$|\tilde{D}_{N,M}(x)| \leqslant \frac{1}{\sin\frac{1}{2}x} \leqslant \frac{\pi}{|x|}, \qquad 0 < |x| \leqslant \pi. \qquad 2)$$

On the other hand,

$$\sum_{n=N}^{M} a_n \sin nx = a_M \tilde{D}_{N,M}(x) + \sum_{n=N}^{M-1} \Delta a_n . \tilde{D}_{N,n}(x), \tag{3}$$

where

$$\Delta a_n - a_n - a_{n+1} \geqslant 0.$$

From this one deduces that

$$\left|\sum_{n=N}^{M} a_n \sin nx\right| \leqslant \frac{\pi a_N}{|x|}, \qquad 0 < |x| \leqslant \pi. \tag{4}$$

SOLUTION: (a): (4) shows that the series (*) is uniformly convergent for $0 < \delta \leqslant |x| \leqslant \pi$. It clearly converges for $x = 0$. Thus its sum is an odd function, continuous for $0 < |x| \leqslant \pi$.

SOLUTION: (b): If the series (*) is uniformly convergent, then

$$\lim_{N\to\infty}\{\sup_{x}|\sum_{\frac{1}{2}N+1\leqslant n\leqslant N} a_n \sin nx|\} = 0.$$

In particular,

$$\lim_{N\to\infty}\sum_{\frac{1}{2}N+1\leqslant n\leqslant N} a_n \sin\frac{n\pi}{2N} = 0.$$

Now,

$$\sum_{\frac{1}{2}N+1\leqslant n\leqslant N} a_n \sin\frac{n\pi}{2N} \geqslant \frac{Na_N}{2\sqrt{2}}, \tag{5}$$

which proves that $Na_N \to 0$, and thus that (i) => (iii).

Let

$$\varepsilon_N = \sup_{n\geqslant N} na_n.$$

If $0 < |x| \leqslant \pi$, there exists an integer p such that

$$\frac{\pi}{p+1} < |x| \leqslant \frac{\pi}{p}.$$

Using (4) with N replaced by $N + p$ and letting $M \to \infty$, one then has

$$\begin{aligned}|\sum_{n\geqslant N} a_n \sin nx| &\leqslant |\sum_{N\leqslant n<N+p} a_n \sin nx| + |\sum_{n\geqslant N+p} a_n \sin nx|\\ &\leqslant |x|\sum_{N\leqslant n<N+p} na_n + \frac{\pi a_{N+p}}{|x|}\\ &\leqslant p|x|\varepsilon_N + (p+1)x_{N+p},\end{aligned}$$

so that, because

$$p|x| \leqslant \pi \quad \text{and} \quad (p+1)a_{N+p} \leqslant (N+p)a_{N+p} \leqslant \varepsilon_N,$$

we have at last

$$\Big|\sum_{n \geqslant N} a_n \sin nx\Big| \leqslant (\pi+1)\varepsilon_N, \tag{6}$$

which proves (iii) =>;(i).

If the series (*) is uniformly convergent, its sum is a continuous function f whose Fourier coefficients are obtained by integrating term by term, which shows that the Fourier series of f is certainly (*). Thus (i) => (ii).

If (*) is the Fourier series of a continuous function, then the Féjer sums $\sigma_N(x)$ of (*) converge uniformly. Since $\sigma_N(0) = 0$, it follows that

$$\lim_{N\to\infty} \sigma_{2N}\left(\frac{\pi}{2N}\right) = 0.$$

If $x = \pi/2N$ then $nx \leqslant \pi$, hence $\sin nx \geqslant 0$, when $n \leqslant 2N$ and $nx \leqslant \pi/2$, hence $\sin nx \geqslant \frac{2nx}{\pi} = \frac{n}{N}$, when $n \leqslant N$. Thus

$$\sigma_{2N}\left(\frac{\pi}{2N}\right) \geqslant \sum_{n\leqslant N}\left(1 - \frac{n}{2N+1}\right)\frac{na_n}{N} \geqslant \frac{1}{2N}\sum_{n\leqslant N} na_n$$

$$\geqslant \frac{a_N}{2N}\sum_{n\leqslant N} n = \frac{1}{4}(N+1)a_N.$$

We have thus proved that (ii) => (iii).

SOLUTION: (c): Inequality (5) proved above again shows that (i) => (iii), and in the same way (iii) => (i) follows from (6). Furthermore, (i) implies that the sum of the series (*) belongs to $L^\infty(T)$, and Lebesgue's Theorem justifies obtaining the Fourier coefficients of this sum by integrating term by term, so that (*) is certainly the Fourier series of its sum. It has thus been proved that (i) => (ii). On the other hand, if (ii) is satisfied, the σ_N are uniformly bounded, and the inequality

$$\sigma_{2N}\left(\frac{\pi}{2N}\right) \geqslant \frac{1}{4}(N+1)a_N$$

proved in part (b) shows that (ii) => (iii).

SOLUTION: (d): Letting $N = 1$ and $M \to \infty$ in (3) yields

$$f(x) = \sum_{n=1}^{\infty} a_n \sin nx = \sum_{n=1}^{\infty} \Delta a_n . \tilde{D}_{1,n}(x).$$

Note that by (1)

$$\tilde{D}_{1,n}(x) = \frac{\cos\frac{1}{2}x - \cos(n+\frac{1}{2})x}{2\sin\frac{1}{2}x} = \frac{\sin^2\frac{1}{2}nx}{\tan\frac{1}{2}x} + \frac{1}{2}\sin nx,$$

so that

$$f = g + h,$$

where

$$g(x) = \sum_{n=1}^{\infty} \Delta a_n \frac{\sin^2\frac{1}{2}nx}{\tan\frac{1}{2}x},$$

$$h(x) = \frac{1}{2}\sum_{n=1}^{\infty} a_n \sin nx.$$

Since

$$\Delta a_n \geqslant 0 \quad \text{and} \quad \sum \Delta a_n = a_0,$$

the function h is continuous (note, furthermore, that as the series which gives h is absolutely convergent, this series is certainly the Fourier series of h). Furthermore, because

$$\frac{\sin^2\frac{1}{2}nx}{\tan\frac{1}{2}x} \geqslant 0 \quad \text{for } 0 < x \leqslant \pi,$$

$g \geqslant 0$ and

$$\int_0^\pi g(x)dx = \sum_{n=1}^\infty \Delta a_n \int_0^\pi \frac{\sin^2\frac{1}{2}nx}{\tan\frac{1}{2}x}\,dx \tag{7}$$

It follows that $f \in L^1(T)$ if and only if the series which appears in the right side of (7) converges. Furthermore, in this case the Fourier series of f certainly is (*), for the integrals of $g(x)\sin nx$ and $h(x)\sin nx$ are obtained by integrating term by term, and therefore

$$\frac{1}{\pi}\int_{-\pi}^{\pi} f(x)\sin nx dx = \sum_{k=1}^\infty \Delta a_k \frac{1}{\pi}\int_{-\pi}^{\pi} \tilde{D}_{1,k}(x)\sin nx dx = \sum_{k=n}^\infty \Delta a_k = a_n.$$

Thus (ii) => (i).

As the fact that (i) => (ii) follows from exercise 11·192, in order to prove that (ii) <=> (iii) it remains to prove that the convergence of the right side of (7) is equivalent to $\sum n^{-1}a_n < \infty$. Now, since

$$\frac{1}{\tan\frac{1}{2}x} - \frac{2}{x}$$

is integrable on $(0,\pi)$, we have

$$\int_0^\pi \frac{\sin^2\frac{1}{2}nx}{\tan\frac{1}{2}x}\,dx = 2\int_1^{n\pi} \frac{\sin^2\frac{1}{2}x}{x}\,dx + O(1) = \int_1^{n\pi}\frac{dx}{x} + O(1)$$

$$\sim \log n,$$

and because

$$\log n \sim 1 + \frac{1}{2} + \cdots + \frac{1}{n},$$

(ii) is therefore equivalent to

$$\sum_{n=1}^{\infty} \left(1 + \frac{1}{2} + \cdots + \frac{1}{n}\right) \Delta a_n < \infty. \tag{8}$$

Since

$$\sum_{n=1}^{N} \frac{a_n}{n} = \left(1 + \frac{1}{2} + \cdots + \frac{1}{N}\right) a_{N+1} + \sum_{n=1}^{N} \left(1 + \frac{1}{2} + \cdots + \frac{1}{n}\right) \Delta a_n,$$

it is clear that (iii) implies (8). The converse is also true, because

$$\left(1 + \frac{1}{2} + \cdots + \frac{1}{N}\right) a_{N+1} = \left(1 + \frac{1}{2} + \cdots + \frac{1}{N}\right) \sum_{n \geqslant N+1} \Delta a_n$$

$$\leqslant \sum_{n \geqslant N+1} \left(1 + \frac{1}{2} + \cdots + \frac{1}{n}\right) \Delta a_n.$$

Finally, if s_N is the n-th partial sum of the series (*), then

$$f(x) - s_N(x) = \sum_{n>N} \Delta a_n . \tilde{D}_{1,n}(x)$$

(cf. (3)), and consequently

$$\|f - s_N\|_1 < \sum_{n>N} \Delta a_n \|\tilde{D}_{1,n}\|_1.$$

But by the preceding

$$\|\tilde{D}_{1,n}\|_1 \sim \frac{1}{\pi} \log n,$$

hence by (8)

$$\|f - s_N\|_1 \to 0.$$

EXERCISE 11·211: (a): If $(a_n)_{n \geqslant 0}$ is a sequence of complex numbers such that

$$a_n \to 0, \qquad \sum_{n=0}^{\infty} |\Delta^2 a_n| < \infty,$$

prove that the series

$$f(x) = \tfrac{1}{2}a_0 + \sum_{n=1}^{\infty} a_n \cos nx \tag{1}$$

converges for $0 < x < 2\pi$, and that

$$f(x) = \tfrac{1}{2} \sum_{n=0}^{\infty} (n + 1)\Delta^2 a_n F_n(x), \tag{2}$$

where F_n denotes the n-th Féjer kernel.

(b): Show that if

$$a_n \to 0, \qquad \sum_{n=0}^{\infty} (n + 1)|\Delta^2 a_n| < \infty,$$

then the function $f \in L^1$, and that if:

$$a_n \to 0, \qquad n\Delta a_n \to 0, \qquad \sum_{n=0}^{\infty} (n + 1)|\Delta^2 a_n| < \infty,$$

then the right side of (1) is the Fourier series of f.

Δ∇Δ = ∇Δ∇ = Δ∇Δ = ∇Δ∇ = Δ∇Δ

SOLUTION: (a): We have (where D_n is the n-th Dirichlet kernel):

$$s_N(x) = \frac{a_0}{2} + \sum_{n=1}^{N} a_n \cos nx = \frac{1}{2}\sum_{n=0}^{N} a_n(D_n(x) - D_{n-1}(x))$$

$$= \frac{1}{2} a_N D_N(x) + \frac{1}{2}\sum_{n=0}^{N-1} \Delta a_n D_n(x).$$

Since

$$D_0 + \cdots + D_n = (n + 1)F_n,$$

this yields

$$s_N(x) = \tfrac{1}{2}a_N D_N(x) + \tfrac{1}{2}N\, a_{N-1}F_{N-1}(x) + \tfrac{1}{2}\sum_{n=0}^{N-1} (n + 1)\Delta^2 a_n . F_n(x).$$

For $0 < x < 2\pi$:

$$|D_N(x)| \leqslant \frac{1}{\sin\frac{1}{2}x}, \qquad |F_n(x)| \leqslant \frac{1}{(n + 1)\sin^2\frac{1}{2}x},$$

which gives the result.

SOLUTION: (b): If

$$a_n \to 0, \qquad \sum (n + 1)|\Delta^2 a_n| < \infty,$$

then (2) shows that $f \in L^1(T)$, since $\|F_n\|_1 = 1$. Moreover, the convergence in (2) holds in $L^1(T)$, so for every integer k:

$$2\hat{f}(k) = \sum_{n=0}^{\infty} (n + 1)\Delta^2 a_n \widehat{F_n}(k)$$

$$= \sum_{n=k}^{\infty} (n + 1 - k)\Delta^2 a_n.$$

Now,

$$\sum_{n=k}^{N} (n + 1 - k)\Delta^2 a_n = a_k - a_{N+1} + (N + 1 - k)\Delta a_{N+1}.$$

Consequently, if, in addition,

$$N\Delta a_n \to 0$$

we obtain that

$$\hat{f}(k) = \tfrac{1}{2}a_k,$$

which proves the last assertion.

EXERCISE 11·212: (a): Let $f \in L^1(T)$, $f \geqslant 0$, and $\hat{f} \geqslant 0$. Show that for $0 < a \leqslant \pi$,

$$\frac{1}{2\pi}\int_{-\pi}^{\pi} f(t)\mathrm{d}t \leqslant \frac{1}{a}\int_{-a}^{a} f(t)\mathrm{d}t.$$

(b): Let $(c_n)_{n\in\mathbb{Z}}$ be a sequence of positive real numbers. Suppose that there exists $\delta > 0$ and $h \in L^2(-\delta,\delta)$ such that

$$\sum_{n=-\infty}^{+\infty} c_n\hat{\varphi}(-n) = \frac{1}{2\pi}\int_{-\delta}^{\delta} h(x)\varphi(x)\mathrm{d}x \tag{*}$$

for every function φ that is infinitely differentiable and has compact support contained in $]-\delta,\delta[$.

Show that

$$\sum_{n=-\infty}^{+\infty} c_n^2 \leqslant \frac{1}{\delta}\int_{-\delta}^{\delta} |h(x)|^2\mathrm{d}x.$$

Δ∇Δ = ∇Δ∇ = Δ∇Δ = ∇Δ∇ = Δ∇Δ

SOLUTION: (a): Let ϑ be the characteristic function of $[-a/2, a/2]$ and $\varphi = \vartheta * \vartheta$. Then

$$0 \leqslant \varphi \leqslant \frac{a}{2\pi} \quad \text{and} \quad \varphi(x) = 0 \text{ if } a \leqslant |x| \leqslant \pi.$$

Furthermore,

$$\hat{\varphi}(n) = |\hat{\vartheta}(n)|^2 \geqslant 0.$$

Therefore

$$\frac{a}{2\pi}\int_{-a}^{a} f(t)dt \geqslant \int_{-\pi}^{\pi} f(t)\varphi(t)dt = 2\pi \sum_{n=-\infty}^{+\infty} \hat{f}(n)\hat{\varphi}(-n)$$

$$\geqslant 2\pi\hat{f}(0)\hat{\varphi}(0) = |\vartheta(0)|^2\int_{-\pi}^{\pi} f(t)dt$$

$$= \left(\frac{a}{2\pi}\right)^2\int_{-\pi}^{\pi} f(t)dt,$$

and the result follows

SOLUTION: (b): Set, as usual

$$(f|g) = \frac{1}{2\pi}\int_{-\pi}^{\pi} f\bar{g}.$$

For $s = 1,2...$ let $\vartheta_s \in C^\infty(T)$, $\vartheta_s \geqslant 0$, $\|\vartheta_s\|_1 = 1$, and $\vartheta_s(x) = 0$ if $1/s \leqslant |x| \leqslant \pi$. Then

$$\lim_s \hat{\vartheta}_s(n) = 1, \tag{1}$$

for all $n \in \mathbb{Z}$. Denote by $C^\infty(-\delta,\delta)$ the set of infinitely differentiable functions with compact support contained in $]-\delta,\delta[$. In the usual manner $C^\infty(-\delta,\delta)$ is identified with a subspace of $C^\infty(T)$.

$\vartheta_s * \check{\vartheta}_s \in C^\infty(-\delta,\delta)$ whenever s is large enough. Applying (*) to this function yields

$$\sum_n c_n |\hat{\vartheta}_s(n)|^2 < \infty,$$

Hence one can define a continuous function

$$g_s(x) = \sum_n c_n |\hat{\vartheta}_s(n)|^2 e^{inx}. \tag{2}$$

Let $0 < \gamma < \delta$ and $\varphi \in C^\infty(-\gamma,\gamma)$. Whenever s is large enough $\varphi * \vartheta_s * \check{\vartheta}_s \in C\ (-\delta,\delta)$, and consequently

$$(g_s|\varphi) = \sum_n c_n |\hat{\vartheta}_s(n)|^2 \overline{\hat{\varphi}(n)} = (h|\varphi * \vartheta_s * \check{\vartheta}_s)$$

$$= (h * \vartheta_s * \vartheta_s|\varphi)$$

Therefore

$$g_s = h * \vartheta_s * \check{\vartheta}_s \quad \text{on} \quad]-\gamma,\gamma].$$

Taking the complex conjugates of both sides of (2) and then multiplying term by term, one sees that the Fourier coefficients of $|g_s|^2$ are all positive. By part (a) one therefore has

$$\|g_s\|_2^2 \leqslant \frac{1}{\gamma}\int_{-\gamma}^{\gamma} |h * \vartheta_s * \check{\vartheta}_s|^2 \leqslant \frac{2\pi}{\gamma} \|h * \vartheta_s * \check{\vartheta}_s\|_2^2 \leqslant \frac{2\pi}{\gamma} \|h\|_2^2,$$

and on making $\gamma \to \delta$:

$$\|g_s\|_2^2 \leqslant \frac{2\pi}{\delta} \|h\|_2^2. \tag{3}$$

Taking (1) and (2) into account, we have $\lim_s \hat{g}_s(n) = c_n$

and, using (3),

$$\sum_n |c_n|^2 = \sup_N \sum_{|n|\leq N} |c_n|^2 = \sup_N(\lim_s \sum_{|n|\leq N} (\hat{g}_s(n))^2)$$

$$\leq \frac{2\pi}{\delta} \|h\|_2^2.$$

EXERCISE 11·213: With every function $f \in L^1(T)$ there is associated its *ADJOINT FOURIER SERIES*:

$$\sum -i\,\text{sign}(n)\hat{f}(n)e^{inx},$$

where

$$\text{sign}(n) = \begin{cases} -1 & \text{if } n \leq -1, \\ 0 & \text{if } n = 0, \\ 1 & \text{if } n \geq 1. \end{cases}$$

The *ADJOINT FOURIER SUMS* of f are the numbers

$$s_N^*(x) = \sum_{|n|\leq N} -i\,\text{sign}(n)\hat{f}(n)e^{inx}.$$

(a): Show that

$$s_N^* = f * D_N^*,$$

where D_N^* is a function to be determined.

(b): Show that

$$\int_0^\pi D_N^*(u)du \sim 2\log N.$$

(c): If ε is an integrable function such that

$$\lim_{u\to 0}\varepsilon(u) = 0,$$

show that

$$\int_0^{\pi} \varepsilon(u)D_N^*(u)du = o(\log N).$$

(Set

$$\Delta_N^*(u) = D_N^*(u) - \sin Nu,$$

and note that

$$\Delta_N^*(u) \geqslant 0 \quad \text{if } 0 \leqslant u \leqslant \pi).$$

(d): Deduce from the preceding that if

$$\ell = \lim_{u\to 0}(f(x + u) - f(x - u))$$

then

$$\lim_N \frac{s_N^*(x)}{\log N} = -\frac{\ell}{\pi}\ .$$

(e): Then prove the following result: If f has left and right hand limits at all points, and if

$$|\hat{f}(n)| = o\left(\frac{1}{n}\right)$$

then f is continuous.

ΔVΔ = VΔV = ΔVΔ = VΔV = ΔVΔ

SOLUTION: (a): It is clear that

$$D_N^*(u) = 2\sum_{n=1}^{N} \sin nu. \tag{1}$$

Multiplying both sides by $\sin\frac{1}{2}u$ we find

$$D_N^*(u) = \frac{\cos\frac{1}{2}u - \cos(N+\frac{1}{2})u}{\sin\frac{1}{2}u}, \qquad 0 < |u| \leqslant \pi. \tag{2}$$

SOLUTION: (b): By (1),

$$\int_0^\pi D_N^*(u)du = 2\sum_{n=0}^{[\frac{1}{2}(N-1)]} \frac{1}{n+\frac{1}{2}}$$

$$\sim 2\log N.$$

SOLUTION: (c): It is easily seen that for $0 < |u| \leqslant \pi$,

$$\Delta_N^*(u) = D_N^*(u) - \sin Nu = (1 - \cos Nu)\cot\tfrac{1}{2}u.$$

By the Riemann-Lebesgue Theorem:

$$\int_0^\pi \varepsilon(u)D_N^*(u)du = \int_0^\pi \varepsilon(u)\Delta_N^*(u)du + o(1). \tag{3}$$

Let $\alpha > 0$, and choose $\delta > 0$ so that

$$|\varepsilon(u)| < \alpha \quad \text{if} \quad 0 \leqslant u \leqslant \delta.$$

The Riemann-Lebesgue Theorem and the fact that

$$\Delta_N^*(u) \geqslant 0 \quad \text{if} \quad 0 \leqslant u \leqslant \pi.$$

show that

$$\left|\int_0^\pi \varepsilon\Delta_N^*\right| \leqslant \alpha\int_0^\pi \Delta_N^* + \int_\delta^\pi \varepsilon(u)\cot\tfrac{1}{2}u du + o(1).$$

Since

$$\int_0^\pi \Delta_N^* = \int_0^\pi D_N^* + o(1)$$

$$\sim 2\log N,$$

from this one concludes that

$$\limsup_N (\log N)^{-1} \left| \int_0^\pi \varepsilon \Delta_N^* \right| \leqslant 2\alpha,$$

and, since α is arbitrary, that

$$\int_0^\pi \varepsilon \Delta_N^* = o(\log N).$$

By (3) we also have:

$$\int_0^\pi \varepsilon D_N^* = o(\log N).$$

SOLUTION: (d): As the function D_N^* is odd, it is easily seen that

$$s_N^*(x) = -\frac{1}{2\pi}\int_0^\pi (f(x+u) - f(x-u))D_N^*(u)du.$$

If

$$f(x+u) - f(x-u) = \ell + \varepsilon(u) \quad \text{with } \lim_{u\to 0}\varepsilon(u),$$

from parts (b) and (c) above one deduces that

$$\lim_N (\log N)^{-1} s_N^*(x) = -\frac{\ell}{\pi}. \tag{4}$$

SOLUTION: (e): If f has left and right hand limits at all points then ℓ exists at every point x and is equal to

$$f(x+0) - f(x-0).$$

Further, if

$$\hat{f}(n) = o\left(\frac{1}{n}\right),$$

then

$$|s_N^*(x)| \leqslant \sum_{n=1}^{N} (|\hat{f}(n) + \hat{f}(-n)) = o\left(\sum_{n=1}^{N} \frac{1}{n}\right) = o(\log N).$$

From (4) it is then deduced that $\ell = 0$, which proves that f is continuous.

EXERCISE 11·214: (a): Let $(u_n)_{n\geqslant 1}$ be a sequence of positive numbers such that

$$u_n \leqslant \frac{A}{n}.$$

Set:

$$a_n = \frac{u_1 + 2u_2 + \cdots + nu_n}{n},$$

$$b_n = \frac{u_1^2 + 4u_2^2 + \cdots + n^2u_n^2}{n},$$

$$\gamma_n = n \sum_{s=1}^{\infty} u_s^2 \sin^2 \frac{s\pi}{2n}.$$

Show that the conditions $a_n \to 0$, $b_n \to 0$, $\gamma_n \to 0$ are equivalent. (Prove that

$$b_n \leqslant Aa_n, \qquad a_n^2 \leqslant b_n \leqslant \gamma_n,$$

and that for every integer ν

$$\gamma_n \leqslant \frac{\pi^2\nu}{4} b_{n\nu} + \frac{A^2}{\nu}.$$

(b): Let f be a real function with left and right hand limits at all points, and with period 2π. For every integer $n \geqslant 1$ set

$$\varphi_n(u) = \sum_{r=0}^{2n-1} |f(u + x_{r+1}) - f(u + x_r)|^2,$$

where $x_r = \pi r/n$.

Show that if

$$|f(\xi + 0) - f(\xi - 0)| > d > 0 \quad \text{at some point } \xi,$$

then whenever n is large enough

$$\varphi_n(u) > d^2 \quad \text{for almost all } u.$$

Show also that if f is continuous and has bounded variation, then $\varphi_n \to 0$ uniformly on R.

(c): Calculate:

$$\int_0^{2\pi} \varphi_n(u)du,$$

using Plancherel's Formula.

(d): With the aid of what has gone before, prove the following theorem (owed to Wiener): A necessary and sufficient condition for a function f with bounded variation to be continuous is that:

$$\lim_n \frac{\hat{f}(1) + 2\hat{f}(2) + \cdots + n\hat{f}(n)}{n} = 0.$$

ΔVΔ = VΔV = ΔVΔ = VΔV = ΔVΔ

SOLUTION: (a): Since $s^2u_s^2 \leqslant Asu_s$, it is clear that $b_n \leqslant Aa_n$. The Cauchy-Schwarz Inequality shows that

$$a_n \leqslant \left(\frac{u_1^2 + 4u_2^2 + \cdots + n^2u_n^2}{n}\right)^{\frac{1}{2}} = b_n^{\frac{1}{2}}.$$

Furthermore, noting that $\sin u \geqslant 2u/\pi$ if $0 \leqslant u \leqslant \pi/2$,

$$\gamma_n \geqslant n \sum_{s=1}^{n} u_s^2\left(\frac{s}{n}\right)^2 = b_n.$$

Finally, if $\nu > 0$ is an integer, then

$$\gamma_n \leqslant n \sum_{s=1}^{n\nu} u_s^2\left(\frac{s\pi}{2n}\right)^2 + A^2n \sum_{s>n\nu} \frac{1}{s^2}$$

$$\leqslant \frac{\pi^2\nu}{4} b_{n\nu} + A^2n\int_{n\nu}^{\infty} \frac{dx}{x^2} = \frac{\pi^2\nu}{4} b_{n\nu} + \frac{A^2}{\nu},$$

All this shows that $a_n \to 0$ is equivalent to $b_n \to 0$, then that $\gamma_n \to 0$ implies $b_n \to 0$, and finally that if $b_n \to 0$ then for every integer $\nu > 0$

$$\limsup_n \gamma_n \leqslant \frac{A^2}{\nu},$$

and consequently $\gamma_n \to 0$.

SOLUTION: (b): If

$$|f(\xi + 0) - f(\xi - 0)| > d > 0$$

there exists an integer n_0 such that:

$$\left|f\left(\alpha + \frac{\pi}{n}\right) - f(\alpha)\right| > d \quad \text{if } n \geqslant n_0 \text{ and } \alpha < \xi < \alpha + \frac{\pi}{n}. \tag{1}$$

If $u \notin \xi + \pi\mathbb{Q}$ there exists an integer k such that

$$u < \xi + 2k\pi < u + 2\pi,$$

and then

$$\xi + 2k\pi \neq u + x_r \quad \text{for all } r.$$

Hence there exists r such that

$$u + x_r < \xi + 2k\pi < u + x_{r+1}, \qquad 0 \leqslant r < 2n.$$

By (1) and the periodicity of f, this implies that

$$|f(u + x_{r+1}) - f(u + x_r)| > d,$$

and we have thus proved that

$$\varphi_n(u) > d^2 \quad \text{if} \quad n \geqslant n_0 \quad \text{and} \quad u \notin \xi + \pi\mathbb{Q}.$$

If it is now assumed that f is continuous and has bounded variation, then for all $\varepsilon > 0$ there exists $\eta > 0$ such that

$$|f(u) - f(v)| < \varepsilon \quad \text{if} \quad |u - v| < \eta,$$

and consequently if $\pi/n < \eta$,

$$\varphi_n(u) \leqslant \varepsilon \sum_{r=0}^{2n-1} |f(u + x_{r+1}) - f(u + x_r)| \leqslant \varepsilon V(f;0,2\pi).$$

This shows that $\varphi_n \to 0$ uniformly on $\mathbb{R}$.

SOLUTION: (c):

$$f(u + x_{r+1}) - f(u + x_r) \sim \sum_s (e^{isx_{r+1}} - e^{isx_r})\hat{f}(s)e^{isu}$$

$$\sim \sum_s 2ie^{is(x_r + \pi/2n)} \sin\left(\frac{s\pi}{2n}\right)\hat{f}(s)e^{isu}.$$

Plancherel's Formula gives

$$\frac{1}{2\pi}\int_0^{2\pi} \varphi_n(u)du = 8n \sum_{s=-\infty}^{+\infty} |\hat{f}(s)|^2 \sin^2\left(\frac{s\pi}{2n}\right),$$

and as f is real, $\hat{f}(-s) = \overline{\hat{f}(s)}$, so that

$$\frac{1}{2\pi}\int_0^{2\pi} \varphi_n(u)du = 16n \sum_{s=1}^{\infty} |\hat{f}(s)|^2 \sin^2\left(\frac{s\pi}{2n}\right).$$

SOLUTION: (d): By the results proved in parts (b) and (c),

$$n \sum_{s=1}^{\infty} |\hat{f}(s)|^2 \sin^2\left(\frac{s\pi}{2n}\right) \to 0$$

if and only if f is continuous. Furthermore, as f has bounded variation,

$$|\hat{f}(n)| \leqslant \frac{A}{n}.$$

Using part (a), Wiener's Theorem is obtained.

EXERCISE 11·215: (a): Show that there exists a sequence such that $0 \leqslant a_s \leqslant 1$ and that for every integer $p \geqslant 1$:

$$(1 + \cos t)(1 + \cos 4t)\cdots(1 + \cos 4^{p-1}t)$$

$$= \sum_{s=0}^{(4^p-1)/3} a_s \cos st. \qquad (1)$$

(b): Denote by γ_p the number of indices s such that:

$$0 \leqslant s \leqslant \frac{1}{3}(4^p - 1), \qquad a_s \neq 0.$$

Show that

$$\gamma_{p+1} - \gamma_p = 3^p.$$

(c): Deduce from this that there exists a continuous function f with period 2π, such that for all x

$$f(x) = -x + \lim_{p\to\infty}\int_0^x (1 + \cos t)(1 + \cos 4t)\cdots(1 + \cos 4^{p-1}t)dt. \qquad (4)$$

(d): Show that f has bounded variation and that its Fourier coefficients are not $o(1/n)$. Calculate $\hat{f}(4^n))$.

Δ∇Δ = ∇Δ∇ = Δ∇Δ = ∇Δ∇ = Δ∇Δ

SOLUTION: (a): First note that

$$(1 + \cos t)(1 + \cos 4t)\cdots(1 + \cos 4^{p-1}t)$$

is a linear combination of functions $\cos st$ for

$$0 \leqslant s \leqslant 1 + 4 + \cdots + 4^{p-1} = \frac{1}{3}(4^p - 1).$$

Multiplying this combination of functions by $\cos 4^p t$, one obtains, for $s \neq 0$, the two terms

$$\tfrac{1}{2}\cos(4^p + s)t, \qquad \tfrac{1}{2}\cos(4^p - s)t.$$

The coefficients $4^p + s$ and $4^p - s$ lie respectively between $4^p + 1$ and $\frac{1}{3}(4^{p+1} - 1)$ and between $\frac{2}{3}4^p + \frac{1}{3}$ and $4^p - 1$. Now

$$\frac{1}{3}(4^p - 1) < \frac{2}{3}4^p + \frac{1}{3},$$

which shows that there exists a sequence (a_s) satisfying (1) for all $p \geq 1$. By what has gone before it is clear that $a_s = 2^{-u}$ with $u \in \mathbb{N}$, for all s such that $a_s \neq 0$; in particular, $0 < a_s \leq 1$.

SOLUTION: (b): $\gamma_{p+1} - \gamma_p$ is equal to the number of terms which appear upon multiplying the right side of (1) by $\cos 4^p t$ and replacing the products of cosines by sums; by the analysis carried out in part (a)

$$\gamma_{p+1} - \gamma_p = 2(\gamma_p - 1) + 1 = 2\gamma_p - 1,$$

and consequently:

$$(\gamma_{p+1} - \gamma_p) = 3(\gamma_p - \gamma_{p-1}).$$

One can show that $\gamma_2 - \gamma_1 = 3$, so

$$\gamma_{p+1} - \gamma_p = 3^p.$$

SOLUTION: (c): Let us denote by $F_p(x)$ the value of the integral which appears in the right side of (2). We have

$$|F_{p+1}(x) - F_p(x)| \leq \sum_{(4^p-1)/3 < s \leq (4^{p+1}-1)/3} \frac{a_s}{s} \leq \qquad \text{(Contd)}$$

(Contd) $$\leqslant \frac{\gamma_{p+1} - \gamma_p}{\frac{1}{3}(4^p - 1)} = \frac{3^{p+1}}{4^p - 1}.$$

Therefore $-x + F_p(x)$ converges uniformly on $\mathbb{R}$ to a continuous function f. Furthermore, the integral over $[0,2\pi]$ of

$$-1 + (1 + \cos t)(1 + \cos 4t)\cdots(1 + \cos 4^{p-1}t)$$

is zero. Hence $-x + F_p(x)$ has period 2π. Therefore so does f.

SOLUTION: (d): For all p, F_p is increasing; thus the same is true for its limit. But then f has bounded variation because it is the difference of two increasing functions. The Fourier coefficients of f are the limit of those of $-x + F_p(x)$, so

$$f(x) = \sum_{s=1}^{\infty} \frac{a_s \sin sx}{s}.$$

It is clear that:

$$2i\hat{f}(4^s) = 4^{-s},$$

which proves that $s\hat{f}(s)$ does not tend to zero.

EXERCISE 11·216: The object of the exercise is to construct a function $f \in \mathrm{Lip}(\alpha)$ whose Fourier coefficients are not $o(n^{-\alpha})$ (cf. exercise 11·194). Let $a > 1$ and $0 < \alpha < 1$. Set

$$f(x) = \sum_{n=0}^{\infty} a^{-n\alpha}\cos a^n x, \qquad -\pi \leqslant x \leqslant \pi.$$

(a): Let $u \geqslant 0$. Show that

$$|f(x + a^{-u}) - f(x - a^{-u})| \leqslant Aa^{(1-\alpha)[u]-u} + Ba^{-\alpha[u]},$$

where A,B are constants depending only upon a,α, and where $[u]$ denotes the integral part of u.

Deduce from this that $f \in \mathrm{Lip}(\alpha)$.

(b): Set

$$F(x) = \frac{\sin\pi x}{\pi x} \quad \text{if } x \neq 0, \qquad F(0) = 1.$$

Prove that

$$\hat{f}(s) = \sum_{n=1}^{\infty} \frac{a^{(1-\alpha)n}}{a^{n} + s} F(a^{n} - s). \tag{*}$$

(c): Let $0 < \lambda < \log a$, and let n_0 be an integer such that

$$a^{n_0} > \frac{1}{2(\log a - \lambda)}.$$

For every integer $\nu > n_0$ denote by s_ν the integer such that

$$s_\nu - \tfrac{1}{2} \leqslant a^{\nu} < s_\nu + \tfrac{1}{2}.$$

Show that

$$|F(a^{n} - s_\nu)| < \frac{1}{\pi(a^{\nu}\log a - \frac{1}{2})} \quad \text{if } n > \nu,$$

and

$$|F(a^{n} - s_\nu)| < \frac{1}{\pi\lambda a^{n}} \quad \text{if } n_0 \leqslant n < \nu.$$

(d): Deduce from this that

$$s_\nu^{\alpha}\hat{f}(s_\nu) \geqslant \delta > 0$$

whenever ν is large enough, and that, in particular, $\hat{f}(s)$ is not $o(s^{-\alpha})$.

REMARK: When $\alpha = 1$ every function $f \in \mathrm{Lip}(1)$ has bounded variation. In this case then see exercise 11·215.

$\Delta\nabla\Delta = \nabla\Delta\nabla = \Delta\nabla\Delta = \nabla\Delta\nabla = \Delta\nabla\Delta$

SOLUTION: (a):

$$f(x + a^{-u}) - f(x - a^{-u}) = -2 \sum_{n=1}^{\infty} a^{-n\alpha} \sin a^{n-u} \sin a^{n} x,$$

so

$$|f(x + a^{-u}) - f(x - a^{-u})| \leqslant 2 \sum_{n=1}^{[u]-1} a^{(1-\alpha)n-u} + 2 \sum_{n=[u]}^{\infty} a^{-n\alpha}$$

$$\leqslant \frac{2a^{(1-\alpha)[u]-u}}{a^{1-\alpha} - 1} + \frac{2a^{-\alpha[u]}}{1 - a^{-\alpha}} .$$

Setting:

$$A = \frac{2}{a^{1-\alpha} - 1}, \qquad B = \frac{2}{1 - a^{-\alpha}},$$

the desired inequality is obtained. Then

$$\frac{|f(x + a^{-u}) - f(x - a^{-u})|}{2a^{-\alpha u}} \leqslant \tfrac{1}{2}[Aa^{(1-\alpha)([u]-u)} + Ba^{\alpha(u-[u])}]$$

$$\leqslant \tfrac{1}{2}(A + Ba^{\alpha}),$$

which shows that $f \in \mathrm{Lip}(\alpha)$.

SOLUTION: (b): The series which defines f is absolutely convergent, and consequently

$$\hat{f}(s) = \sum_{n=0}^{\infty} a^{-n\alpha} \frac{1}{\pi}\int_0^{\pi} (\cos a^n x)(\cos sx)\,dx.$$

On observing that

$$(a^n + s)F(a^n + s) = (a^n - s)F(a^n - s),$$

and that consequently

$$F(a^n + s) + F(a^n - s) = \frac{2a^n}{(a^n + s)} F(a^n - s),$$

this yields

$$\hat{f}(s) = \sum_{n=0}^{\infty} \frac{a^{(1-\alpha)n}}{a^n + s} F(a^n - s).$$

SOLUTION: (c): Note that

$$|F(x)| \leqslant \frac{1}{\pi|x|} \text{ if } x \neq 0,$$

and that if $n > \nu$

$$a^n - s_\nu \geqslant a^n - a^\nu - \tfrac{1}{2} > (n - \nu)a^\nu \log a - \tfrac{1}{2} > a^\nu \log a - \tfrac{1}{2}.$$

Since

$$a^\nu \log a - \tfrac{1}{2} > a^{n_0} \log a - \tfrac{1}{2} > \lambda a^{n_0} > 0,$$

it follows that

$$|F(a^n - s_\nu)| < \frac{1}{\pi(a^\nu \log a - \frac{1}{2})} \quad \text{if } n > \nu.$$

If it is now assumed that $n_0 \leqslant n < \nu$ then

$$s_\nu - a^n > a^\nu - a^n - \tfrac{1}{2} > (\nu - n)a^n \log a - \tfrac{1}{2} \geqslant a^n \log a - \tfrac{1}{2} > \lambda a^n,$$

because

$$a^n \geqslant a^{n_0} > \frac{1}{2(\log a - \lambda)}.$$

Thus

$$|F(a^n - s_\nu)| < \frac{1}{\pi\lambda a^n} \quad \text{if } n_0 \leqslant n < \nu.$$

SOLUTION: (d): Taking into account part (c) and the inequalities $|F(x)| \leqslant 1$ for all x and $|F(x)| \geqslant 2/\pi$ if $|x| \leqslant \frac{1}{2}$, we obtain

$$\left|\sum_{n=0}^{n_0-1} \frac{a^{(1-\alpha)n}}{a^n + s_\nu} F(a^n - s_\nu)\right| \leqslant \frac{1}{s_\nu} \sum_{n=0}^{n_0-1} a^{(1-\alpha)n} = O(a^{-\nu}),$$

$$\left|\sum_{n=n_0}^{\nu-1}\right| \leqslant \frac{1}{\pi\lambda s_\nu} \sum_{n=0}^{\infty} a^{-\alpha n} = O(a^{-\nu}),$$

$$\left|\sum_{n=\nu+1}^{\infty}\right| \leqslant \frac{1}{\pi(a^\nu \log a - \frac{1}{2})} \sum_{n=0}^{\infty} a^{-\alpha n} = O(a^{-\nu}),$$

and

$$\frac{a^{(1-\alpha)\nu}}{a^\nu + s_\nu} F(a^\nu - s_\nu) \geqslant \frac{2}{\pi} \frac{a^{(1-\alpha)\nu}}{2a^\nu + \frac{1}{2}}$$

$$\sim \frac{1}{\pi} a^{-\alpha\nu}.$$

It follows that if $0 < \delta < 1/\pi$ then

$$a^{\alpha\nu}|\hat{f}(s_\nu)| > \delta$$

when ν is large enough. Since $s_\nu \sim a^\nu$ we also have

$$s_\nu^\alpha|\hat{f}(s_\nu)| > \delta$$

for ν large enough, which proves that the $\hat{f}(s)$'s are not $o(s^{-\alpha})$.

EXERCISE 11·217: (a): If $-1 < r < 1$ set

$$P_r(x) = \sum_{-\infty}^{+\infty} r^{|n|}e^{inx}.$$

Prove that

$$P_r(x) = \frac{1 - r^2}{1 - 2r\cos x + r^2}.$$

(b): For every function $f \in L^1(T)$ set

$$A_r(f;x) = \sum_{-\infty}^{+\infty} r^{|n|}\hat{f}(n)e^{inx}.$$

Prove that if $f(x + 0)$ and $f(x - 0)$ exist at a point x, then

$$\lim_{r\to 1^-} A_r(f;x) = \tfrac{1}{2}(f(x + 0) + f(x - 0)).$$

(c): Set

$$F(x) = \int_{-\pi}^{x} f(t)dt.$$

Prove that if $F(\pi) = 0$, then

$$A_r(f;x) = \frac{1}{2\pi}\int_{-\pi}^{\pi} \frac{F(x+t) - F(x-t)}{2\sin t} Q_r(t)dt,$$

where Q_r is a function to be determined.

(d): Prove that as $r \to 1$ $(r < 1)$ the Q_r's form an approximate identity in $L^1(T)$.

(e): Prove that if at some point x the "symmetric derivative of F", $D_s f(x) = \lim_{t\to 0}[F(x+t) - F(x-t)]/2t$, exists and is finite, then

$$\lim_{r\to 1^-} A_r(f;x) = D_s F(x).$$

(f): Compare this result with the Lebesgue-Féjer theorem.

Δ∇Δ = ∇Δ∇ = ΔVΔ = ∇Δ∇ = Δ∇Δ

SOLUTION: (a):

$$P_r(x) = 1 + \frac{re^{ix}}{1 - re^{ix}} + \frac{re^{-ix}}{1 - re^{ix}} = \frac{1 - r^2}{1 - 2r\cos x + r^2}$$

SOLUTION: (b): Replacing the $\hat{f}(n)$ by their defining expressions in the formula for $A_r(f;x)$ shows that

$$A_r(f;x) = (f*P_r)(x).$$

Note that the P_r form an approximate identity in $L^1(T)$ as $r \to 1$, since

$$P_r \geqslant 0, \qquad \frac{1}{2\pi}\int_{-\pi}^{\pi} P_r(x)dx = 1,$$

and for all $\delta > 0$

$$\frac{1}{2\pi}\int_{\delta\leqslant|x|\leqslant\pi} P_r(x)dx \leqslant P_r(\delta), \qquad \lim_{r\to 1} P_r(\delta) = 0.$$

Set

$$s = \tfrac{1}{2}(f(x + 0) - f(x - 0)),$$

$$Q(t) = f(x + t) - f(x - t) - 2s.$$

Taking into account that $P_r(-x) = P_r(x)$, one easily sees that

$$A_r(f;x) - s = \frac{1}{2\pi}\int_0^{\pi} Q(t)P_r(t)dt.$$

Let $\varepsilon > 0$; if δ is chosen so that

$$|Q(t)| < \varepsilon \quad \text{if } 0 < t \leqslant \delta,$$

then

$$|A_r(f;\sigma) - s| \leqslant \varepsilon + P_r(\delta)\|Q\|_1,$$

which proves that

$$\lim_{r\to 1^-} A_r(f;x) = s.$$

SOLUTIONS: (c);(d): Integrating by parts and taking into account that $P_r'(-t) = -P_r'(t)$ yields

$$A_r(f;x) = \frac{1}{2\pi}\int_{-\pi}^{\pi} f(t)P_r(x - t)dt = \frac{1}{2\pi}\int_{-\pi}^{\pi} F(t)P_r'(x - t)dt$$

$$= \frac{1}{2\pi}\int_{-\pi}^{\pi} F(x - t)P_r'(t)dt = -\frac{1}{2\pi}\int_{-\pi}^{\pi} F(x + t)P_r'(t)dt,$$

so

$$A_r(f;x) = -\frac{1}{2\pi}\int_{-\pi}^{\pi} \tfrac{1}{2}(F(x+t) - F(x-t))P_r'(t)dt$$

setting

$$Q_r(t) = -\sin t . P_r'(t),$$

we have

$$A_r(f;x) = \frac{1}{2\pi}\int_{-\pi}^{\pi} \frac{F(x+t) - F(x-t)}{2\sin t} Q_r(t)dt. \qquad (1)$$

It is clear that

$$Q_r(-t) = Q_r(t) \quad \text{and} \quad Q_r \geqslant 0.$$

Also, if $f(x) = \cos x$ then $F(x) = \sin x$; therefore taking $f(x) = \cos x$ and $\gamma = 0$ in (1) yields

$$\frac{1}{2\pi}\int_{-\pi}^{\pi} Q_r(t)dt = r.$$

Finally an elementary calculation shows that

$$Q_r(t) = 2r(1 - r^2)\left(\frac{\sin t}{1 - 2r\cos t + r^2}\right)^2$$

attains its maximum when

$$\cos t = \frac{2r}{1 + r^2} .$$

Therefore, if $0 < \delta < \pi$ then

$$\sup_{\delta \leqslant t \leqslant \pi} Q_r(t) \leqslant Q_r(\delta),$$

as long as

$$\cos\delta \leqslant 2r(1 + r^2)^{-1}.$$

Since $Q_r(\delta) \to 0$ as $r \to 1$, the Q_r form an approximate identity in $L^1(T)$.

SOLUTION: (e): By adding a constant to f, $F(\pi)$ may be assumed to 0. Setting

$$\psi(t) = \frac{F(x + t) - F(x - t)}{2\sin t} - D_s F(x),$$

we have

$$A_r(f;x) - D_s(x) = \frac{1}{2\pi}\int_{-\pi}^{\pi} \psi(t) Q_r(t) dt + (r - 1) D_s F(x),$$

since $\psi(t) \to 0$ as $t \to 0$, one can show as in (b) that the integral on the right side tends to 0 as $r \to 1$, from which the result follows. (Note that there is no problem at $\pm\pi$ since at these points

$$Q_r(t) \sim (\text{constant})\sin^2 t).$$

SOLUTION: (f): In particular

$$A_r(f;x) \to f(x)$$

at all points where $f(x) = F'(x)$, i.e. almost everywhere. Now, the condition $f(x) = F'(x)$ can be written

$$\lim_{h\to 0} h^{-1}\int_0^h (f(x + u) - f(x)) du = 0.$$

Therefore this result is better than that of Lebesgue-Féjer, which says $\sigma_N(f;x) \to f(x)$ if

$$\lim_{h\to 0} h^{-1}\int_0^h |f(x+u) - f(x)|\,du = 0.$$

EXERCISE 11·218: The series $\sum_{-\infty}^{\infty} c_n$ converges to s in the sense of Borel (written

$$\sum_{-\infty}^{\infty} c_n \overset{(B)}{=} s)$$

if

$$\lim_{r\to\infty} e^{-r} \sum_{n=0} \frac{s_n r^n}{n!} = s$$

where

$$s_n = \sum_{|p|\leqslant n} c_p.$$

(a): Show that if

$$\sum_{-\infty}^{+\infty} c_n = s$$

in the ordinary sense, it also converges to s in the Borel sense (use Exercise 10·177).

(b): Show that

$$\sum_{n=0} z^n \overset{(B)}{=} \frac{1}{1-z} \quad \text{if} \quad \mathrm{Re}(z) < 1.$$

(c): For every function $f \in L^1(T)$ set:

$$B_r(f;x) = e^{-r} \sum_{n\geq 0} s_n(f;x) \frac{r^n}{n!} .$$

Show that

$$B_r(f;x) = (f*B_r)(x),$$

where

$$B_r(x) = e^{-2r\sin^2\frac{1}{2}x} \frac{\sin(\frac{1}{2}x + r\sin x)}{\sin\frac{1}{2}x} .$$

(d): Prove that

$$J(x) = \int_0^x \frac{|\sin u|}{u}\, du \sim \frac{2}{\pi} \log x \quad \text{as } x \to \infty.$$

(e): Prove that

$$\|B_r\|_1 = \frac{2}{\pi}\int_0^{\pi/2} e^{-2ru^2} \frac{|\sin(2r + 1)u|}{u}\, du + o(1) \quad \text{as } r \to \infty.$$

Deduce from this that

$$\|B_r\|_1 \sim \frac{2}{\pi^2} \log r, \quad r \to \infty.$$

(f): Using (e), show that there exist continuous functions whose Fourier series does not converge in the Borel sense at certain points.

(g): Prove that if $f \in L^1(T)$, and if

$$\lim_{t\to 0}(f(x + t) - f(x))\log \frac{1}{|t|} = 0$$

at a point x, then

$$\sum_{-\infty}^{+\infty} \hat{f}(n)e^{inx} \overset{(B)}{=} f(x).$$

ΔVΔ = VΔV = ΔVΔ = VΔV = ΔVΔ

SOLUTION: (a): If (r_i) is a sequence of real numbers that tends to $+\infty$, and if:

$$u_{n,i} = e^{-r_i}\frac{r_i^n}{n!},$$

then:

(i): $u_{n,i} \geqslant 0$;

(ii): $\sum_{n=0}^{\infty} u_{n,i} = 1$;

(iii): $\lim_{i\to\infty} u_{n,i} = 0$.

By exercise 10·177 this implies

$$\lim_{r\to\infty} e^{-r}\sum_{n=0}^{\infty}\frac{s_n r^n}{n!} = s \quad \text{if } s_n \to s.$$

SOLUTION: (b): In this case

$$s_n = \frac{1 - z^{n+1}}{1 - z},$$

and consequently

$$e^{-r} \sum_{n=0}^{\infty} \frac{s_n r^n}{n!} = \frac{1}{1 - z}(1 - ze^{r(z-1)}).$$

from which the result follows.

SOLUTION: (c): Because $s_n(f;\cdot) = f*D_n$ it is clear that

$$B_r(f;\cdot) = f*B_r,$$

where

$$B_r = e^{-r} \sum_{n=0}^{\infty} \frac{D_n r^n}{n!}.$$

Recall that

$$D_n(x) = \mathrm{Im}\left(\frac{e^{i(n+\frac{1}{2})x}}{\sin\frac{1}{2}x}\right), \qquad 0 < |x| \leqslant \pi,$$

and consequently

$$B_r(x) = \frac{e^{-r}}{\sin\frac{1}{2}x} \mathrm{Im}(e^{i\frac{1}{2}x + re^{ix}})$$

$$= e^{-r(1-\cos x)} \frac{\sin(\frac{1}{2}x + r\sin x)}{\sin\frac{1}{2}x}.$$

SOLUTION: (d): First of all, if $n > 0$ is an integer, then

$$J(\pi n) = \sum_{s=0}^{n-1} \int_0^{\pi} \frac{\sin x}{x + s\pi}\, dx.$$

The integral that appears in this expression is equivalent to $2/\pi s$ are $s \to \infty$, which proves that

$$J(\pi n) \sim \frac{2}{\pi}\log n.$$

Therefore, when $x \to \infty$:

$$0 \leqslant J(x) - J(\pi[x/\pi]) \leqslant \frac{1}{[x/\pi]},$$

$$J(\pi[x/\pi]) \sim \frac{2}{\pi}\log[x/\pi] \sim \frac{2}{\pi}\log x,$$

and consequently

$$J(x) \sim \frac{2}{\pi}\log x.$$

SOLUTION: (e):

$$\|B_r\|_1 = \frac{2}{\pi}\int_0^{\pi/2} \frac{e^{-2r\sin^2 u}}{\sin u}\,|\sin(u + r\sin 2u)|\,du.$$

As the function $\left|\frac{1}{u} - \frac{1}{\sin u}\right|$ is bounded on $[0,\pi/2]$, it is clear that as

$$\|B_r\|_1 = \frac{2}{\pi}\int_0^{\pi/2} \frac{e^{-2r\sin^2 u}}{u}\,|\sin(u + r\sin 2u)|\,du + o(1).$$

Using the mean value theorem and the inequalities

$$\sin u \geqslant \frac{2u}{\pi}, \qquad 0 \leqslant u \leqslant \frac{\pi}{2},$$

$$u^2 - \sin^2 u \leqslant \frac{u^4}{3},$$

this yields

$$\int_0^{\pi/2}(e^{-2r\sin^2 u} - e^{-2ru^2})\,\frac{du}{u} \leqslant \frac{2r}{3}\int_0^{\pi/2} u^3 e^{-8ru^2/\pi^2}\,du = O\left(\frac{1}{r}\right) .$$

Consequently

$$\|B_r\|_1 = \frac{2}{\pi}\int_0^{\pi/2} \frac{e^{-2ru^2}}{u}\,|\sin(u + r\sin 2u)|\,du + o(1).$$

Applying the mean value theorem once again, and using the inequality

$$u - \sin u \leqslant \frac{u^3}{6} , \qquad u \geqslant 0,$$

it follows that

$$|\sin(u + 2ru) - \sin(u + r\sin 2u)| \leqslant \frac{4ru^3}{3} ,$$

and since

$$\frac{4r}{3}\int_0^{\pi/2} u^2 e^{-2ru^2}\,du = O\left(\frac{1}{\sqrt{r}}\right) ,$$

one deduces that

$$\|B_r\|_1 = \frac{2}{\pi}\int_0^{\pi/2} \frac{e^{-2ru^2}}{u}\,|\sin(2r + 1)u|\,du + o(1).$$

To evaluate the integral which appears in the preceding formula, choose a $\delta > 0$ and write that

$$e^{-2\delta^2} J((2r + 1)r^{-\frac{1}{2}}\delta) \leqslant \int_0^{\delta/\sqrt{r}} \leqslant J((2r + 1)r^{-\frac{1}{2}}\delta) \sim \frac{1}{\pi}\log r,$$

$$\int_{\delta/\sqrt{r}}^{\pi/2} \leqslant \int_{\delta}^{\infty} \frac{e^{-2u^2}}{u}\, du.$$

Then

$$\frac{2}{\pi^2} e^{-2\delta^2} \leqslant \liminf_{r\to\infty} \frac{\|B_r\|_1}{\log r} \leqslant \limsup_{r\to\infty} \frac{\|B_r\|_1}{\log r} \leqslant \frac{2}{\pi^2},$$

and making $\delta \to 0$ yields

$$\lim_{r\to\infty} \frac{\|B_r\|_1}{\log r} = \frac{2}{\pi^2}.$$

SOLUTION: (f): The linear functional on $C(T)$ defined by

$$f \mapsto B_r(f;x)$$

have $\|B_r\|_1$ as norm. By the principle of condensation of singularities, for all x there exists a G_δ everywhere dense in $C(T)$ such that

$$\sup_r |B_r(f;x)| = \infty$$

if $f \in G_\delta$.

SOLUTION: (g): Note that

$$\frac{1}{2\pi}\int_{-\pi}^{\pi} B_r = e^{-r} \sum_{n\geqslant 0} \frac{r^n}{n!} \frac{1}{2\pi}\int_{-\pi}^{\pi} D_n$$

$$= e^{-r} \sum_{n\geqslant 0} \frac{r^n}{n!} = 1.$$

If, for $0 < |t| \leqslant \pi$, we set

$$\varphi(t) = (f(x + t) - f(x))\log\left|\frac{2\pi}{t}\right| ,$$

$$Q_r(t) = \frac{B_r(t)}{\log\left|\frac{2\pi}{t}\right|} ,$$

then

$$B_r(f;x) - f(x) = \frac{1}{2\pi}\int_{-\pi}^{\pi} \varphi(t)Q_r(t)dt. \tag{1}$$

Note that for $0 < \delta \leqslant |t| \leqslant \pi$:

$$|Q_r(t)| \leqslant \frac{e^{-2r\sin^2\frac{1}{2}\delta}}{(\log 2)\sin\frac{1}{2}\delta} .$$

On the other hand, some calculations similar in every way to those carried out in (e) show that

$$\|Q_r\|_1 = \frac{2}{\pi}\int_0^{\pi/2} \frac{e^{-2ru^2}}{u\log\frac{\pi}{u}} |\sin(2r + 1)u|du + o(1).$$

Decomposing the integral above into an integral taken between 0 and $1/\sqrt{r}$, and another taken between $1/\sqrt{r}$ and $\pi/2$, it is seen to be bounded by:

$$\frac{J((2r + 1)r^{-\frac{1}{2}})}{\log\pi\sqrt{r}} + \frac{1}{\log 2}\int_1^{\infty} \frac{e^{-2u^2}}{u} du.$$

By (d) we therefore have

$$\|Q_r\|_1 \leqslant A, \qquad A = \text{constant}.$$

But then if $\varepsilon > 0$ and if $\delta > 0$ is such that $|\varphi(t)| < \varepsilon$ for $|t| \leqslant \delta$, by (1) we shall have

$$|B_r(f;x) - f(x)| \leq \varepsilon A + \frac{e^{-2r\sin^2\frac{1}{2}\delta}}{(\log 2)\sin\frac{1}{2}\delta}\int_{\delta\leq|t|\leq\pi}|\varphi(t)|\,dt,$$

which proves that

$$\lim_{r\to\infty} B_r(f;x) = f(x).$$

EXERCISE 11·219: Let (a_n) be a sequence of numbers $a_n \geq 0$, and

$$A_n = a_0 + \cdots + a_n.$$

Suppose that

$$A_n \to \infty, \qquad \frac{a_n}{A_n} \to 0.$$

The series

$$\sum_{-\infty}^{+\infty} c_n$$

is said to converge to s in the sense of Nörlund, written

$$\sum_{-\infty}^{+\infty} c_n \overset{(N)}{=} s,$$

if

$$\nu_n = \frac{a_n s_0 + \cdots + a_0 s_n}{A_n} \to s,$$

where

$$s_n = \sum_{|p|\leqslant n} c_p.$$

(I): (a): Show that if a series converges in the ordinary sense it converges in the Nörlund sense. (Use exercise 10·177).

(I): (b): Prove that if $0 < a_0 \leqslant a_1 \leqslant \cdots$ Cesaro convergence implies Nörlund convergence.

(II): In the whole of this Part (II) assume that:

$$a_0 \geqslant a_1 \geqslant \cdots \geqslant a_n \geqslant \cdots, \qquad a_n \to 0. \tag{*}$$

For $f \in L^1$ set

$$N_n(f) = \frac{1}{A_n} \sum_{p=0}^{n} a_p s_{n-p}(f),$$

where $s_n(f)$ denotes the n-th Fourier sum of f.

(II): (a): Show that

$$N_n(f) = f * (U_n - V_n),$$

where

$$U_n(x) = \frac{1}{A_n} \frac{\sin(n+1)x}{\sin\frac{1}{2}x} \sum_{p=0}^{n} a_p \cos(p + \tfrac{1}{2})x, \tag{1}$$

$$V_n(x) = \frac{1}{A_n} \frac{\cos(n+1)x}{\sin\frac{1}{2}x} \sum_{p=0}^{n} a_p \sin(p + \tfrac{1}{2})x. \tag{2}$$

Show that

$$\int_0^{1/n} |U_n(x)|\,dx$$

is bounded for $n \geqslant 1$.

(II): (b): Show that there exist constants B, C, D such that:

$$\int_{1/n}^{\pi} \frac{|\sin(p+1)x|}{2\sin^2\frac{1}{2}x}\,dx \leqslant B(p+1)\log\frac{n}{p} + C(p+1), \quad 1 \leqslant p < n,$$

$$\int_{1/n}^{\pi} \frac{|\sin x|}{2\sin^2\frac{1}{2}x}\,dx \leqslant D\log n, \qquad n \geqslant 2.$$

(II): (c): By carrying out some suitable Abel transformations on (1) and (2), prove that

$$\|V_n\|_1 \leqslant 1,$$

and:

$$\int_{1/n}^{\pi} |U_n(x)|\,dx < C + \frac{1}{A_n}\left(E\log n + B\sum_{p=1}^{n-1} A_p \log\left(1 + \frac{1}{p}\right)\right),$$

where E is a constant.

(II): (d): Now assume that

$$\sup_n \|U_n\|_1 < \infty.$$

By carrying out another Abel transformation on (1) for U_n, and observing that

$$\sin^2(n+1)x \leqslant |\sin(n+1)x|,$$

show first of all that the quantity

$$\int_0^{\pi} (\sin^2(n+1)x)\left\{\frac{\cos(n+\frac{1}{2})x}{\sin\frac{1}{2}x} + \frac{2}{A_n}\sum_{p=0}^{n-1} A_p \sin(p+1)x\right\}dx$$

stays bounded, then by studying the behaviour of the integrals:

$$\int_0^\pi \frac{\sin^2(n+1)x.\cos(n+\frac{1}{2})x}{\sin\frac{1}{2}x}\,dx,$$

$$\int_0^\pi \cos(2n+2)x.\sin(p+1)x\,dx,$$

deduce that

$$\sup_n \frac{1}{A_n}\sum_{p=1}^{n}\frac{A_p}{p} < \infty.$$

(II): (e): Conclude from the preceding that when (*) is satisfied the following conditions are equivalent:

(i): For all $f \in L^1(T)$

$$\sum_{-\infty}^{+\infty} \hat{f}(n)e^{inx} \overset{(N)}{=} f(x)$$

at every point x where f is continuous;

(ii):

$$\sup_n \frac{1}{A_n}\sum_{p=1}^{n}\frac{A_p}{p} < \infty.$$

(III): For every real number f set

$$\Lambda_n^\alpha = \frac{(\alpha+1)\cdots(\alpha+n)}{n!},$$

and assume that

$$a_n = \Lambda_n^{\alpha-1}.$$

(III): (a): Show that

$$A_n = \Lambda_n^{\alpha}$$

and that if $\alpha > 0$ one obtains a summation method in the style of Nörlund.

Next show that if $0 < \alpha < 1$ one is in the case studied in part (II) above, if $\alpha > 1$, in that of part (I)(b); and that finally, if $\alpha = 1$ Cesaro's method is recovered.

(III): (b): It is said that:

$$\sigma_n^{\alpha} = \frac{s_0\Lambda_n^{\alpha-1} + \cdots + s_n\Lambda_0^{\alpha-1}}{\Lambda_n^{\alpha}}$$

is the *CESARO SUM OF ORDER* α of the series $\sum c_n$, and that the latter is (C,α)-*CONVERGENT* if σ_n^{α} is $(\alpha > 0)$.

Prove that for every $\alpha > 0$ the Fourier series of a function $f \in L^1(T)$ is (C,α)-convergent to f at every point where f is continuous.

ΔVΔ = VΔV = ΔVΔ = VΔV = ΔVΔ

SOLUTION: (I): (a):

$$\nu_n = \sum_{p=0}^{\infty} u_{n,p}s_p,$$

where

$$u_{n,p} = \begin{cases} \dfrac{a_{n-p}}{A_n} & \text{if } 0 \leqslant p \leqslant n, \\ \\ 0 & \text{if } p > n. \end{cases}$$

It is clear that

$$\sum_{p\geqslant 0} u_{n,p} = 1.$$

Moreover,

$$u_{n,p} \leqslant \frac{a_{n-p}}{A_{n-p}},$$

and consequently

$$\lim_n u_{n,p} = 0.$$

By exercise 10·177 $s_n \to s$ implies $\nu_n \to s$.

SOLUTION: (I): (b): Writing σ_n for the n-th Cesaro sum of the series $\sum c_n$,

$$\nu_n = \frac{1}{A_n}\sum_{p=0}^{n} a_{n-p}((p + 1)\sigma_p - p\sigma_{p-1})$$

$$= \frac{(n + 1)a_0}{A_n}\sigma_n + \frac{1}{A_n}\sum_{p=0}^{n-1} (p + 1)(a_{n-p} - a_{n-p-1})\sigma_p$$

$$= \sum_{p=0}^{\infty} \lambda_{n,p}\sigma_p,$$

where

$$\lambda_{n,p} = \begin{cases} \dfrac{(p + 1)}{A_n}(a_{n-p} - a_{n-p-1}) & \text{if } 0 \leqslant p \leqslant n - 1, \\ \dfrac{(n + 1)a_0}{A_n} & \text{if } p = n, \\ 0 & \text{if } p > n. \end{cases}$$

As the sequence (a_n) is increasing, all the $\lambda_{n,p} \geqslant 0$. It is clear that

$$\lim_n \lambda_{n,p} = 0.$$

Furthermore, when all the s_n's are equal to one, the same holds for all the σ_n's and ν_n's. so

$$\sum_{p=0}^{\infty} \lambda_{n,p} = 1.$$

By Exercise 10·177 $\sigma_n \to s$ implies $\nu_n \to s$.

SOLUTION: (II): (a): $N_n(f)$ is equal to the convolution product of f with

$$\frac{1}{A_n}\sum_{p=0}^{n} a_p D_{n-p}(x) = \frac{1}{A_n}\sum_{p=0}^{n} a_p \frac{\sin(n - p + \frac{1}{2})x}{\sin\frac{1}{2}x}$$

$$- \frac{1}{A_n}\frac{\sin(n + 1)x}{\sin\frac{1}{2}x}\sum_{p=0}^{n} a_p \cos(p + \tfrac{1}{2})x$$

$$- \frac{1}{A_n}\frac{\cos(n + 1)x}{\sin\frac{1}{2}x}\sum_{p=0}^{n} a_p \sin(p + \tfrac{1}{2})x,$$

which implies the first result desired.

Note that

$$\frac{\sin(n + 1)x}{\sin\frac{1}{2}x} \leqslant 2(n + 1),$$

and consequently

$$|U_n| \leqslant 2(n + 1),$$

so

$$\int_0^{1/n} |U_n(x)|\,dx \leqslant 2\left(1 + \frac{1}{n}\right) \leqslant 4.$$

SOLUTIONS: (II): (b);(c): Denoting by F_n the n-th Féjer kernel and setting

$$\Delta_p = a_p - a_{p+1},$$

we obtain

$$V_n(x) = \frac{1}{A_n}\cos(n+1)x\{(n+1)a_nF_n(x) + \sum_{p=0}^{n-1}(p+1)\Delta_pF_p(x)\}. \tag{1}$$

Since

$$\Delta_p \geqslant 0 \quad \text{and} \quad \|F_p\|_1 = 1,$$

it follows that

$$\|V_n\| \leqslant \frac{1}{A_n}\{(n+1)a_n + \sum_{p=0}^{n-1} p + 1)\Delta_p\} = 1.$$

Similarly, noting that

$$\sum_{p=0}^{n}\cos(p+\tfrac{1}{2})x = \frac{\sin(n+1)x}{2\sin\frac{1}{2}x}$$

we obtain

$$U_n(x) = \frac{1}{A_n}\sin(n+1)x \tag{2}$$

$$\times\left\{a_n\frac{\sin(n+1)x}{2\sin^2\frac{1}{2}x} + \sum_{p=0}^{n-1}\Delta_p\frac{\sin(p+1)x}{2\sin^2\frac{1}{2}x}\right\}.$$

Now note that

$$\int_{1/n}^{\pi} \frac{\sin x}{2\sin^2\frac{1}{2}x}\,dx \leqslant \int_{1/n}^{\pi} \frac{x\,dx}{\left(\frac{2x^2}{\pi^2}\right)} = \frac{\pi^2}{2}\log(\pi n) \leqslant D\log n.$$

for all $n \geqslant 2$ if D is chosen large enough.

When $1 \leqslant p \leqslant n$,

$$\int_{1/n}^{\pi} \frac{|\sin(p+1)x|}{2\sin^2\frac{1}{2}x}\,dx \leqslant (p+1)\int_{1/n}^{1/p} \frac{x\,dx}{\left(\frac{2x^2}{\pi^2}\right)} + \int_{1/p}^{\pi} \frac{dx}{\left(\frac{2x^2}{\pi^2}\right)}$$

$$\leqslant B(p+1)\log\frac{n}{p} + C(p+1),$$

with $B = C = \pi^2/2$. Therefore

$$\int_{1/n}^{\pi} |U_n| \leqslant \frac{C}{A_n}\left[(n+1)a_n + \sum_{p=1}^{n-1}(p+1)\Delta_p\right]$$

$$+ \frac{1}{A_n}\left[D\Delta_0\log n + B\sum_{p=1}^{n-1}(p+1)\Delta_p\log\frac{n}{p}\right].$$

Notice that the first bracket is equal to $A_n - \Delta_0$, and that

$$\sum_{p=1}^{n-1}(p+1)\Delta_p\log\frac{n}{p} = 2a_1\log n + \sum_{p=2}^{n} a_p\log\frac{n}{p}$$

$$+ \sum_{p=2}^{n} pa_p\log\left(1-\frac{1}{p}\right).$$

The last term of the right side is negative, and

$$\sum_{p=2}^{n} a_p\log\frac{n}{p} = \sum_{p=1}^{n-1} A_p\log\left(1+\frac{1}{p}\right) - A_1\log n.$$

Therefore, writing

$$E = D\Delta_0 + 2Ba_1,$$

we have

$$\int_{1/n}^{\pi} |U_n| \leq C + \frac{1}{A_n}\left(E\log n + B\sum_{p=1}^{n-1} A_p \log\left(1 + \frac{1}{p}\right)\right).$$

SOLUTION: (II): (d): With the help of an Abel transformation one obtains

$$U_n(x) = \frac{1}{A_n}\frac{\sin(n+1)x}{\sin\frac{1}{2}x}$$

$$\times\left\{A_n\cos(n+\tfrac{1}{2})x + 2\sin\tfrac{1}{2}x \sum_{p=0}^{n-1} A_p \sin(p+1)x\right\}$$

$$= \sin(n+1)x . G_n(x),$$

where

$$G_n(x) = \frac{\cos(n+\frac{1}{2})x}{\sin\frac{1}{2}x} + \frac{2}{A_n}\sum_{p=0}^{n-1} A_p \sin(p+1)x.$$

Since

$$\sin^2(n+1)x|G_n(x)| \leq |\sin(n+1)x . G_n(x)| = |U_n(x)|,$$

we have

$$\sup_n \int_0^{\pi} \sin^2(n+1)x|G_n(x)|\,dx < \infty.$$

Let

$$M = \sup_{0 \leqslant x \leqslant \pi} \left|\cot\tfrac{1}{2}x - \frac{2}{x}\right| .$$

Since

$$\int_0^\pi \frac{\sin^2(n + 1)x\cos(n + \frac{1}{2})x}{\sin\frac{1}{2}x}\,dx$$

$$= \int_0^\pi \sin^3(n + 1)x\,dx + \int_0^\pi \sin^2(n + 1)x\cos(n + 1)x\cot\tfrac{1}{2}x\,dx,$$

this integral remains bounded in absolute value by

$$\pi(M + 1) + 2\left|\int_0^{\pi(n+1)} \sin^2 u.\cos u\,\frac{du}{u}\right| .$$

Now, the generalized integral:

$$\int_0^\infty \sin^2 u.\cos u\,\frac{du}{u}$$

is convergent, from which it follows that

$$\frac{1}{A_n}\sum_{p=0}^{n-1} A_p\int_0^\pi \sin^2(n + 1)x.\sin(p + 1)x\,dx$$

remains bounded. Also, if $0 \leqslant p \leqslant n - 1$, then

$$\left|\int_0^\pi \cos(2n + 2)x.\sin(p + 1)x\,dx\right| \leqslant \frac{1}{2n + p + 3} + \frac{1}{2n - p + 1}$$

$$\leqslant \frac{2}{n} .$$

Since

Since

$$\frac{1}{A_n}\sum_{p=0}^{n-1} A_p \leqslant n,$$

it follows that:

$$\frac{1}{A_n}\sum_{p=0}^{n-1} A_p\int_0^{\pi}\sin(p+1)x\,dx = \frac{2}{A_n}\sum_{0\leqslant 2p\leqslant n-1}\frac{A_{2p}}{(2p+1)}$$

remains bounded. Writing

$$\frac{1}{A_n}\sum_{p=1}^{n}\frac{A_p}{p} = \frac{1}{A_n}\sum_{1\leqslant 2p\leqslant n}\left(1+\frac{1}{2p}\right)\frac{A_{2p}}{(2p+1)}$$

$$+\frac{1}{A_n}\sum_{2\leqslant 2p\leqslant n+1}\left(1+\frac{2}{2p-1}\right)\frac{A_{2p-1}}{(2p+1)},$$

because $A_{2p-1} \leqslant A_{2p}$ one obtains

$$\frac{1}{A_n}\sum_{p=1}^{n}\frac{A_p}{p} \leqslant 5\,\frac{A_{n+2}}{A_n}\,\frac{1}{A_{n+2}}\sum_{0\leqslant 2p\leqslant n+1}\frac{A_{2p}}{(2p+1)}.$$

Since

$$\frac{A_{n+2}}{A_n} = 1 + \frac{a_{n+1}+a_{n+2}}{A_n} \leqslant 1 + 2\,\frac{a_n}{A_n} \to 1,$$

at last we obtain

$$\sup_n \frac{1}{A_n}\sum_{p=1}^{n}\frac{A_p}{p} < \infty.$$

SOLUTION: (II): (e): Let

$$\phi_n = U_n - V_n = \frac{1}{A_n}\sum_{p=0}^{n} a_p D_{n-p}.$$

It is clear that

$$\frac{1}{2\pi}\int_{-\pi}^{\pi} \phi_n = 1.$$

Since the n-th Féjer kernel is

$$F_n(x) = \frac{1}{n+1}\left(\frac{\sin(n+1)\frac{1}{2}x}{\sin\frac{1}{2}x}\right)^2,$$

by using the expression (1) for V_n found in (II)(b)(c), one obtains for all $\delta > 0$

$$\sup_{\delta \leqslant |x| \leqslant \pi} |V_n(x)| \leqslant \frac{1}{A_n \sin^2\frac{1}{2}\delta}\left(a_n + \sum_{p=0}^{n-1} \Delta_p\right) = \frac{a_n}{A_n}\frac{1}{\sin^2\frac{1}{2}\delta} .$$

Similarly, the expression (2) for U_n provides the bound

$$\sup_{\delta \leqslant |x| \leqslant \pi} U_n(x) \leqslant \frac{a_n}{A_n}\frac{1}{2\sin^2\frac{1}{2}\delta} .$$

Thus

$$M_n(\delta) = \sup_{\delta < |x| < \pi} |\phi_n(x)| \to 0 \quad \text{as } n \to \infty.$$

If

$$\sup_n \frac{1}{A_n}\sum_{p=1}^{n} \frac{A_p}{p} < \infty, \tag{3}$$

then

$$\sup_n \frac{1}{A_n} \sum_{p=1}^{n} \frac{1}{p} < \infty,$$

i.e.

$$\sup_n \frac{1}{A_n} \log n < \infty.$$

By the inequalities

$$\|V_n\|_1 \leqslant 1, \qquad \int_0^{1/n} |U_n| \leqslant 4,$$

$$\int_{1/n}^{\pi} |U_n| \leqslant C + \frac{1}{A_n} \left(E \log n + B \sum_{p=1}^{n-1} A_p \log\left(1 + \frac{1}{p}\right)\right),$$

$$\log\left(1 + \frac{1}{p}\right) \leqslant \frac{1}{p},$$

it is then seen that (3) implies that

$$\sup_n \|\phi_n\|_1 << \infty. \tag{4}$$

In particular, in this case (ϕ_n) is an approximate identity in $L^1(T)$. If $f \in L^1(T)$ is continuous at x, the bound

$$|N_n(f;x) - f(x)| \leqslant \|\phi_n\|_1 \sup_{|t| \leqslant \delta} |f(x + t) - f(t)| + M_n(\delta)\|f\|_1$$

shows that

$$\lim_n N_n(f;x) = f(x). \tag{5}$$

If it is now assumed that (5) is satisfied for every continuous function f and all x, then in particular

$$\sup_n |N_n(f;0)| < \infty, \qquad f \in C(T).$$

By the Banach-Steinhaus Theorem, (4) then holds, and therefore

$$\sup_n \|U_n\|_1 < \infty,$$

because

$$\|V_n\|_1 \leqslant 1.$$

But the (II)(d) this implies (3).

SOLUTION: (III): (a):

$$\sum_{n=0}^{\infty} \Lambda_n^{\alpha} x^n = (1 - x)^{-\alpha-1}, \qquad -1 < x < 1,$$

from which it is immediately deduced that

$$\sum_{p=0}^{n} \Lambda_p^{\alpha-1} = \Lambda_n^{\alpha}.$$

Thus by setting $a_n = \Lambda_n^{\alpha-1}$ we have

$$A_n = \Lambda_n^{\alpha}, \qquad \lim_n A_n = \infty, \qquad \lim_n \frac{a_n}{A_n} = \lim_n \frac{\alpha}{\alpha + n} = 0.$$

Furthermore,

$$\frac{a_{n+1}}{a_n} = 1 + \frac{(\alpha - 1)}{(n + 1)},$$

which proves that a_n increases with n if $\alpha > 1$, and decreases with n if $0 < \alpha < 1$; when $\alpha = 1$ we have $a_n = 1$ for all n.

SOLUTION: (III): (b): If $\alpha \geqslant 1$ the (C,α)-convergence of $\sum \hat{f}(n)e^{inx}$ to $f(x)$ at every point where f is continuous results from Féjer's Theorem and from part (I)(b) above. If $0 < \alpha < 1$ note that

$$\Lambda_n^\alpha \sim \frac{n^\alpha}{\Gamma(\alpha + 1)},$$

so

$$\sum_{p=1}^{n} \frac{\Lambda_p^\alpha}{p} \sim \frac{n^\alpha}{\alpha\Gamma(\alpha + 1)} \sim \frac{1}{\alpha} \Lambda_n^\alpha.$$

Thus, (3) is satisfied, and by part (II)(e) this implies (C,α)-convergence at every point of continuity.

EXERCISE 11·220: Let $1 \leqslant p \leqslant 2$ and $\frac{1}{p} + \frac{1}{q} = 1$.

(a): Show that if $f \in L^p(T)$, then

$$\sum_{n=-\infty}^{+\infty} |\hat{f}(n)|^q \leqslant \|f\|_p^p.$$

(b): Let $(c_n)_{n\in\mathbb{Z}}$ be a sequence of complex numbers such that

$$\sum_{n=-\infty}^{+\infty} |c_n|^p < \infty.$$

Show that the series

$$\sum_{n=-\infty}^{+\infty} c_n e^{inx}$$

converges in $L^q(T)$ to a function f, and that

$$\|f\|_q^q \leqslant \sum_{n=-\infty}^{+\infty} |c_n|^p.$$

(Use the Riesz-Thorin Theorem, cf. exercise 6•123).

ΔVΔ = VΔV = ΔVΔ = VΔV = ΔVΔ

SOLUTION: (a): Let T be the mapping which associates with $f \in L^\infty(T)$ the sequence of its Fourier coefficients. Then

$$\|T(f)\|_\infty \leqslant \|f\|_1, \tag{1}$$

as

$$|\hat{f}(n)| \leqslant \|f\|_1 \quad \text{for all } n,$$

and by Plancherel's Theorem

$$\|T(f)\|_2 = \|f\|_2. \tag{2}$$

If $0 \leqslant t \leqslant 1$ and

$$\frac{1}{p} = \frac{1-t}{1} + \frac{t}{2} = 1 - \frac{t}{2},$$

$$\frac{1}{q} = \frac{1-t}{\infty} + \frac{t}{2} = \frac{t}{2},$$

then by the Riesz-Thorin Theorem

$$\|T(f)\|_q \leqslant \|f\|_p \tag{3}$$

Note that p runs over the segment [1,2] when t varies from 0 to 1, and that $\frac{1}{p} + \frac{1}{q} = 1$. Formula (3), proved for $f \in L^\infty(T)$, is still valied if $f \in L^p(T)$, since $L^\infty(T)$ is dense in $L^p(T)$.

SOLUTION: (b): Consider the space E of sequences $c = (c_n)_{n \in \mathbb{Z}}$, all the terms of which are zero except for a finite number, and let

$$\phi(c)(x) = \sum_{n=-\infty}^{+\infty} c_n e^{inx}, \qquad x \in \mathbb{R}.$$

Then

$$\left.\begin{aligned} \|\phi(c)\|_\infty &\leqslant \|c\|_1, \\ \|\phi(c)\|_2 &= \|c\|_2, \end{aligned}\right\} \tag{4}$$

so

$$\|\phi(c)\|_q \leqslant \|c\|_p. \tag{5}$$

We finish by observing that E_0 is dense in $\ell^p(\mathbb{Z})$.

Erratum to Exercise 3 • 45

THE PROOF THAT (iii) => (ii) IS VALID ONLY IF $f(0) = f(1)$

If this condition is not fulfilled, consider a continuous function ϑ_ε such that

$$0 < \vartheta_\varepsilon < 1, \quad \vartheta_\varepsilon(0) = \vartheta_\varepsilon(1) = 0, \quad \vartheta_\varepsilon(x) = 0 \text{ if } \varepsilon < x < 1 - \varepsilon ,$$

and write

$$n^{-1} \sum_{i=1}^{n} f(x_i) - \int_0^1 f$$

$$= n^{-1} \sum_{i=1}^{n} \vartheta_\varepsilon(x_i) f(x_i) - \int_0^1 \vartheta_\varepsilon f - \int_0^1 (1 - \vartheta_\varepsilon) f$$

$$+ n^{-1} \sum_{i=1}^{n} (1 - \vartheta_\varepsilon(x_i)) f(x_i).$$

Assuming that $|f| < M$, we have

$$|n^{-1} \sum_{i=1}^{n} f(x_i) - \int_0^1 f| < |n^{-1} \sum_{i=1}^{n} \vartheta_\varepsilon(x_i) f(x_i) - \int_0^1 \vartheta_\varepsilon f| +$$

(Contd)

(Contd) $$+ M\int_0^1 (1 - \vartheta_\varepsilon) + Mn^{-1} \sum_{i=1}^{n} (1 - \vartheta_\varepsilon(x_i)),$$

and therefore

$$\limsup_n \left|n^{-1} \sum_{i=1}^{n} f(x_i) - \int_0^1 f\right| < 2M\int_0^1 (1 - \vartheta_\varepsilon) < 4\varepsilon M.$$

Bibliography

N.A. BARY: *A Treatise on Trigonometric Series, Vol.s I, II*, (Pergamon Press, Inc., New York), (1964).

J. DIEUDONNÉ: *Eléments d'Analyse, Tome II*, (Gauthier-Villars, Paris), (1969).

R.E. EDWARDS: *Fourier Series, Vol.s I, II*, (Holt, Rinehart and Winston, Inc., New York), (1967).

C. GEORGE: *Cours d'Intégration* (École des Mines, Nancy), (1977).

P.R. HALMOS: *Measure Theory*, (Van Nostrand, New York), (1950).

Y. KATZNELSON: *An Introduction to Harmonic Analysis*, (John Wiley and Sons, Inc., New York), (1968).

A. KOLMOGOROV and S. FOMIN: *Elements of the Theory of Functions and Functional Analysis*, (Mir, Moscow), (1974).

L.H. LOOMIS: *An Introduction to Abstract Harmonic Analysis*, (Van Nostrand, New York), (1953).

F. RIESZ and B. NAGY: *Lecons d'Analyse Fonctionelle*, (4th edn.), (Budapest—Paris), (1965).

W. RUDIN: *Real and Complex Analysis*, (McGraw-Hill, New York), (1966).

K. YOSIDA: *Functional Analysis*, (Springer-Verlag, Berlin), (1965).

A. ZYGMUND: *Trigonometric Series, Vol.s I, II*, (Cambridge University Press), (1959).

Name Index

The notions introduced in Chapter 0 do not appear in this Index. The numbers below refer to the exercises.